NATURE'S MAGIC

Nature's Magic presents a bold new vision of the evolutionary process – from the Big Bang to the 21st century. Synergy of various kinds is not only a ubiquitous aspect of the natural world but it has also been a wellspring of creativity and the "driver" of the broad evolutionary trend toward increased complexity, in nature and human societies alike. But in contrast with the many theories of emergence or complexity that rely on some underlying force or "law," the "Synergism Hypothesis," as Peter Corning calls it, is in essence an economic theory of biological complexity; it is fully consistent with mainstream evolutionary biology. Corning refers to it as Holistic Darwinism. Among the many important insights that are provided by this new paradigm, Corning presents a scenario in which the human species invented itself; synergistic behavioral and technological innovations were the "pacemakers" of our biological evolution. Synergy has also been the key to the evolution of complex modern societies, he concludes. The final chapter addresses our current challenges and future prospects.

Peter Corning's varied career has included a tour as a naval aviator, a stint in journalism as a science writer for *Newsweek*, a Ph.D. in the social sciences, post-doctoral training and research in biology and behavioral genetics, several years of teaching in Stanford University's interdisciplinary Human Biology Program, and broad private-sector experience as a senior partner in a Silicon Valley consulting firm. Currently director of the Institute for the Study of Complex Systems, Dr. Corning is a member of several scientific organizations and a past president of the International Society for the Systems Sciences. He has also published more than 150 professional articles and three previous books and is widely known for his work on the role of synergy as a causal agency in evolution.

Nature's Magic

Synergy in Evolution and the Fate of Humankind

Peter Corning

Institute for the Study of Complex Systems
Palo Alto, California

CAMBRIDGE
UNIVERSITY PRESS

PUBLISHED BY THE PRESS SYNDICATE OF THE UNIVERSITY OF CAMBRIDGE
The Pitt Building, Trumpington Street, Cambridge, United Kingdom

CAMBRIDGE UNIVERSITY PRESS
The Edinburgh Building, Cambridge CB2 2RU, UK
40 West 20th Street, New York, NY 10011-4211, USA
477 Williamstown Road, Port Melbourne, VIC 3207, Australia
Ruiz de Alarcón 13, 28014 Madrid, Spain
Dock House, The Waterfront, Cape Town 8001, South Africa

http://www.cambridge.org

First published 2003

Printed in the United States of America

Typefaces Sabon 10/13 pt. and ITC Symbol   *System* LaTeX 2_ε [TB]

A catalog record for this book is available from the British Library.

Library of Congress Cataloging in Publication data

Corning, Peter A., 1935–
    Nature's magic : synergy in evolution and the fate of humankind / Peter A. Corning.
        p. cm.
    Includes bibliographical references (p.).
    ISBN 0-521-82547-4 (HB)
        1. Evolution (Biology) 2. Human evolution. I. Title.

QH366.2. C73 2003
576.8–dc21   2002031552

ISBN 0 521 82547 4 hardback

TO JOHN MAYNARD SMITH

A GUIDING LIGHT

Two are better than one,
Because they have good return for their work:
If one falls down,
his friend can help him up.
But pity the man who falls
and has no one to help him up.
Also, if two lie down together,
they will keep warm.
But how can one keep warm alone?
Though one may be overpowered,
two can defend themselves.
A cord of three strands
is not quickly broken.

Ecclesiastes 4:9–12

Contents

1 Prologue: The New Evolutionary Paradigm

> True innovation occurs when things are put together
> for the first time that had been separate.
>
> Arthur Koestler
> *Beyond Reductionism*

When Arthur Koestler, the famed novelist and respected polymath, penned those words more than 30 years ago, he was seeking to draw our attention to a phenomenon that is greatly underrated and vastly more important even than Koestler imagined. I call it nature's magic.

Grand theories are commonplace these days. It seems that new ideas must shout to be heard. So the claims for this book may sound like hyperbole as usual. The thesis, in brief, is that synergy – a vaguely familiar term to many of us – is actually one of the great governing principles of the natural world. It has been a wellspring of creativity in the evolution of the universe, and it has greatly influenced the overall trajectory of life on Earth. It has played a decisive role in the emergence of humankind. It is vital to the workings of every modern society. And it is no exaggeration to say that our ultimate fate depends upon it.

All this may sound like so much dust-jacket rhetoric, but the Synergism Hypothesis (as I call it) is a serious scientific theory that is fully consistent with Darwin's theory, and with the canons of the physical, biological, and social sciences, not to mention the new science of complexity. The theory, in a nutshell, is that synergy is not only a ubiquitous effect in nature; it has also played a key *causal* role in the evolutionary process. It has been at once the fountainhead and the *raison d'être* for the progressive increase in complexity over the broad span of evolutionary history. Far from being law-like and predictable, however, this trend has always involved an open-ended, creative, historically constrained experiment in which economic criteria (broadly defined) have predominated. Complexity – in nature and human societies alike – is not the product of some inexorable force, or mechanism, or "law." It has been shaped by the immediate functional advantages – the "payoffs" – arising from various forms of synergy.

What Is Synergy?

How do I define synergy? Very broadly, the term refers to the combined, or cooperative, effects produced by the relationships among various forces, particles, elements, parts, or individuals in a given context – effects that are not otherwise possible. The term is derived from the Greek word *synergos*, meaning "working together" or, literally, "co-operating." Synergy is often associated with the cliché, "the whole is greater than the sum of its parts" (which dates back to Aristotle, in the *Metaphysics*), but this is actually a rather narrow and even misleading characterization. In fact, synergy comes in many different forms; sometimes wholes are not greater than the sum of their parts, just different. We will examine the phenomenon of synergy in greater depth in the next few chapters. Here are just a few brief examples, starting with some of the basic forces of nature:

* The center of gravity of an object, say an automobile, is actually a synergistic effect. It depends upon how the combined weight of all its parts is distributed, as we learned in school. But if we were to disassemble the car, its center of gravity would disappear; it would be parceled out (so to speak) among each of the 15,000 or so individual parts.
* The vortex, or whirlpool, that occurs when your bath water flows down the drain is actually a complex effect produced by the combined actions of several different forces – gravity, water pressure, air pressure, rotational forces, centrifugal forces, even the initial state of the bath water.
* "Supermolecules" of 50 atoms or more may take on wholly new collective properties that their lightweight cousins lack – greater stability, better binding capabilities, a different geometry, less energy dissipation (entropy), and the like.
* Chlorine and sodium are both toxic to humans by themselves, but when they are combined they produce a totally new substance that is positively beneficial (in moderate amounts) – ordinary table salt.

* Chrome–nickel–steel, an alloy synthesized from three natural elements, may be stronger by 35% than all of its constituents added together. In the bargain, chrome–nickel–steel has rust-free properties, another synergistic effect. (The nickel adds strength to the steel and the chromium reduces its tendency to oxidize.)

* Synergy is commonplace in medicine and health care. One example is the effect produced by using atropine and prednisone together to treat eye inflammations. The atropine serves to dilate the eyes so that the prednisone, an anti-inflammatory drug, can work more effectively.

* Our alphabet is also highly synergistic. Take the words "rat," "cat," and "bat." Each combination of letters produces a different image in the reader's mind. But imagine what would happen if the vowels were removed. Like the coins that magically disappear into a prestidigitator's folded handkerchief, the synergy would vanish and we would be left with the two-letter nonsense combinations rt, ct, bt.

* One cup of beans, eaten by itself, provides the nutritional equivalent of 2 ounces of steak. Three cups of whole-grain flour consumed alone provides the equivalent of 5 ounces of steak. But when they are ingested together, they provide the equivalent of 9.33 ounces of steak, or 33% more usable protein. The reason is that their constituent amino acids are highly complementary. Grains are low in lysine, while legumes are low in methionine. When combined they compensate for each other's deficiencies. In other words, the whole taco is truly greater, nutritionally, than the sum of its parts.

* Lichen, patchy growths that are found on tree trunks, rocks, and even bare ground in many woodland areas, are legendary for their ability to colonize barren environments as well. The key to their success as "nature's pioneers" lies in their complementary talents. Lichens actually consist of symbiotic partnerships between various kinds of green algae, or cyanobacteria, and fungi. (There are more than 20,000 different lichen species all told.) The algae or cyanobacteria are photosynthesizers. They provide energy-capturing services, while the fungi bring to the partnership both surface-gripping and water-storage capabilities – talents that are especially useful in a harsh environment. The partners may even join forces to create a specialized reproductive organ called a thallus that produces combined, symbiotic spores. Together, the "team" can do what neither partner can do alone.[1]

* "Tensegrity" (tensional integrity) refers to the way in which the counteracting effects of compression and tension can be used synergistically

to achieve structural "integrity" in certain self-stabilizing physical structures. The term tensegrity was coined by the well-known engineer–inventor Buckminster Fuller (who, incidentally, also promoted the concept of synergy) to characterize his most famous invention, the remarkable geodesic domes that today number in the hundreds of thousands world-wide.[2] We now know that many kinds of tensegrity structures also exist in nature. One example is the appropriately named "buckminsterfullerene" – carbon-60 and several variants. The great stability and remarkable binding properties achieved by these recently synthesized "supermolecules" of pure carbon (affectionately known as "Bucky Balls") derive from their physical resemblance to geodesic domes and soccer balls. Another example of tensegrity, closer to home, is the human body. The interaction between our bones, muscles, tendons, and ligaments gives our bodies their distinctive combination of structural stability and mobility. Likewise, every one of the ten trillion or so cells in each of our bodies is supported by an internal scaffolding, called a cytoskeleton, which is composed of actin filaments and microtubules. The actin filaments counteract pulling forces that are exerted on the cell and the microtubules resist compression forces. We are totally dependent on these and many other kinds of synergy. (Indeed, Harvard pathology professor Donald E. Ingber sees tensegrity as one of the basic organizing principles in life.[3])

The Causal Role of Synergy in Evolution

Accordingly, I will argue that synergy ranks up there with such heavyweight concepts as gravity, energy, entropy, and information as one of the keys to understanding how the world works and how we got here – not to mention where we are going. Moreover, synergy has been a creative dynamo and a prolific source of innovation in evolution, as we shall see. Synergy was present at the "Big Bang." It has been deeply involved in the evolution of our physical universe. Some time after the Earth first evolved, some 4.5 billion years ago, synergy provided the payoffs (the emergent functional effects) that arose in the still-mysterious process by which networks of complex prebiotic molecules joined together to catalyze the first living systems. It also provided the "benefits" which, over time, produced the awesome complexity of photosynthesis. (Entire books have been devoted to describing our as-yet-imperfect understanding of how photosynthesis works.) Synergy is found also in the intricate combination of labor in complex eukaryotic cells and in the "enchanted loom" of the human mind – to use the soaring image of neurobiologist Charles Sherrington – where wondrous new synergies are invented and actualized every day. In other words, the unique cooperative effects produced by various combinations of "parts" in a given context are themselves distinct, partially independent causes of subsequent evolutionary events.

The universe can be portrayed as a vast structure of synergies, a many-leveled edifice in which the synergies produced at one level serve as the building blocks for the next level. Moreover, unpredictable new forms of synergy, and even new principles, emerge at each level of organization. I like to call it a "Magic Castle" (with a nod to Walt Disney), because there is something truly magical about this creative aspect of nature. In the course of providing a guided tour of this Magic Castle (in Chapters 2, 3, and 4), I will show that synergy is of central importance in virtually every scientific discipline, though it very often travels incognito under various aliases (mutualism, cooperativity, symbiosis, win–win, emergent effects, a critical mass, coevolution, interactions, threshold effects, even non-zero-sumness).

According to the reigning dogma of evolutionary biology – commonly known as Neo-Darwinism – "random" gene mutations (and related molecular-level phenomena) are said to be the underlying source of creativity in evolution. It is said that the course of biological evolution has been shaped over time by relentless competition among "selfish genes." I will argue that the Neo-Darwinists have got things skewed. In fact, it is the functional benefits – the survival advantages – produced by novelties of *various* kinds and at various levels (including even behavioral innovations, as we shall see) that have defined the trajectory of evolution. Contrary to the popular misconception, natural selection does not (literally) select genes. It differentially rewards (or disfavors) different genes, and gene combinations, based on the *effects* they produce in a given environment. It is the functional payoffs that matter.

In this light, it is novel forms of functional synergy (cooperative effects) that have been responsible, over time, for shaping the progressive evolution of complexity in nature through a process that can be characterized (after biologist John Maynard Smith) as "synergistic selection." I call this new paradigm "Holistic Darwinism," and I side with the growing number of contemporary biologists who hold that evolution must be viewed as a multi-leveled process in which selfish genes are most often subordinated to the dictates of "selfish genomes" – synergistic systems. Outlaw genes are the exception rather than the rule. The Synergism Hypothesis and the theory of Holistic Darwinism will be developed in some detail in Chapters 5 and 6.

The "Synergistic Ape"

Many different theories of human evolution have been proposed over the years. (I will briefly describe some of them in Chapter 7.) Humans have been variously characterized as the "killer ape," the "naked ape," and the "talking ape." We have been called "man the hunter," "woman the gatherer," and even the "selfish ape" (looking out primarily for ourselves and our kin).

However, I will propose a radically different scenario for human evolution. I will develop the theory that, in effect, we invented ourselves through a process that I have dubbed "Neo-Lamarckian Selection." We are uniquely the "inventive ape." Moreover, the many new kinds of synergy that our ancestors invented over the course of perhaps 6 million years played a starring role; we are also, quintessentially, the "synergistic ape." Finally, the Synergism Hypothesis also applies to the explosive rise of complex human societies during the past few thousand years (as described in Chapter 8). Indeed, the mostly unrecognized common denominator in every one of the recent game theory models (so-called) of cultural evolution is synergy. It is synergy that has been responsible for the evolution of cooperation in nature and humankind, not the other way around.

The Perils of Prediction

It is a common misconception that synergy always refers to positive effects; synergy is presumed always to be a good thing. But this is not so. Every day, in a thousand different ways, our lives are shaped, and re-shaped, by synergy. Yet our attitude toward it – our judgment about whether it is a good thing or a bad thing – depends on our values and where we stand (or perhaps which side we're on). In fact, there is a mirror image on the "dark side" for every one of the different categories of positive synergy that I will describe in Chapter 2. I will discuss "negative synergy," or sometimes "dysergy," in some detail in Chapter 4 ("Black Magic"). I will also highlight some special categories of positive and negative synergy – what I call the "Bingo Effect" (when some new combination crystallizes, often unexpectedly), as well as the twin phenomena of "synergy plus one" and especially "synergy minus one." As we shall see, both kinds of disruptions may represent a potentially serious threat to any complex system.

A colleague, the science writer Connie Barlow, has pointed out that the Synergism Hypothesis is more than a hypothesis, or a theory. It also provides a worldview that focuses on the effects produced by the relationships between things. It highlights a fundamental property of the universe and, more relevant for humankind, a fundamental property of human societies. One of the most important implications of this worldview, in fact, is developed in the penultimate chapter, where it is argued that the enduring search for some hidden "law" of history – some deterministic force or mechanism – that will allow us to predict the future course of the "human career" (in anthropologist Richard Klein's term) is fundamentally flawed. The "Neo-Pythagoreans" – as I call them – exclude a priori (by the very nature of their quest for universal "laws") the contingent, historical, synergistic phenomena that have shaped the overall course of the evolutionary process. As a result, these theorists are blind to a major causal agency in evolution. What is required instead, I will argue, is a "science of history."

The implications of this worldview are discussed in the final chapter: "Conjuring the Future: What Can We Predict?" The synergy paradigm provides an answer to this "ultimate question" that is at once challenging, empowering, and threatening. If we should choose to ignore these implications, we will do so at our peril.

2   The "Enchanted Loom"

The brain is waking and with it the mind is returning ...
Swiftly the head-mass becomes an enchanted loom,
where millions of flashing shuttles weave a dissolving pattern ...
a shifting harmony of sub-patterns.

Charles S. Sherrington

Sherrington's famous metaphor for the human mind could be applied to the rest of nature as well. The natural world could be likened to an enchanted loom that weaves a golden tapestry of synergy. For synergy is all around us, and within us; we are completely dependent on it. Yet we often take it for granted or fail to appreciate its *gravitas*, its weightiness. It's profoundly paradoxical. And potentially dangerous. (We'll come back to this point in the final chapter.)

A Golden Tapestry of Synergies

Let's start our survey with something as ordinary as the humble clay brick, one of humankind's oldest and least honored technologies. Without a plan for how to use them and a supply of mortar (itself a synergistic combination of cement, lime, sand, and water), a pile of bricks will be ... well, a pile of bricks. But when the bricks are arranged in precisely ordered patterns and then bonded together they "collaborate" to form a great variety of structures: factories, fireplaces, canals, churches, prisons, watchtowers, garden walls, roads, sidewalks, even kilns for making more bricks.[1]

When the Toronto Maple Leafs recently moved out of their famed Maple Leaf Gardens after 67 years of making hockey history (every game had been sold out since 1949), a sentimental remembrance in a Toronto newspaper noted that the Gardens in its heyday was one of the largest brick buildings in the world. Its 750,000 bricks, if laid end to end, would stretch for 28 miles. Of course, in that case the synergy would disappear; you might then have only the longest line of bricks in the world. And so it is with any building that is demolished. In a fraction of the time that it took to build it, the synergy is gone forever – sometimes to our regret. There is an entire book filled with photographs, drawings, and descriptions of architectural masterpieces by Frank Lloyd Wright, perhaps America's greatest architect, entitled *Lost Wright*.

Synergy is such a commonplace aspect of the way we prepare our food that we don't even notice it most of the time. A particularly tasteful example is a lemon pie – a combination of butter, eggs, lemon juice, sugar, flour, and salt. How do we know it's synergistic? Imagine what would happen if an absent-minded baker switched the prescribed quantities of sugar (one and one-quarter cups in our recipe) and salt (one-eighth teaspoon). Or imagine what would happen if a slightly different combination of ingredients – say butter, egg yolks, lemon juice, salt, pepper, and vinegar – were prepared in a slightly different way. We would call it Hollandaise sauce. In each of these cases, what our taste buds respond to are the synergies – the combined effects.

Written language is also synergistic. The 26 letters of our English alphabet make meaningful words only in precise combinations (with some obvious exceptions). To illustrate, it takes only an alteration in the order of the same combination of letters to magically transform the word "being" to "begin" or "note" to "tone." Or consider "unite" and "untie," two words which have almost opposite meanings. And we can add more levels of synergy by stringing words together into phrases, sentences, paragraphs, etc. We can see how the synergy works by changing the word order in the newspaper headline "dog bites man" to "man bites dog" – the classic journalism school example of what is, and is not, a newsworthy story. (Palindromes like "Madam I'm Adam" are based on the fact that certain meaningful letter combinations will read the same in reverse.)

Note, however, that the synergy in these examples is not located in the words themselves. The words are essentially arbitrary two-dimensional patterns. The synergy is what happens in the reader's mind. The words will evoke no synergy at all for an infant or an adult who can't read English. Moreover, the very same word can produce different synergies in different languages. We all know what the word "gift" means in English. In German it means "poison." Even within the same language, a word can have very different meanings, depending on the context. Take the following sentence: "I am at present present to present a present."

Technology as Synergy

Human technology is also rife with synergy – needless to say. Look at duralumin, a compound of aluminum, copper, manganese, and magnesium that combines the light weight of aluminum with the strength of steel. There is also synergy in the so-called superalloys composed of nickel, cobalt, and various other elements. Superalloys are favored for jet engines and spacecraft because they can resist very high temperatures, high pressures, and oxidation. Then there are the superconductors – crystalline compounds like yttrium–barium–copper oxide, or bismuth–strontium–calcium–copper oxide that, in ways still unknown, allow for the flow of electricity through the material with little or no resistance and at much higher temperatures than had previously been possible (near absolute zero).

"Two Plus Two Equals Five" reads an ad for GE's co-generation equipment. If an industrial plant needs both electricity to power its machinery and steam heat or hot water for various other needs, a co-generation system can do both jobs at once with results that are synergistic. An electrical power plant alone has an efficiency that rarely exceeds 40%. A conventional hot-water heater has an efficiency of about 65%. In both cases, the unused energy goes to waste (entropy). By combining the two processes in one system, energy efficiencies of 95% can be achieved at a much lower overall cost. Co-generation systems typically pay for themselves in three to five years.

Consider also a commonplace consumer product like an automobile. Actually, an automobile is a technological wonder that our not-so-remote ancestors of, say, 200 years ago would surely have marveled at. It represents an assemblage of (depending on the car and how you count) some 15,000–20,000 precisely designed and manufactured parts, comprising some 60 different materials. It also embodies many different technologies, from weaving to glassmaking, metallurgy, ceramics, hydraulics, rubber-vulcanizing, electricity, paints, plastics, and the latest in electronics. And it also incorporates literally thousands of different human inventions: threaded screws, articulated gears, springs, hinges, clamps, cotter pins, bolts, chains, filters, locks, lock washers, Velcro fasteners, ball bearings, fans, pumps, valves, storage batteries, electric motors, and, of course, internal combustion engines.

Furthermore, the synergy produced by these self-propelled machines will occur only if all, or almost all, of the parts work together harmoniously. Most of us are oblivious to how the fuel injectors, timing belts, intake valves, exhaust manifolds, and a plethora of other parts "cooperate"; we notice only a few of the details and pay close attention only when something malfunctions. In fact, one test for the presence of synergy is that the "whole" may not work if a major part is removed or breaks down – a wheel, the alternator, the ignition key, or the driver for that matter. We will make more use of this "synergy test" as we go along. (As an aside, this method of testing for synergy was originally suggested by Aristotle more than 2300 years ago in his classic study of first principles, later renamed the *Metaphysics*.)

An automobile also represents something truly novel in the history of life on Earth, if not in the universe. It is inexplicable in terms of the laws of physics, or the dynamical attractors of chaos theory, or the quarks of quantum theory, or even the laws of thermodynamics – although each of these disciplines has something useful to say about automobiles and how they work. Equally important, you cannot "explain" an automobile simply by listing all of its parts. Nor can you throw all of those parts into a disorganized heap and still get the synergy. It takes a very particular arrangement of the parts to make the magic happen.

We also tend to forget that the synergy produced by an automobile depends on an enormous number of supporting industries and a vast human enterprise involving literally millions of workers, from the manufacturers of oil drilling bits to the producers of electronic sensors and our 500,000 auto mechanics. An automobile is a collective miracle, and there are about 215 million of these complex, synergistic machines – including trucks and buses – currently operating on the streets and highways of the United States alone.[2]

Human organizations also regularly produce synergistic effects. The much-maligned U.S. Postal Service may still lag in customer service, courtesy, and responsiveness to complaints. But consider what it does accomplish. In 1900, the total volume of mail was 7.1 billion pieces for a population of 76 million. Today the Postal Service is handling a mail volume that is 28 times larger for a population that is about four times as large. The total volume in 2000 was more than 200 billion pieces, or about 548 million a day. (The U.S. Postal Service delivers 46% of the world's total letter and card volume. Japan is second with 6%.[3])

To handle this mountain of mail (more than 12 million tons), the Postal Service in 2000 employed 798,000 workers. All told, there are more than

38,000 postal facilities and some 134 million "delivery points" in the United States. And the postal system is there when you need it (its on-time delivery record in 1998 was 93%), because the total cost of supporting it is shared by its many millions of daily users. It represents an immense synergistic effect. How do we know it's synergistic? Just remove, say, the fuel supply from the postal system's huge fleet of some 202,000 vehicles – and imagine the consequences.

Synergies 'R' Us

Synergy is also vital to the workings of the human body, in a myriad of ways. Recall the phenomenon of "tensegrity" in the Prologue; we are completely dependent on this fullerine form of synergy. Another example, among the many that can be found in any biology textbook, involves muscle action. Say you want to lift this book from your lap to eye level and then put it back down again. In order to bend your arm first upward and then downward, there must be a closely coordinated set of actions involving both your biceps (on the front side of your upper arm) and your triceps (on the back side). During the upward movement, the biceps will contract while the triceps relaxes. During the downward movement, the reverse will occur. When you hold your arm steady in mid-air, the two muscles will be in a dynamic tension. The actions of these two "striated" or voluntary muscles are synergistic; by coordinating their actions they produce combined effects that would not otherwise be possible.

The human immune system is another miracle of synergy. The highly evolved system that defends our bodies against the enormous number of potentially harmful microbes in our environment consists of nine different "sub-systems," all told. Perhaps the most impressive part is our "acquired" immune system, actually a widely dispersed network of primary and secondary organs (the thymus and bone marrow, the lymph nodes, spleen, and tonsils) which can orchestrate a highly coordinated defense of the body using an array of specialized cells and molecules that range throughout the body via the lymphatic and circulatory systems. Together these highly specialized parts can do things that none of them could do alone. It's a synergistic system.[4]

The most awesome example of synergy in the biological realm, by far, is Sherrington's "enchanted loom" – the human mind.[5] For starters, the brain is an intricate network consisting of some 100 billion neurons with an estimated 100 trillion connections between them. This means that each axon (the neuron's main output trunk, or cable) has an average of about 1,000 linkages to other neurons, while some giant pyramidal cells may have as many as 20,000 connections – a mind-boggling concept. If all of the microscopic neurons in your brain were laid out end to end, like the bricks in the Maple Leaf Gardens, the path would stretch for several hundred thousand miles – far enough, probably, to reach the moon and back. Of course, in that case your brain would no longer be able to produce its own special kind of magic.

Many aspects of the brain's *modus operandi* – particularly the "binding" process by which the activities of the various parts are integrated into the flow of our conscious experience – still elude us. However, we do know that our brains and nervous systems exhibit an intricate division (and synthesis) of labor among numerous functionally specialized areas. The thalamus, hypothalamus, hippocampus, amygdala, corpus callosum, medulla, cerebellum, motor cortex, and all the rest are interconnected in complex ways. And what we do know affirms that the workings of the brain are synergistic. The various parts of the system are in constant communication with one another through a neuronal network of staggering complexity. It was once thought that the brain's functional elements were localized and could be "mapped," but recent work has shown that the mind more nearly resembles a symphony orchestra than a collection of individual performers.

For instance, what we "see" – the visual images that we continuously sense and process in real time as we go about our daily lives – are actually synergistic effects. They are artifacts of the way our sensory apparatus works and how our brain processes and interprets the inputs that it receives from our photoreceptors. In fact, most other species see things very differently, if at all. Some animals see only in black and white, or infrared, or ultraviolet. Others can detect mostly things in motion and little else. Bats "see" only sonar reverberations. Vipers rely on their heat-sensing capabilities. And many species monitor two different fields of vision simultaneously. We still have only a partial understanding of how our brains perform these routine miracles.

What still lies well beyond our understanding at this point is how the brain is able to do its many-faceted analyses, in real time and at great speed, and then put the pieces back together again into a consciously perceived and meaningful whole – how it produces the synergy. (It appears to involve the parietal lobe and the temporal lobes on either side of the brain.) Even more amazing is how the brain is able to direct the eyes to be discriminating (technically it's called "selective attention") and filter the raw data it receives, millisecond by millisecond. For "seeing" is really an interactive process that relies heavily on past experience and interpretation; images and our "deductions" about their meanings are tightly coupled. Indeed, it has been discovered that the perception of meaningful "wholes" generally precedes our recognition of the "parts," even though our receptors are specialized to collect the raw data in parts.[6]

Among its many tricks, the brain automatically integrates the overlapping but different images from our two eyes, so that ordinarily we "see" a unified picture of the outside world. The brain also fills in our notorious "blind spot," where the optic nerve intrudes into the retina, so that we do not as a rule know it's there. And perhaps most remarkable, when a driver looks into the car's rearview mirror, his/her brain has learned to convert a small mirror image located overhead and about 30° to the right of the car's centerline – or to the left in England – into a picture of what is happening behind the car. It is rather like shooting a movie, processing it, editing it, distributing it, and viewing it all within a fraction of a second. And another new movie is being produced each fraction of a second.

Synergistic Neurons

Even the activity of a single neuron turns out to be incredibly complex. Briefly, the electrochemical "spike" associated with the "firing" of a neuron involves what the Nobel geneticist and neuroscientist Francis Crick characterizes as a "complex dynamic sum" of both excitatory and inhibitory inputs from all of the many other neurons with which it is in contact via the "synapses" (or junctions) between its own cell body and the axon endings, or "knobs," of its neighboring neurons. A neuronal signal is a synergistic effect.

Moreover, the transmission process within each neuron is, in Crick's words, a "chemical miracle." It is not at all like electricity flowing through a wire but an intricate electrochemical dance that involves balance shifts in the concentrations of sodium and potassium ions inside and immediately outside the membrane of the neuron. This process is controlled by an elaborate system of molecular "gates" and metabolic "pumps" that move the ions through the neuron's membrane in such a way as to propagate a localized (and quickly reversed) change in the internal ion balance (a change in polarity measured in millivolts) that is called an "action potential." It is this localized shift in the concentration of ions that cascades through the axon in a fraction of a millisecond.[7]

Unbeknownst to you, all of this and much more is happening inside your brain many millions of times each second. As Crick concludes: "A neuron, then, is tantalizingly simple. . . . It is only when we try to figure out exactly how it responds . . . that we are overwhelmed by the inherent complexity of its behavior. . . . All this shows, if nothing else, that we cannot just consider one neuron at a time. It is the combined effect of many neurons that we have to consider."[8] Although Crick doesn't use the term, he is talking about synergy.

It should be evident that synergy is not just related to corporate mergers, or drug interactions. Synergistic effects are ubiquitous. They are of central importance to such fundamental sciences as physics, chemistry, and biology. They are deeply embedded in the phenomena studied by economists, psychologists, and sociologists. They are the absorbing preoccupation – and stock-in-trade – of engineers, architects, business entrepreneurs, clothing designers, writers, and artists, as we shall see.

The synergy principle also plays a key part in such familiar institutions as airlines, banks, hospitals, libraries, newspapers, railroads, telephone systems, and much more. It is utilized in our poetry, music, movies, painting, and photography. It is widely employed in politics and government: in legislatures, political parties, reform movements, and revolutionary conspiracies. It affects how we organize our businesses, how we design our homes, how we plan and cultivate our gardens, how we decorate the interiors of our homes, and how we dress ourselves. It is even a major facet of our sports, our military, and our family relationships.

Almost daily I see media reports of new, or newly documented, forms of synergy. Here are just a few recent examples:[9]

* The brand new Akashi Kaikyo bridge in Japan, the world's longest suspension span, is supported by two immense cables, each composed of 290 strands. Each strand, in turn, contains 127 massive, high-strength steel wires, anchored in such a way that each one carries an equal share of the total load. Another way of putting it is that the weight of the bridge is divided into 73,660 equal parts.
* Carlsbad, California, boasts a "Legoland" theme park, including a scaled-down version of the New York City skyline, constructed from 20 million Lego blocks.

* Literally millions of desktop computer owners are being yoked together these days into remote networks that provide researchers with
the equivalent of massive supercomputer power. The computers are
commandeered at times when they would otherwise be idling with
screen savers. The networks are utilized for scientific data-crunching
projects that would otherwise be impossible, or prohibitively expensive.
SETI@Home, for instance, has obtained 1 million years of computer time
from some 3.5 million computers over the last three years, enabling astronomers to sift through enormous quantities of astronomical data for
hints of extraterrestrial intelligence.
* An 11-minute TV cartoon may appear to the viewer as an unbroken flow
of action, but in fact it consists of almost 6,000 separate pictures that
enter your brain at the rate of some 16 images per second.
* "Process engineering" has been transforming various industries by inte-
grating the information stream, control and feedback processes among
the various elements, from raw materials suppliers to parts subcontrac-
tors, manufacturing units, sales and marketing departments, distribution
services, and point-of-sale inventory and sales activities. Indeed, some
companies, like Dell Computers, have eliminated the middleman altogether. Dell allows individual purchasers to place customized orders by
phone or on the Internet. Each computer is then built to the individual's
specifications and is shipped directly from the factory to the purchaser.
Dell's prices are aggressively low, yet its profit margins are enviable.
* The Toyota Prius and Honda Insight are "parallel hybrids," autos that
utilize two different power trains in a collaborative, synergistic relationship. These cars have much smaller, more efficient gas engines than
usual, because they are assisted by electric power from a relatively small,
efficient battery pack that can be quickly recharged by the engine. The
Prius can also use some of the car's braking energy to help recharge the
battery. The EPA's gas mileage rating for the Prius is a stunning 52 miles
per gallon in city driving.

To reiterate, the word synergy means "working together" or, literally,
"cooperating" – though we are more interested here in the functional *effects* produced by cooperation. To reiterate, synergistic effects need not be
"more" than what the parts can produce alone, only different. Look at
some of the English words that have the prefix "syn," meaning "with,"
"together," or "at the same time" – syndicate, synagogue, synthesis, syn-
dicalism, synchronize, synthetic, syntax, synod, syngamy, and synoecism.

There are also a number of words that have the equivalent prefix "sym" – symmetry, symphony, symbiosis, sympathy, symposium, symptoms, and others. These syn/sym words are indicative of synergy's many dimensions.

The Varieties of Synergy

All of nature and every complex human society – indeed, every one of us – represents a multi-leveled structure of synergistic effects. We are surrounded by it. It's inescapable. Let's briefly survey a few of the many different kinds of synergy. We will be referring to them frequently in the chapters that follow.

SYNERGIES OF SCALE Many forms of synergy arise from adding (or multiplying) more of the same thing. A bigger molecule, a bigger organism, a bigger group, or a bigger organization may be able to do things that smaller ones cannot. And when it comes to competition between groups, the odds are that the bigger unit will prevail. God may not always be on the side of the big battalions, as Voltaire claimed, but it certainly helps. I call it a "synergy of scale." Economists prefer the term "economies of scale," of course, but some synergies are not more economical or cost-efficient, just more potent.

This point was demonstrated in a famous set of experiments many years ago by the well-known biologist of the 1930s and 1940s, Warder C. Allee. Allee showed that a large number of flatworms (planaria) could collectively detoxify a solution that would otherwise be lethal to any one of them alone. When a single planarian was placed in a silver-rich solution, its head began to degenerate within 10 hours. But when groups of 10 or more planaria were put into the same solution at once, they were able to survive without ill effects. Each planarian absorbed a portion of the silver colloid, and their collective efforts reduced the ambient silver concentration below the toxic level.[10]

There are many other synergies of scale in nature. Large colonies of the predatory bacterium *Myxococcus xanthus* are jointly able to engulf much larger prey than any one or a few of them could do and, equally important, are collectively able to secrete digestive enzymes in concentrations that would otherwise be dissipated in the surrounding medium.[11] Similarly, large coalitions of male lions are generally able to take over and hold a pride of females against smaller groups, or individual males. And in the Japanese ant *Colobopsis nipponicus*, there is a caste of super-sized workers that can ward off raiders by plugging up the colony's entrances with their heads.[12]

Synergies of scale also occur with bee stings. We know that a single bee sting is not likely to be fatal to a human, except for those rare individuals who are highly allergic. But 100 bee stings can be deadly even to those who are not prone to toxic shock.

One of the most dramatic examples of a synergy of scale in nature, perhaps, is the strategy used by Ridley's sea turtles for reducing the loss of their eggs to avian predators. When the turtles come ashore to lay their eggs on the beaches of Costa Rica and elsewhere in Central America, they do so in a synchronized action called an arribada. As many as 400,000 of them will storm the beaches together at the rate of about 5,000 per hour and may lay 40 million eggs in the sand within a few days, far more than their predators can consume before the eggs are hatched.

There are so many synergies of scale in human societies that one hardly knows where to begin – or end. One that economists and planners must deal with every day goes under the heading of "aggregating demand." A fundamental problem in designing any public facility – a mass transit system, a parking garage, a restaurant, a private club – is how to generate a sufficient number of users to support the system. Likewise, the "limited partnerships" that create pools of capital for various private development projects often depend on many individual investors (sometimes hundreds or thousands) who purchase "shares" in the venture. But perhaps the ultimate synergy of scale in our society is the TV audience for a Superbowl. With some 130 million domestic viewers alone watching the 2001 Superbowl game, the network could command $2.3 million per minute for a commercial, more than many of us will earn in our lifetime.

Synergies of scale can be greatly accelerated with multiplicative processes (exponential growth). An elegant example is the famous riddle used with French schoolchildren. Suppose you have a pond on which a waterlily is growing, and the plant doubles in size each day. If it is allowed to grow unchecked, the lily pond will be completely covered in 30 days, choking off all other forms of life in the pond. If you decide to cut the lily back when it covers one-half of the pond, on what day will that be? The answer, of course, is the twenty-ninth day. You will have only one day left to save the pond.

The lily pond is often invoked as a metaphor for our population problem. As the Reverend Thomas Malthus, one of the pioneer economists, put it in his famous monograph, *An Essay on the Principle of Population*

(1798): "Population, when unchecked ... increases in geometrical ratio ... [while] the means of subsistence, under circumstances the most favourable to industry, could not possibly increase faster than in an arithmetic ratio." Accordingly, Malthus envisioned relentless population pressures that could only be checked by "the ruthless agencies of hunger and poverty, vice and crime, pestilence and famine, revolution and war."[13] (Perhaps it was Malthus's gloomy vision that inspired the writer Thomas Carlyle to characterize economics the "dismal science.")

THRESHOLD EFFECTS These are special cases of a synergy of scale; I call it "synergy plus one." Threshold effects occur when a critical point is reached that precipitates an abrupt change of state. A familiar example is the old saw about "the straw that broke the camel's back." A more up-to-date variation on this theme is the children's story by Pamela Allen, *Who Sank the Boat?* It was not the mouse that caused the rowboat to sink, of course, but the combined weight of all the animals in the boat – their synergy. Another commonplace example is a tug-of-war. The two sides may be perfectly matched, but if you add one more player to either side (synergy plus one) the war may soon be over.

Threshold effects may also travel incognito, disguised as a "critical mass," an "optimum number," and (in ecological circles) "density dependence" and "frequency dependence." In fact, we are surrounded by threshold effects: rush-hour traffic tie-ups, overloaded telephone circuits, break-even business revenues, "standing-room only" events, and many others. Economic markets also provide innumerable examples. When the oil-producing countries banded together and agreed to reduce daily oil production by a mere 3%, this was enough to change the demand–supply balance (as the economists put it) and boost oil prices by more than 20% initially, and much more later on.

Nature is laced with such threshold effects. Indeed, natural selection is often produced by the aggregate effects of a population, not individual characteristics alone. For example, many species respond to population increases beyond a critical level with various adaptive responses – emigration, reduced fertility, inhibited development, even infanticide and cannibalism. A subtle variation on this theme is the phenomenon of brood parasitism in birds. Some 80 species of birds, including especially various cuckoos and cowbirds, make it a regular practice to lay their eggs in the nests of other species. However, it has been shown that the effectiveness of this strategy is very much influenced by such variables as the number of nest sites available, the relative number of eggs laid by the parasite birds and their "hosts" and even the number of other parasites that the hosts may be subject to (like botflies).[14]

In short, threshold effects are typically context-sensitive. This was illustrated by biologist Garrett Hardin some years ago in his classic (and much-reprinted) *Science* article, "The Tragedy of the Commons."[15] In the Middle Ages in England, many villages provided a "common" pasture where local farmers could graze their excess cattle, much like public parks today provide common recreational areas for humans. The problem was that there were no limits set and no charges to the user. So every farmer had an incentive to use the common pasture as much as possible. The inevitable – synergistic – result was frequent overgrazing and destruction of the ground cover when the "carrying capacity" of the pasture was exceeded. The solution, of course, was to enclose the commons and regulate its use, a solution we are now applying for similar reasons to such overcrowded national parks as Yosemite, Glacier, and the Grand Tetons.

PHASE TRANSITIONS These are related to threshold phenomena. They involve an abrupt and radical change of state in many physical and biological systems under certain conditions. Physicists often use as examples the crystallization of water into ice, the loss of magnetic properties in ferromagnets at extremely high temperatures, or the onset of superconductivity in various materials at extremely low temperatures.

The physicist Herman Haken, who has spent more than 20 years developing a science of cooperative phenomena called "synergetics," likes to use the laser as an illustration.[16] Lasers have a magical quality. Although they employ the energy in light photons, lasers can do many things that ordinary light cannot. In industry, lasers are used to cut metal, weld joints, burn holes in diamonds, provide carriers for communications signals, and serve as optical scanners in supermarkets, among many other uses. And in medicine their uses range from delicate eye surgery to the precision zapping of tumors with minimal damage to the surrounding tissue.

How do lasers do it? Ordinary light is chaotic (incoherent), both in terms of its directionality and its mix of frequencies. Laser magic arises when great quantities of light photons of the same frequency are generated through "stimulated emission" – synergistic interactions between excited atoms in a light-amplifying substance like a ruby. The energized atoms are then marshaled with mirrors into a tightly focused directional beam. The result is a very concentrated, intense energy stream with little or no interference (energy-dissipation) between the photons. In this highly "co-herent" form, the photons are able to produce powerful combined effects that would otherwise be impossible. A common ruby laser, for instance, can create surface temperatures of about 5,500° Celsius (10,000° Fahrenheit) and can burn a hole through a 1/16-inch sheet of steel with a single burst.

Many phase transitions are simulated these days with computer models, using non-linear mathematics. The so-called "chaos" models, for instance, illuminate how the interactions in unpredictable, disordered processes may collectively lead to a stable state – a dynamical attractor. Conversely, the phenomenon of "self-organized criticality" associated with physicist Per Bak and others involves a "global dynamic" – a synergistic process in which a system may self-evolve to a critical state, which can lead to a "catastrophic" change. Bak's favorite example is a sand pile.[17] (In biology, phase transitions often have a functional basis. They can be observed in such phenomena as seed germination, seasonal changes in deciduous trees, metamorphosis, and the birthing process, although there are innumerable examples at the cellular level as well.[18])

GESTALT EFFECTS This term derives from a branch of psychology known as "Gestalt theory." Founded in Germany before World War One, Gestalt psychology is concerned with the ability of the human mind to see patterns, relationships, or "wholes" composed of many parts. Our Gestalt capabilities are especially apparent when some of the parts in a visual pattern are missing or garbled. For instance, anyone who can read this book should be able to recognize the term "m_ddle-class" and should know that the missing letter is different from the one that is missing in the expression "m_ddle-headed." I had a personal Gestalt experience in my car recently when I approached a road construction site and saw ahead of me a large flashing sign that directed the traffic to change lanes. Two of the twelve lights that composed the arrow were burned out, but I could still correctly interpret the meaning of the sign. (Our Gestalt capabilities are also regularly put to the test by cartoonists, poor spellers and, these days, our e-mail correspondents.) The Gestalt theorists have also identified certain "laws" of visual perception (proximity, similarity, good continuation, and closure).[19]

The term "Gestalt effects" can also be used in a broader sense, though. It could refer to any synergistic effect that arises from the pattern of physical/spatial relationships among different parts – its form or structure. Recall how the synergy would disappear if the bricks in the Maple Leaf Gardens, or the neurons in the human brain, or the parts of an automobile were laid out end-to-end. Or imagine the result if the many miles of thread in the clothes you are (presumably) wearing were unraveled and laid out in a single line. Or suppose you were asked to use a bath towel that had been unraveled.

An especially illuminating example was provided by biologist John H. Campbell. If you mix charcoal, sulfur, and saltpeter in a ratio of 2:3:15 and sprinkle it on the ground, it will burn with a sputter. But if you pack it into a cardboard container, it will make a firecracker. And if it is packed inside a hollow cylindrical steel casing together with a metal projectile and an opening at one end, it will become a cannon. Another example comes from the hot new field of nanotechnology, where the synergistic combination of super-strong chemical bonds and a honeycomb architecture yields a substance (nanotubes) that is stronger than steel.

These and many other forms of synergy are dependent on how the parts are physically ordered. Indeed, artists, landscape architects, interior decorators, and many other professional designers daily create Gestalt effects – patterned arrangements of various elements or parts that produce synergistic visual, spatial, or even auditory effects.

FUNCTIONAL COMPLEMENTARITIES The water molecules that collectively produce the Mississippi River are all identical. So (more or less) are the grains of sand that make up a beach, or the vesicles of glutamate that trigger a neuronal spike. But many other forms of synergy depend on different properties or capabilities that join forces to give the combination new functional characteristics. Velcro fasteners are one example. Their remarkable gripping ability depends on the interaction between the tiny hooks and loops on the two opposing strips. Another example is the tacit partnership between computer hardware and software. IBM, one of the few companies that develops and manufactures both components, has been able to exploit this form of synergy in various ways.[20] Likewise, aspirin and opium have analgesic properties in combination that exceed the sum of their separate effects, for reasons that are now much better understood.[21] Still other examples were mentioned in Chapter 1: the combined action of atropine and predisone, lichen symbioses, and the complementary amino acid contents of beans and corn. And when bricks are combined with mortar, the two can do things together that neither one can do alone. How do we know that these complementarities are synergistic? Just apply the "synergy test."

EMERGENT PHENOMENA The term emergence has a venerable history. It dates to a school of evolutionary theorists in the late nineteenth and early twentieth centuries (especially G. H. Lewes and Conwy Lloyd Morgan) who focused on what they called the "qualitative novelties" produced by things of "unlike kind."[22] Their preeminent example was the human mind. Today, unfortunately, the term emergence is used in a bewildering variety of ways, often as a synonym for synergy. However, I side with the early theorists; emergence should properly be confined to those forms of synergy in which different parts merge, lose their identity, and take on new physical or functional properties. Thus table salt has emergent properties. Chrome–nickel–steel also has emergent properties. And so do the color combinations that are produced by melding the primary color pigments in various ways. However, a golf putting green composed of a hundred thousand blades of close-cropped Bent grass is not an emergent phenomenon. Nor is a sandpile, or a rain puddle, or a lake. The human body, on the other hand, could be said to be an emergent phenomenon. Our many trillions of cells are interdependent and form a unified "whole" of many parts that produces combined, synergistic effects. And so do the 20,000 parts in an automobile.

One problem with using the term emergence as a general-purpose synonym for synergy is that it began life as a word that was equivalent to "appearance," "flowering," "growth," or "coming into view." One of the dictionaries I consulted illustrates the word emergence with "the sun emerged from behind a cloud." And once a week an on-line journal search service provides me with about two dozen references under the keyword "emergence," almost all of which refer to such subjects as the emergence of some cult, the emergence of Mexico's democracy, the emergence of the euro, the emergence of cooperation among schools, the emergence of the Internet, the emergence of the private sector in the Russian economy, and many more. Such semantic vagary invites confusion. A more serious objection, though, is that all forms of emergent phenomena involve synergy, but there are many forms of synergy that do not have emergent properties, by any reasonable definition. Accordingly, the term should be confined to organized "wholes" composed of functionally distinct "parts" that produce irreducible combined effects – like the human mind.

AUGMENTATION OR FACILITATION This involves combined, synergistic effects that enhance a dynamic process, or in some cases make it possible. One example is catalysts, substances that decrease the activation energy required for various chemical reactions while themselves remaining unchanged. The enzymes that serve as catalysts in biochemical processes are especially important. (We will have more to say about catalysts in Chapter 3.) Another example is the enhancement of mutation rates by exposing an organism to a combination of gamma rays and metallic salts.

An important biological example involves hemoglobin, one of the most vital molecules in our body. Hemoglobin is a complex protein whose four distinct strands (or monomers) enfold a ring-shaped "heme group" containing an iron atom. There is a functional partnership (a complementarity) between the scaffolding provided by each monomer, the so-called "globin chain," and its "heme group." Together the two distinct parts are able to reversibly bind and then transport molecules of oxygen, carbon dioxide, and (we have recently learned) nitric oxide, all of which are involved in the workings of complex biological systems. However, it also happens that the four globin chains are linked together and have "allosteric" properties; when any one chain binds, say an oxygen molecule, it increases the binding affinity of the other three, so that each hemoglobin molecule will be more likely to load up and transport four oxygen molecules on each trip through the bloodstream. Our lives literally depend on such synergies.[23]

JOINT ENVIRONMENTAL CONDITIONING Through joint action individual organisms can often achieve significant economies, or efficiencies that would not otherwise be possible. One of my favorite examples is the emperor penguin. During the brutally cold Antarctic winter, when temperatures can fall to $-15.5°$ Celsius ($-60°$ Fahrenheit) and the winds can reach hurricane force, the hardy penguins that live in this desolate, snow-swept environment huddle together in tightly packed colonies, sometimes numbering 10,000 or more, for several months at a time. By doing so, they are able to share precious body heat, which would otherwise go to waste. Equally important, they also provide insulation for one another. A scientist who studied them extensively, Yvonne Le Maho, found that the huddling behavior reduces the penguins' overall energy expenditures by 20–50%.[24]

There are many variations on this theme in the natural world. When elephant seals come ashore on the California coast each spring to reproduce, the males engage in sometimes bloody contests for dominance and preferential (often exclusive) mating privileges with the "harems" of females. But once the fighting is over, the losers huddle together in peaceable "rookeries" to facilitate mutual defense and, again, to provide insulation for one another and share heat.[25]

Temperature regulation in honeybee hives is especially impressive. Honeybee workers routinely share body heat or join forces to engage in fanning activities, as the need arises, in an effort to keep the core temperature of the hive within about ±1° Celsius (1.8° Fahrenheit).[26] Mexican desert spiders utilize a very different kind of joint environmental conditioning. By clustering together in the thousands during the heat of the dry season, they are able to reduce individual water loss and thus decrease thus risk of dehydration.[27]

Humans also utilize this form of synergy, needless to say. As the quote from Ecclesiastes on the frontispiece reminds us, heat-sharing has been practiced by humans for many centuries. And a recent newspaper photograph confirmed that heat-sharing is still a common practice in many parts of the world. In the photograph, a dozen Russian soldiers in the field during the winter months were shown huddling tightly together for warmth while they slept inside a tent.

Of course, animals often collectively re-shape their environments in more active ways. Nests, dams, underground burrows, prepared sleeping sites, even woodland animal trails may be the product of joint efforts. And the same is true in humankind, needless to say. From the communal fires and crude shelters of our remote ancestors to the collectively irrigated fields and rice paddies of contemporary agriculturalists and the well-worn footpaths that exist even today in many rural and woodlands areas, we often shape our environments together to suit our needs. Indeed, many of our most important modern technologies – from furnaces to air conditioners – are designed to assist us with a vital basic need: thermoregulation.

Unfortunately, we also collaborate in producing negative environmental impacts, often without meaning to do so. Although none of us, presumably, wants global warming to occur, most of us are unintentionally contributing to it (and many other forms of environmental damage) as we use the goods and services provided by a modern society. (More on negative synergy in Chapter 4.)

RISK- AND COST-SHARING One of the pillars of social life, in both nature and human societies, is the ability to "economize" by sharing with others the costs and risks inherent in living. There are innumerable examples in the natural world: fish schools, migratory bird formations, synchronized breeding, joint nest-building, collective foraging, and many more. Thus many birds and some animals divide up the job of lookout duty; they take turns scanning the environment for potential predators. Other species benefit from having "helpers at the nest," or surrogate "allomothers," or "baby sitters."[28]

One of the most dramatic (and well-documented) examples of co-operative risk-reduction in nature involves vampire bats, which subsist entirely on animal blood (though not human blood, as a rule). In a classic set of studies over a ten-year period, biologist Gerald Wilkinson found that, when vampire bats return to their communal nests from a successful night's foraging, they frequently regurgitate blood and share it with other nest-mates, including even non-relatives.[29]

The reason, it turns out, is that blood-sharing greatly improves each bat's chances of survival. A bat that fails to feed for two nights is likely to die. (No wonder they're bloodthirsty.) Wilkinson showed that the blood donors are typically sharing their surpluses and, in so doing, are saving unsuccessful foragers that are close to starvation. So the costs are relatively low and the benefits are relatively high. Since no bat can be certain of success on any given night, it is likely that the donor will itself eventually need help from some nest-mate. In effect, the vampire bats have created a kind of mutual insurance system.

Risk-sharing is also a widespread practice in human societies, needless to say, and it represents a potent source of synergy. Our highly institution-alized insurance industry provides a prime example. As Winston Churchill put it, insurance brings "the magic of averages to the rescue of millions." It is a concept that dates back to the so-called "funeral societies" of ancient Greece. Since it is obviously not possible to predict exactly when a per-son will die, the people of Periclean Athens conceived the idea of banding together and making small annual contributions to cover the funeral ex-penses for whoever happened to die during the succeeding year, rather than requiring each person to be "self-insured" for the entire amount. Today, we use the same basic principle to insure ourselves at relatively low cost against every conceivable risk, from earthquakes to dental cavities. Indeed, the "futures" markets were developed to insure investors against such di-verse economic risks as weather, commodity prices, monetary exchange rates, and a host of other things.

Cost-sharing represents an even more important form of synergy in a complex modern economy. As we learned in Economics 101, almost every business that sells goods and services to the general public distributes its "fixed costs" and "overhead costs" among its customers or clients (with some exceptions). This has the effect of spreading, reducing, and equalizing the proportion that any one customer must bear. A variation on this theme occurred recently when AT&T lined up a $40 billion loan, which was shared among 39 different banks.

Another form of cost-sharing involves "shared-use" goods and services that are not sold outright but are "rented" or leased for varying periods of time. Use-sharing is so pervasive in modern economies that we don't think of it as being synergistic – a type of tacit cooperation that is mutually advantageous. Yet it represents an important way of achieving efficiencies and reducing individual user costs. In fact, many of our most visible and successful businesses are based on the shared-use principle: hotels, rental cars, movies, restaurants, ski resorts, airlines, cruise liners, shared office suites, time-share condos, Disney World, and many more. Although we seldom stop to think about it, we are implicitly cooperating with other users of these services when we pay for and use a proportionate share of the total available time (or space) and then promptly vacate the premises to make room for others.

A COMBINATION OF LABOR One of the most important sources of synergy – in nature and human societies alike – involves what the economists call a "division of labor." Plato was perhaps the first social theorist to appreciate that synergy lies at the very foundation of human societies; the division of labor produces mutually beneficial results because different people have different aptitudes, and specialization increases a person's skill and efficiency. In his great philosophical dialogue, the *Republic*, Plato wrote: "Things are produced more plentifully and easily and of better quality when one man does one thing which is natural to him and does it in the right way, and leaves other things."[30]

The classical economist Adam Smith, in *The Wealth of Nations*, provided us with one of the textbook examples. At a pin factory that Smith had personally observed, ten workers performing ten different tasks were able to manufacture about 48,000 pins per day. But if each of the laborers were to work alone, attempting to perform all of the tasks associated with making pins rather than working cooperatively, Smith doubted that on any given day they would be able to produce even a single pin per man.[31]

The writers of modern-day economics textbooks are fond of using Adam Smith's pin factory as an illustration of the division of labor, but this characterization downplays the synergy. Another way of looking at the pin factory is in terms of how various specialized skills, tools, and production operations were *combined* into an organized "system." It should really be called a "combination of labor." The system included not only the roles played by each of the workers, which had to be precisely coordinated, but also the appropriate machinery, energy to run the machinery, sources of raw materials, a supporting transportation system, and (not least) markets where the pins could be sold to recover production costs.

Not only that but the pin factory required "management" – one or more persons responsible for hiring and training workers, for planning, for production decisions, for marketing and selling the pins, for payroll and book-keeping, and so forth. In other words, the economic benefits (the synergies) realized by the pin factory were the result of the total system, including a complex network of production tasks and cooperative relationships. How do we know it was synergistic? Just imagine what might happen if one of the key pin-making machines broke down, or if its highly skilled operator called in sick.

The division/combination of labor is also widespread in nature. There are, for instance, the orb-web spiders that collaborate in building immense collective webs to span the woodland streams where their insect prey are especially abundant. There are also the many carnivores that engage in collaborative hunting behavior – wild dogs, wolves, chimpanzees, lions, and others. In a careful study of 486 group-hunting episodes by a pride of lions in Namibia, ethologist Paul Stander identified a clear-cut division of labor. (Stander's terms, "wings" and "centres," "drivers" and "catchers," were suggestive of soccer players.) Stander also observed consistent differences among the animals in terms of their preferred stalking roles.[32]

Social insects are also renowned for exploiting the division/combination of labor, and recent work by biologists Nigel Franks, Carl Anderson, and others has established that the partitioning of the work effort in insect colonies is far more elaborate than had previously been supposed. There are, in fact many instances of closely coordinated "team-work" related to various tasks and subtasks. For instance, in colonies of *Pheidole pallidula*, worker ants will coordinate their efforts to pin down or immobilize an intruder ant, while a larger "major" is called upon to be the executioner. Similarly, in the termite *Hodotermes mossambicus*, one group of workers will climb grass stems and cut off pieces, which drop to the ground below, while a second group is responsible for hauling the booty back to the nest.[33]

One of the most impressive practitioners of a division/combination of labor in nature, though, is *Eciton burchelli*, a species of army ants found in Central and South America.[34] These creatures form highly organized colonies of about 500,000 members, with four distinct "castes" that divide up the responsibilities for colony defense, foraging, transport, nest-making, and care of the brood. The big "sub-majors" (or porters), for instance, team up to carry sometimes very large prey which, if split up into pieces, would be more than each individual sub-major could carry alone.

The army ants' highly coordinated foraging system is also legendary. In a single day, a raiding party of up to 200,000 workers – armed with potent stingers and marching in a dense phalanx – might reap some 30,000 prey items, many of which are then split up and hauled back to the nest for all to share. Because they forage *en masse*, army ants can also collectively subdue much larger prey than would otherwise be possible – even lizards, snakes, and nestling birds. It's the ultimate sting operation, and another impressive example of a synergy of scale.

Perhaps the most remarkable form of synergy in army ants, though, involves the way the colony builds its nests. The workers form the nest out of many thousands of their own interlinked bodies. Not only are these living nests quick and efficient to construct but, most impressive, they are able to maintain a constant internal nest temperature that varies no more than ±1° Celsius. (These nests are also ideally suited for a tropical species that must frequently relocate its home base when the local food supply is exhausted.)

ANIMAL–TOOL "SYMBIOSIS" The many functional relationships that exist in nature between organisms and various "tools" amount to a form of symbiosis. The animal–tool relationship produces otherwise unattainable synergistic effects, many of which spell the difference between life and death. Thus, some birds use rocks to break open egg shells while others deploy thorns to dig for grubs under the bark of trees. Some chimpanzees use "wands" to fish for buried insects while others use stone anvils and hammers to crack open the proverbial tough nuts. California sea otters are legendary for using rocks that rest on their bellies while they float on their backs as a "tool" for breaking open mussels and other hard-shelled prey. I've seen it done. Elephants are especially impressive tool-users. Among other things, they scratch or clean their ear cavities with grass or other vegetation; they scratch their bodies with sticks held in their trunks; they wipe cuts with clumps of grass held in their trunks; they reach toward inaccessible food and hit humans with sticks held in their trunks; they throw objects at other animals with great accuracy, including humans and their vehicles; they brandish or wave branches, apparently to chase away flies or to threaten other animals; they lay mats of grass over their backs to keep biting flies away; they may pile up branches or push a large tree down onto a fence, forcing it to sag so that they can walk over it; they even use sticks and stones for making spontaneous "drawings" in the dirt.[35]

In short, animal–tool symbiosis is widespread in nature, and the difference between humans and other tool-using species, as Darwin noted, is a matter of degree; there is no difference in kind. Moreover, the synergies that result from these symbiotic relationships have helped to shape the very course of biological evolution. We will return to this subject in Chapters 3 and 6 and, especially, when we consider human evolution in Chapter 7.

INFORMATION SHARING AND COLLECTIVE INTELLIGENCE Information sharing is one of the more common forms of synergy, both in nature and in human societies. Indeed, all socially organized species absolutely depend on it. Very often it is a service that can be provided to others at no cost to the possessor, or at an incremental additional cost, while the benefits can be multiplied many times over. The benefits can range from more efficient foraging efforts to life itself. Although there is a vast and growing research literature on the subject, some of the underlying principles were first developed by Darwin himself in one of his lesser known but now more widely appreciated books, *The Expression of the Emotions in Man and Animals* (1872).

Information can take many different forms, from the chemical trails laid down by bacteria and social insects to the vocal signals of many mammals and birds and the visual displays of foraging honeybees. (Indeed, the honeybees' famous "waggle dance" turns out to be loaded with subtly variable information.) In socially organized species, alarm calling is especially well documented, and there is also considerable support for the famous "information center" hypothesis of Peter Ward and Amotz Zahavi; communal nesting may provide a venue for gathering and sharing information about widely scattered, patchy food sources – though there are obviously other important reasons for joint nesting as well.[36] By the same token, much of the vocalization that occurs in various species of birds, carnivores, dolphins, and primates relates to making group-level decisions about migration and foraging.[37]

Information sharing in nature often involves imitation, or undirected "cultural" transmission. For instance, young songbirds in many species must learn all or part of their distinctive song pattern from a parent. Many other species learn what is edible, or dangerous, from observing experienced animals. This was demonstrated in a classic 1970s experiment. Ethologist

Eberhard Curio and his colleagues were able to trick naive young blackbirds into mobbing a small stuffed Australian honeyeater after they observed the illusion (contrived with mirrors) that adult birds were doing so. (The adult "teachers" were in fact reacting to an unseen decoy predator – a stuffed owl.) Likewise, Japanese rhesus macaques learned to wash potatoes before eating them by observing the monkey genius (Imo) that invented the technique, and Israeli black rat pups learn from their mothers how to strip pine cones to get at the edible seeds inside.[38]

As suggested earlier, animals not only share information with each other, they frequently engage in pooling "data" and making collective decisions. This too represents a form of synergy. Indeed, some of the most remarkable examples of joint decision making in nature can be found in bacterial colonies, such as myxobacteria, that migrate and forage for food cooperatively and form various joint structures, as the occasion demands, including specialized reproductive "fruiting bodies." All this is accomplished via an array of individual chemical signals that produce combined, threshold effects.[39]

Social insects also benefit from what has been called "swarm intelligence." For instance, entomologist Nigel Franks has shown that army ant colonies use information to engage in joint problem-solving behaviors regarding such things as the allocation of effort to various tasks, which produces synergistic effects for the colony as a whole.[40] Honeybees are even more impressive in this regard, thanks in part to their unique forms of symbolic communications. This was documented in detail by entomologist Thomas Seeley in his classic book *The Wisdom of the Hive* (1995). In a recent review article on the subject, Seeley and Royce Levien wrote: "It is not too much to say that a bee colony is capable of cognition, in much the same sense that a human being is. The colony gathers and continually updates diverse information about its surroundings, combines this with information about its internal state, and makes decisions that reconcile its well-being with its environment... What is remarkable about this collective intelligence is that it arises from fundamentally decentralized information processing." The authors, inspired by Adam Smith's famous metaphor for economic markets, call it an "invisible brain."[41]

CONVERGENT (HISTORICAL) EFFECTS Last, and perhaps least appreciated, is one of the most pervasive and important forms of synergy in nature and human societies alike – the daily assault of fortuitous, often unexpected convergent effects that shape the evolutionary process. Here synergy and history join hands. Many of the synergies that surround us and impact upon our daily lives are unplanned, causally unconnected and highly context-dependent. They may be partly a product of the laws and principles of physics and chemistry, but they also entail many contingent historical influences that combine to produce important functional consequences. (The late Stephen Jay Gould was a crusader for this idea; we will return to it in Chapter 9.)

One example, among many, might serve to illustrate this point. Jared Diamond, in his landmark study, *Guns, Germs and Steel* (1997), takes up the challenge of accounting for the rise of large complex civilizations in humankind over the past 13,000 years or so – not simply the reasons why this trend occurred but also why it happened where and when it did and why it did not happen elsewhere, or at other times. A key aspect of Diamond's approach, one that directly contradicts some of the deepest metatheoretical assumptions of the sciences, is that it is not possible to explain such fundamentally historical phenomena in terms of some context-free, deterministic (law-like) "mechanism." Context-dependent factors have played a crucial role in the process. Each major "breakthrough" in the evolution of complex human societies, as well as each "replication" of the process in some other geographic "venue," was the result of a site-specific nexus – a convergence of many "ultimate" and "proximate" factors (terms Diamond borrows from evolutionary biology but uses in a different sense). Diamond does not use the term synergy. He refers instead to a "package" of contributing factors. But the meaning is the same; each instantiation involved a combination of necessary and sufficient elements that worked together.

In the agricultural revolution, the development of food production and the resulting food surpluses was a key factor, Diamond argues, but this in turn depended on many other factors. One important precursor was the prior emergence of anatomically modern humans, including our language skills and sophisticated cultural resources, perhaps 100,000–150,000 years ago. Another factor was the decline and mass extinction of many of the large "megafauna" upon which evolving humans had come to depend, coupled with an increase in human population levels. This demand–supply imbalance created a growing pressure to find suitable supplements to the standard hunter–gatherer diet. The fortuitous co-location only in the Fertile Crescent of key "founder crops," especially emmer wheat (which could be domesticated with a single gene mutation), together with legumes and animal husbandry (which allowed for a balanced diet), meant that this was the most likely location for a "technological breakthrough" that could provide food for a large, sedentary, concentrated population. Equally vital, however, were such cultural inventions as fire, tools, food storage, draft animals, record-keeping, and complex political organization. (Needless to say, this brief summary can hardly do justice to a much more elaborate synthesis. We'll come back to this subject in Chapter 8.)

One of the most important properties of nature has to do with the fact that, very often, synergy is not simply an object, or a "thing," but a process – a dynamic in which many things, and people, may "work together" (whether wittingly or not) to produce a new form of synergy. Indeed, we are often witnesses to, or participants in, one of those historically rooted moments when some new synergy emerges – when things come together for the first time that had been separate (to paraphrase Koestler). These special moments of creation deserve to have a label, or a name, so that they can be highlighted. For lack of a better term (I'm open to suggestions), I call it the "Bingo Effect." It's rather like the moment when the final, winning number is called out in a bingo game.[42] The Bingo Effect occurs when just the right combination of ingredients, or parts, or people come together to "realize" a new recipe, or a new technology, or a new joint business venture for that matter. However, the "triggering" event is never the primary cause; it only completes a synergistic "package." As we shall see, the Bingo Effect has been associated with some of the major turning points in evolution – and human history. However, we will endeavor to keep our focus on the entire "package" – on the synergies.

Synergy Is Everywhere

In sum, synergy is the very stuff of history. We must look upon the historical process not as the potentially predictable working out of still-hidden laws but as an inescapably creative process, full of challenges and surprises. Historical contingencies for better or worse produce many unique combined effects. And these "vicissitudes" have themselves played a major causal role in the ongoing Epic of Evolution – to use the current buzzword. Furthermore, we must learn to think not in terms of the "prime movers" or principal "causes" of this ongoing saga but in terms of historically de-termined synergies (positive or negative), whether deliberately organized, or "self-organized," or fortuitous. Much of the recent work related to the modeling of complex systems is designed precisely to help us understand these synergies more clearly.

Several years ago I was invited to a small roundtable meeting on complexity, where I gave a short presentation on the role of synergy in the natural world. After the talk, a mathematically oriented complexity theorist commented: "What can I do with something that's everywhere?" My startled response was: "Just because synergy is everywhere doesn't mean it's not important." In retrospect, that's still a good answer. In the next few chapters, we'll see why.

3   The Magic Castle

The whole is something over and above its parts, and not just the sum of them all.

Aristotle

The power and majesty of nature in all its aspects is lost on one who contemplates it merely in the detail of its parts and not as a whole.

Pliny (The Elder)

In the new cosmology, the entire universe is viewed as the product of an evolutionary process. The distinguished physicist Timothy Ferris, in his recent *tour de force* on the subject, *The Whole Shebang* (1997), speaks of a universal phylogeny: "If, as many theorists suspect, the constants of nature were decided by random 'phase transitions' that took place during the first moment of time, then the laws of nature, too, are evidence of historical events. Evolution is creative: In an evolving universe, all events could not be predicted even if we knew the precise state of the early universe. Cosmology is an ongoing *story*" [his emphasis].[1] Ferris notes that the universe may be as much a product of chance as of necessity. However, it is also a product of synergy; the "Epic of Evolution" is quintessentially a story about synergy and history.

The Big Whoosh

Let's begin the story at the Great Beginning.[2] Matter – the basic "stuff" of the universe – would not exist without synergy. Indeed, the synergy started

35

with the "Big Bang" – or what should really be called the "Big Whoosh."
(There are no sound effects in space; it was more akin to an ultra-fast silent
movie.) According to the current mainstream scenario, at the instant of
creation, perhaps 14 billion years ago, there was no time and no space.
There was only an enormous concentration of energy and heat – an almost
infinitely hot, infinitely dense fireball that lasted an infinitesimal ten trillion,
trillion, trillionths of a second (that's ten with 36 zeros after it). One of the
world's leading astronomers, Allan Sandage, has called it truly a "miracle."

In the next fraction of a picosecond, this Almighty "singularity" (in
the physicists' jargon) became a plurality. An incomprehensibly vast and
complex burst of creativity produced the basic raw materials for all of
nature's subsequent magic tricks. At this point, an almost instantaneous
exponential expansion may have occurred, according to the "inflation"
theory developed by physicists Alan Guth, Andrei Linde, and others, but
this part of the story is less certain.

In any case, as the universe rapidly expanded and began to cool, it
became a "seething cauldron" of radiation energy, along with many particles of matter and antimatter. These minute particles materialized spontaneously out of fleeting concentrations of energy but quickly disappeared
again as they collided with one another and were mutually annihilated.
Some physicists have characterized this matter–antimatter clash as a "war."
Fortunately for us, matter was victorious, but it was also a carnage.
The survivors represent perhaps one-half of one percent of the original
combatants.[3]

Then something inexplicable happened. The original "superforce"
accompanying the Big Whoosh differentiated into four distinct kinds of
forces – now known as *gravity*, the *strong nuclear binding force*, the *weak
nuclear force* (which controls solar energy), and the *electromagnetic force*
associated with massless photons and other forms of radiation. (It is now
widely believed that the weak force and the electromagnetic force are different forms of the same "electroweak" force.) Why these four different
forces arose, and what they really are, remains one of the unsolved mysteries of physics, but they have played a major role in cosmic evolution.[4]
They immediately began to construct the material universe, and in the next
fraction of an instant matter got real; the war between opposing particles
gave way to a vast cosmic experiment in which a great variety of exotic
new particles emerged and were given "tryouts" (so to speak).

Most of the particles that were generated in this primordial soup rapidly
decayed or evolved into different forms in what amounted to a cosmic

analogue of natural selection. All of the more massive particles were radically unstable. Only the two lightest ones, now known as quarks and leptons, survived in large quantities, and these minute fragments of matter became the subatomic bricks that were used to construct our material universe.

Enter Synergy

Then the synergy began in earnest. Within the first 3-plus minutes after the onset of the Big Whoosh, protons and neutrons began to appear. However, these familiar constituents of matter – they're the particles that give matter most of its mass – are actually compounds of three quarks "glued" together by gluons ("packets" of the strong nuclear force).[5] Quarks come in six different sizes (or "flavors"), whimsically labeled top, bottom, charm, strange, down, and up. Quarks can't survive independently, but they can (with the help of gluons) survive in threesomes, and together they can produce various interdependent (synergistic) effects. Thus the so-called "up–down–up" combination results in a positively charged proton, while the slightly heavier "down–up–down" arrangement yields an electrically inert neutron. Without this elemental form of synergy (an example of a functional complementarity), the material universe as we know it would not exist.[6]

At this point we can also observe the historical nature of the evolutionary epic. When the universe was only a few seconds old its temperature exceeded 10 billion degrees – 1,000 times the temperature at the center of our Sun. In this super-hot cosmic soup, subatomic particles like protons and neutrons were "cooked" and remained unstable. So the process of creating matter was "frozen" (pardon the pun) by the ambient heat. But after the first few minutes, when the universe had cooled to perhaps 1 billion degrees, it became possible for protons and neutrons to stabilize and begin forming partnerships.

Here again, the synergy principle played a critical role. Neutrons by themselves are unstable; they need to pair up with protons in order to survive. But, in so doing, the neutron–proton pairs also acquire joint capabilities that neither particle possesses alone. If the emerging material universe had consisted only of protons, the cosmic epic would have stopped with the creation of hydrogen, which has only a single proton in its nucleus; there would be no heavier elements. The reason is that, unless the positively charged protons are paired with a neutron, they will repel one another. Not only do neutrons give atoms additional mass without adding to their nuclear charge but they enable proton–neutron pairs (collectively referred to as nucleons, or baryons) to join up with other pairs and create heavier elements. So the proton–neutron marriage is synergistic (another functional complementarity).

The next big step in the cosmic epic was also synergistic – and historically conditioned. When the universe had further expanded and finally cooled to about 3,000° Celsius (5,400° Fahrenheit), the dense fog of free electrons slowed down to the point where they could no longer resist being captured and brought into orbit by the clouds of positively charged hydrogen and helium ions. As these ions acquired electron "shells," it became possible to create stable atoms of matter, thanks to their synergy. This was a turning point in cosmic evolution.

You probably remember learning in your high-school physics or chemistry class that the number of electrons associated with any stable element is always exactly equal to the number of protons in the nucleus (with some important qualifiers), so that the overall charge of the atom is neutralized. The significance of this important synergistic effect is that the individual atoms of matter are able to band together, instead of being subject to mutual repulsion. Just as the quark–gluon triads create stable nuclear material and the proton–neutron partnerships underpin the larger, more complex elements, so the proton–electron pairings allow atoms to aggregate and interact in various ways. Indeed, the captive electrons also play an important role, along with the electromagnetic force, in binding atoms of various kinds together. Thus an ordinary atom of helium, say, is really a conglomerate consisting of three "levels" of synergistic effects.

Synergy and Gravity

And that is only the beginning of the synergy story. The next giant leap in cosmic evolution – from clouds of hydrogen and helium to the more massive elements – required yet another important form of synergy – gravity. We are taught to think of gravity as a mysterious and pervasive cosmic "force." Physicists have even proposed that there exist discrete gravity "packets," similar to gluons and photons, called "gravitons" (naturally). Gravitons are still hypothetical; they have not yet been observed. But gravity is not really a separate agency out there in the ether that acts on matter. It involves a relationship of mutual attraction between material substances, and its influence is always a function of the masses of various substances and their proximity to each other (as we know from Newton's law).[7]

In fact, there is gravitational attraction between every object in the universe (you and this book for instance). But gravity is also the weakest of the four cosmic forces – much too weak for you to notice its influence between very small objects. (Gravity is 10^{43} times *weaker* than the strong nuclear force.) So if the atoms of matter were not electrically neutralized by their electrons, gravity would not be able to overcome the repulsive force of their protons. In other words, the influence of gravity is dependent on the synergy between protons and electrons.

On the other hand, with gravity, as with many other forms of synergy, there is "strength in numbers," as the old saying goes. The combined influence of the Earth's great mass of atoms is strong enough to exercise a very powerful gravitational effect even upon gas molecules and dust particles, so long as they are fairly close to the Earth's surface. And our Sun, which is roughly one million times the mass of the Earth, is powerful enough to hold entire planets in orbit, like so many captive electrons. So the gravity we can observe is really a synergistic phenomenon – a synergy of scale on a cosmic scale. And its effects on material substances in turn depend entirely on the multiple layers of synergy that create stable, electrically neutral atoms.

However, gravitational attraction could not have played its important role in cosmic evolution without another important historical circumstance. It happened that the distribution of matter in the early universe was "clumpy" – it was not evenly spread out. This fortuitous situation (a result, perhaps, of quantum-level "fluctuations" on a vast scale during the Big Whoosh) meant that gravitational attraction was able to exert a strong influence over the subsequent course of events. Once the immense clouds of electrically charged hydrogen and helium ions that littered the expanding universe had been converted into vast patchy clouds of stable, electrically neutral atoms, a process of gradual accretion arose which, in time, led to you and me.

We can imagine that the process was somewhat like what the biophysicist Stuart Kauffman proposed as a "scenario" for the origins of life on Earth. Kauffman characterized it as a collectively autocatalytic "phase transition," and he used the analogy of 10,000 buttons scattered on a hardwood floor that are gradually tied together in pairs until, at some point, the pairs begin to be tied to one another. Then, at some critical point in the continuing process, the buttons will all become connected together in a single interconnected web.[8] In the cosmic version of this scenario, it was gravity that did the binding, and the buttons were the many individual atoms of hydrogen and helium that were being attracted to one another. These pairs in turn combined their gravitational power (their gravitons) and used their synergy to attract larger and more distant aggregates of atoms. Gradually, over many hundreds of millions of years, the clouds of primordial gas became more and more concentrated. As these huge clumps were relentlessly compressed, they began to generate intense internal heat and outward thermal pressure, which eventually became a counter force that stabilized the evolving proto-stars and proto-galaxies (another synergistic effect) and set the stage for the next historical development.

Fusion and Synergy

From this point on, the process of stellar evolution followed various scenarios, depending on many historical circumstances, including the size and density (the mass) of a given gas cloud and its interactions with other clouds. In the case of a typical "main sequence star," the first stage most likely involved a concentration of the helium into an inner core, while the more numerous hydrogen atoms formed an outer "shell." As the internal pressure and temperature continued to rise over some tens of millions of years, the hydrogen eventually "ignited"; the intense heat and compression inside the proto-star "stripped" the hydrogen atoms of their electrons. The evolving star began to throw off a shower of photons and other forms of radiation energy – "sunlight" – that were liberated from the solar pressure-cooker. The naked hydrogen ions then began merging with one another.

This is the process known as "fusion," and it is highly synergistic. In fact, "stellar nucleosynthesis" (as the physicists term it) is the magical means by which all of the other 89 naturally occurring elements in the universe were fabricated. Under intense pressure and heat (1.1 billion degrees Celsius or more), the hydrogen ions in an evolving star will fuse together to form helium while producing some 600 trillion joules of free energy. (No wonder physicists have been trying so hard to re-create the fusion process here on Earth.)

In the larger stars, the fusion process can go much further. Three helium nuclei, each having two protons and two neutrons, can combine to form carbon. Under very high temperatures and pressures, two carbon ions will fuse to form magnesium. Two magnesium ions will combine to make silicon. And two nuclei of silicon can produce iron. All of the remaining, heavier elements are created when a very large star collapses and undergoes a supernova, a "little bang" that has been likened to a monstrous thermonuclear bomb. In fact, our Earth, including its natural elements and chemical compounds, is a "second-generation" conglomerate that coalesced from the remnants of one or more supernovas. (Applying the categories described in Chapter 2, the process involved many phase transitions that produced emergent new effects.)

Chemistry and Synergy

Our dry, sometimes dull textbooks in physics and chemistry treat fusion in a matter-of-fact way. But nucleosynthesis is really a form of sorcery on a grand scale – an alchemist's dream come true. Under the right conditions, even gold is fabricated from the quarks and leptons that are the ultimate raw material of the universe. Different combinations of the same atomic building blocks produce substances with very different physical and chemical properties. Thus chemistry is a bottomless well of creativity in nature and a prodigious source of synergy.

One obvious example is water – that miraculously versatile molecule with properties that are totally different from either of its two constituent gases. No other substance has its fluidity, its responsiveness to heat and cold, or its remarkable ability to dissolve and transport other substances. These properties are a product of the distinctive way in which water molecules network with each other, which in turn derives from the way in which atoms of oxygen and hydrogen interact.[9]

Another obvious example, which we often take for granted, is a wood fire. When atmospheric oxygen and the carbon in wood fiber combine at a high enough temperature, they produce the gaseous compound carbon dioxide, along with heat, light, and a residue of ashes.

Less well known but equally significant are the synergies produced by the many other kinds of chemical compounds that occur naturally, or with a little human coaxing. For instance, copper and zinc oxide together produce greater physical activity than do the two of them separately.[10] Crystals that are unstable alone may be stabilized by "twinning" – being paired up.[11] Polymers represent an especially important class of chemical synergies. A polymer is a giant molecule comprising a long chain with perhaps thousands of repeat units (monomers). Many polymers are produced by living organisms – including wool, silk, rubber, cotton, spider webs, and other substances. Many more have been synthesized artificially by polymer chemists, including plastics, synthetic fibers, and synthetic rubber. The significance of polymer chemistry lies in the synergistic properties that can be produced, such as low density in relation to strength, or toughness, elasticity, compressibility, etc.[12]

And these are only a few of nature's many parlor tricks. There are also the so-called "covalent bonds," where electrons are shared between two atoms. Covalent bonds produce molecules that have greater stability than either of the constituent atoms do alone. The textbook example, again, is water. To simplify the physics a bit, oxygen normally has eight electrons, two in an inner shell and six more in an outer shell that also has space for two additional electrons. When hydrogen and oxygen atoms are combined, the electrons from two hydrogen atoms bond with the outer shell of an oxygen atom, which results in a highly stable molecule of water.

There is also synergy when "ionic bonds" form between atoms. This occurs when an electron is actually transferred between the constituent elements. For instance, the light metal sodium and fluorine gas alone are unstable. Sodium is "electropositive," meaning that it has a tendency to give up an electron from its shell. Fluorine, on the other hand, is "electronegative" – it tends to gain electrons. So when the two elements are combined, an electron is transferred from the sodium to the fluorine. The synergistic result is the stable compound sodium fluoride.

We are so accustomed to chemistry's mundane miracles – its endless synergies – that we take them for granted. When liquid hydrogen and liquid helium are combined in very large quantities, the energy released may be enough to propel a spacecraft to the moon. In its wake, the launch pad will be bathed in a waste product – water.

Likewise, the batteries in your car, your computer, and your cell phone rely on "oxidation–reduction" chemical reactions in which energy (in the form of an electrical voltage) is extracted from a chemical mixture and then returned when the battery is recharged. Recharging reverses the chemical reaction and restores the original energy state. Or consider the catalysts that produce so many magical effects in chemistry without themselves being changed in the process. For example, if you simply mix atmospheric oxygen and hydrogen gases together, nothing will happen (fortunately). But if you add a little platinum to the mixture, the explosive result will be water.[13]

Even more synergy can be found in the myriad of things that atoms and molecules of various substances can do collectively. A large pond of water, for example, is actually a confluence of many trillions upon trillions of water molecules in a loosely bonded lattice structure. (Even a single water droplet can contain many millions of molecules.) Individually, these molecules are not even visible. But together they may be able to float a boat, support an ice skater, drown a swimmer, or provide a home for a goldfish.

The Anthropic Principle

Before we turn to the role of synergy in biological evolution, let's briefly look at how the evolutionary process has been shaped by the historical context. This is highlighted by the so-called "anthropic principle" – the controversial idea that the universe is somehow tailored (or even "designed") for living systems in general and humankind in particular. Chemist James Lovelock, principal author of the much-debated "Gaia Hypothesis" (the theory that the Earth as a whole is a single living system), has called this the "Goldilocks Effect." In order for life on Earth to evolve, the conditions had to be "not too hot, not too cold, but just right."[14]

And much more besides! Many of the different scenarios that have been proposed for the origins of life focus on the biochemistry – how the right combination of ingredients came together (perhaps from the ocean floor, or deep underground, or even from outer space) with the additional stimulus of free energy inputs. In these scenarios, "internal" processes of biochemical autocatalysis, variation, differential survival, and replication are seen as the main themes.

From a synergy perspective, however, the "external" context is equally important. If the emergence of living systems was "inevitable," as some theorists claim, it was also highly improbable. The mathematician/physicist Roger Penrose pegs the odds at about one in 10^{300} – a number that is greater than all the particles in the universe. In fact, an essential prerequisite for the appearance of life on Earth, or anywhere else in the universe, was a fortuitous and highly synergistic combination of preconditions that occurred at a very specific time and place (or places) in the cosmic epic. The list of these preconditions is awesome; numerous books and articles have been published on the subject.[15] Here are just a few of them:

* If the values for any of the four fundamental forces were significantly different, the material universe could not have evolved.

* The electrical charges of protons and electrons are exactly equal. We do not know why, but if this were not the case the procreative role of gravity would have been thwarted. Even if the difference between the two electrical charges were very small, the enormous clouds of hydrogen and helium gas that littered the early universe would never have coalesced into stars and galaxies.
* It is highly significant that the mass of an atom is concentrated in the nucleus – in its "nucleons." This provides basic stability for the atom while allowing its electrons to move about and interact with the environment and other atoms in various ways.
* There are no other stars close to our Sun. If there were, the orbits of the Earth and the other planets would be very unstable.
* The mass and gravitational attraction between the Earth and Sun are also "just right." This keeps the Earth at a comfortable distance from the Sun and allows the Earth to attract and hold light, gaseous compounds. If the Earth's gravity were much stronger, the atmosphere would retain toxic levels of ammonia and methane and would be too dense to allow sunlight to penetrate. If it were much weaker, the atmosphere would lose water and would not be dense enough to block harmful solar radiation. In either case the temperature range on Earth would also be more extreme.
* Solar radiation would not be able to support life if its properties were any different. If the mix of wavelengths in sunlight were even slightly higher, the additional energy would break the weak molecular bonds in organic compounds and even strip atoms of their electrons; the structure and functioning of complex organic molecules would break down. Conversely, if the energy level were significantly lower, sunlight would be absorbed by the water in living organisms (our most important constituent by weight). This would prevent sunlight from being able to play its vital role in photosynthesis – boosting the energy levels of electrons in "antenna" pigment molecules.
* It is the unique chemical properties of the elements hydrogen, carbon, nitrogen, and oxygen that make life possible. Happily, these elements all exist in abundance on (or immediately above and below) the surface of the Earth, and in the oceans.
* Oxygen, a waste product of photosynthesis, is maintained in the atmosphere at a constant 21%. If the percentage were much higher, the Earth would be a much more highly combustible place. If the percentage were much lower, the surface of the Earth might resemble an airliner without cabin pressurization – and no oxygen masks.

✳ The distinctive properties of water are also important to life on Earth, in many ways. For instance, the dynamics of the biosphere depend on the fact that ice floats. If instead it sank to the bottom, the process of freezing would begin at the bottom. The oceans would long ago have turned into solid, lifeless blocks of ice.

There are literally dozens, perhaps hundreds, of other "necessary but not sufficient" preconditions for the evolution of life on Earth, which is why the "anthropic principle" evokes reactions ranging from scientific wonder to religious awe. To quote the eminent biologist George Wald: "It is as though, starting from the Big Bang, the universe pursued the intention to breed life, such is the subtlety with which difficulties in the way are got around, such are the singular choices in the values of key properties that could potentially have taken any value."[16]

The improbability of it all does not prove the existence of an intelligent Designer, of course. But it can't be ruled out. On the other hand, the origin of life on Earth cannot be reduced to the mechanical operation of preexisting "laws of physics" – much less any grand "final theory." The many necessary preconditions were historically convergent and synergistic; they combined to produce a historically unique (or at least rare) window of opportunity for the emergence of living systems. It was a cosmic example of the Bingo Effect – a "winning" combination of circumstances.

Architectonics in Evolution

If you are conversant with cosmology, you will note that we have skipped over many important subjects: black holes and the formation of galaxies; supernovas and super-clusters; the formation of planetary systems; quasars; neutrino showers; cosmological constants; so-called dark matter; vacuum energy, etc. – not to mention the synergies that are found at the mega-level in the cosmos, from binary stars to binary galaxies. Nevertheless, what we have outlined here provides a context and foundation for understanding the process of biological evolution and the emergence of humankind. As we have seen, this foundation is built on synergy. And so is the superstructure; living systems represent a complex hierarchy of synergistic effects.

"Architectonics" is the key to understanding the evolution of this magic castle. The biosphere that we inhabit is a multileveled high-rise structure that has been laboriously constructed, step by step (though also in fits and starts), over the course of the past 3.5 billion years or so. It is not the result either of a historical accident – a "drunkard's walk"– or a deterministic law. It is an intelligible, functionally coherent edifice in which the studs, the stringers, the floor joists, and the underlayment at each of the lower levels have provided the framework and supporting structure for the next higher level. Although this architectonic process has been shaped and constrained by various laws and physical constraints – including, among other things, gravity, the physicochemical properties of the natural world, and the local conditions in different environments – new forms of functional synergy have been a major source of nature's creativity.

The laws of physics may explain why covalent bonds exist but not how living organisms may make use of this phenomenon to solve bioengineering problems. Fractal geometry may illuminate the process of pattern formation in tree leaves but not why leaves have evolved or why some trees have leaves and others deploy needles. Likewise, humans cannot cancel out the Earth's gravity, but with our awesome ingenuity we can devise ways to break free from its hold over us; we can literally create new degrees of freedom. This is the transcendent power that all those multiple levels of synergy have given us. We are more, after all, than a zillion quarks packed together. As the distinguished twentieth-century evolutionary biologist Theodosius Dobzhansky expressed it:

> A man consists of some seven octillian (7×10^{27}) atoms, grouped in about ten trillion (10^{13}) cells. This agglomeration of cells and atoms has some astounding properties; it is alive, feels joy and suffering, discriminates between beauty and ugliness, and distinguishes good from evil. There are many other living agglomerations of atoms, belonging to at least two million, possibly twice that many [or even many more] biological species ... How has this come about?[17]

Nature's Wizardry

How life on Earth began is still a deep mystery, of course. There is no lack of imaginative hypotheses, but none is as yet convincing. As the chairman of a National Academy of Sciences panel on this issue, Harold P. Klein, commented: "The simplest bacterium is so damn complicated from the point of view of a chemist that it is almost impossible to imagine how it happened."[18] And the scientific journal *Nature*, in a recent editorial on "Pursuing Arrogant Simplicities," commented that: "Even after one absorbs a thousand or more pages of text, one would still be unlikely to have a feel for the variability and complexity of even the simplest microbe."[19] Or, for that matter, an appreciation for the centrality of synergy. Though it is not fashionable these days to be awed by the natural world, the wizardry of nature at least deserves our profound respect. And synergy has had a major hand in it.

Some of the most important examples of biological synergy involve the basic machinery of reproduction. In the age of biotechnology and recombinant DNA, the renowned three-letter acronym for deoxyribonucleic acid has become a household word. Most schoolchildren now learn about the double helix and the four nucleotide "letters" that make up the genetic code (not counting uracil, which substitutes for thymine in RNA). What is often glossed over is the fact that the properties and functions of DNA are highly synergistic.

There is, first, the molecular structure and chemistry of DNA itself. The basic scaffolding of these unique molecules consists of reduced carbon surrounded by hydrogen atoms, which allows them to combine easily with the other key elements involved in constructing nucleic acids. There is also the double-stranded antiparallel "backbone" (or ladder) of the DNA molecule, in which phosphate groups alternate with a sugar, deoxyribose, to form covalently linked chains. The structure hangs together because its atoms share pairs of electrons.

Likewise, the four nucleotide "bases" that form the individual "rungs" on the ladder – each a complex synthesis of carbon, nitrogen, hydrogen, and oxygen (except for adenine) – can only perform their vital informational function because of their specific complementarities: Adenine only pairs with thymine and guanine with cytosine. The fortuitous result is that each side of the DNA ladder perfectly complements the other, and each can serve as a template for reproducing its partner. Moreover, it is the precise order in which these base pairs are arrayed in various three-letter "codons" (or "words") that determines which of 20 different amino acids will be produced, another functional complementarity.

Still another level of synergy is added to the process when long chains consisting of hundreds of precisely ordered amino acids are assembled and folded in intricate ways to create any one of several thousand different proteins – each of which in turn produces different synergistic effects. Moreover, as the new science of proteomics (the study of proteins) is learning, proteins typically do not act alone; they form networks akin to electrical circuits. Indeed, the so-called "proteome" is proving to be a far more formidable challenge to biologists than was the genome.[20]

And that's not all. The vital role of DNA in biosynthesis is possible only because of the highly coordinated role that three distinct forms of RNA play in "transcribing" and utilizing the nucleotide bases (a combination of labor). The so-called *messenger RNA* makes a copy of the relevant DNA sequence from a given chromosome in the cell nucleus and transports it to a "manufacturing" site, a ribosome in the surrounding cytoplasm (or cell body); *transfer RNA* brings the appropriate amino acids from elsewhere in the cytoplasm to the ribosome and links them to the RNA "template"; and *ribosomal RNA* plays a key role in fabricating the protein by lining up the amino acids in the proper order. Without its synergy-producing RNA helpers, DNA would be impotent. Biologist James Darnell and his co-authors, in their well-known textbook *Molecular Cell Biology*, conclude that "the development of three distinct functions of RNA was probably the molecular key to the origin of life."[21]

Catch-22 and the Synergy of RNA

Biologists John Maynard Smith and Eörs Szathmáry, in their recent book, *The Origins of Life* (1999), point out that there was a Catch-22 (from Joseph Heller's famous novel) associated with getting life started in the first place. Without enzymes, the faithful replication of long strands of DNA would be impossible. But without long strands of DNA, the production of enzymes could not have occurred. The escape from this dilemma, *mirabile dictu*, was that RNA can form both templates and enzymes; RNA produced the necessary complementarity on its own.

It has also been proposed that the chemistry of DNA requires a protective envelope, in order to segregate the process from its surrounding environment. And, in the earliest stage of biological evolution, substances similar to "liposomes" – fatty bubbles that form whenever clusters of naturally occurring lipid molecules are immersed in water – may have played such a sheltering role. Lipid molecules are hydrophobic at one end and hydrophilic at the other end, so they can combine forces to create an effective non-aqueous "enclosure." Indeed, a similar role is played in cell membranes today by the so-called phospholipids. Not only is this yet another example of nature's magic (a Gestalt effect), but it illustrates how nature uses happenstance opportunities to solve biological problems.[22]

Living systems are also dependent on the synergy produced by enzymes – the organic catalysts that control virtually all biochemical processes.

Altogether, there are about 2,000 enzymes found in living systems, and each one is an awesome example of nature's wizardry. By lowering the activation energy required for a biochemical reaction, enzymes speed up the process and allow it to go forward at a lower temperature. Moreover, the properties of an enzyme may be shaped by synergies with various cofactors, coenzymes, allosteric (cooperative) interactions, and even the properties of the target substrate. Without enzymes, life would be much slower than molasses, as the old saying goes. In fact, it might not be able to go forward at all. Just to illustrate, the conversion of carbon dioxide and water to carbonic acid occurs some 10 million times faster when it is catalyzed by the enzyme carbonic anhydrase. The facilitative effects of enzymes demonstrate again the tremendous potency of synergy.

Another important form of synergy can be found in the reproductive process itself. Biological reproduction involves a complex, frequently multi-stage fabrication effort in which many genes participate cooperatively (a combination of labor). For instance, the production of collagen (the chief constituent of connective tissue like skin and tendons) requires the cooperative efforts of more than a dozen enzymes. Even a simple virus, like the much-studied SV40 in monkeys, has five genes consisting of 5,243 base pairs of DNA. And bacteria, arguably the first truly "autopoietic" (self-maintaining) organisms, may have 2,000–5,000 genes which code for a like number of proteins.

All of this goes on inside a precisely engineered organic system that cannot even be seen with the naked eye. Bacteria, moreover, can perform their reproductive feats with phenomenal speed, in about 20 minutes. A single bacterium is capable (in theory) of producing a billion copies of itself in the course of a day. Humans, by contrast, have some 40,000 genes, which orchestrate the development not just of a single cell but of an organism with at least 10 trillion cells, each of which is far more complex. No wonder it takes us a little longer to reproduce ourselves.

Much evolutionary innovation and "progress" can be credited to single-celled bacteria. Thanks to the many discoveries in microbiology over the past 20 years or so, we now appreciate more fully the important role these microscopic creatures have played in the evolution of life on Earth. In fact, bacteria invented many of the most important biochemical and biomechanical processes in living systems, including photosynthesis, nitrogen fixing, glycolysis, cellular damage repair, mobility, and a gene-sharing conjugal process analogous to sex. Moreover, complex organisms remain dependent on the services provided by bacteria – either directly as internal "endosymbionts" (as they are called) or indirectly as important players in virtually every ecosystem.[23]

Bacteria can also manufacture at least 27 different kinds of inorganic minerals – including magnetite, which they use as an aid to navigation. And they can survive in many extreme environments that are hostile to humankind – from termite intestines to the specialized stomachs of ruminant animals (cattle, horses, deer, etc.), as well as the piping hot seafloor hydrothermal vents, the frigid wastes of the Antarctic tundra, the rich fossil fuel deposits that are found more than a mile underground, and even in toxic pools of sulfuric acid. Bacteria are also remarkably skilled at coping with new environmental threats, like pharmaceuticals, as we have learned to our chagrin.

But more to the point, bacteria also pioneered in exploiting the synergy principle – utilizing cooperative solutions to solve the problems of earning a living in an often hostile environment. They invented communal living, multicellular organization, even the division of labor.

This bacterial "togetherness" has very important theoretical implications. As we noted earlier, evolutionary thinking in recent years has been dominated by the notion that natural selection and evolutionary change have been driven primarily by competition. While competition is undeniably an important aspect of the evolutionary process, our rapidly increasing knowledge of the "microcosmos" suggests a more balanced picture. Contrary to the traditional assumption that primitive life forms are solitary and that cooperation is a rare and relatively recent development, the overwhelming majority of bacteria today live in large communities, and so did their ancestors. The fossil evidence suggests that microbial cooperation may have originated as far back as the Archean Eon, more than 3.5 billion years ago. The basic survival strategy of evolving bacteria has long included cooperation. And synergy. Bacteria began to exploit the synergy principle at a very early date.

The Evidence from Stromatolites

Important evidence for this can be seen in the so-called stromatolites – rocky domes of various sizes that litter seashores from the Bahama Islands to Shark Bay, Australia. Often referred to as "living rocks," the stromatolites are actually composed of layers of bacteria (or algae) and inorganic materials (sand, gravel, mud, and other particles) deposited by wave action over many years and cemented together by bacterial activity. Some of these remarkable structures are still living, but many others may be ancient fossils.[24]

What makes these primordial palaces an important example of synergy is that the resident communities use collective action to create a structure that can resist the destructive effects of wind, water, waves, and ultraviolet (UV) radiation. The top layer of the stromatolite typically consists of dead cells that serve the community by shielding the next layer – the energy-producing photosynthesizers – from the damaging effects of UV radiation. Below that are layers of producers and consumers, along with various channels to allow for the circulation of nutrients, enzymes, and wastes. And below that are limestone and the hardened corpses of many previous generations.

The stromatolite community also has a sophisticated chemistry, utilizing both anaerobic (oxygen-free) and aerobic processes. And when the sediment at the crown becomes too thick, the next layer below will migrate upward, adding a new level to the microbial castle. (The thick bacterial carpets and scums known as "microbial mats" or "biofilms," which can be found almost everywhere in the world today, have similar characteristics but lack the hard structure of the stromatolites.)

Many of these "slime cities," as science writer Andy Coghlan has called them, include a mix of different strains that collaborate in various ways and share in various communal benefits.[25] For instance, the bacterial symbionts that perform the difficult feat of breaking down cellulose in a cow's rumen achieve their microscopic magic through the joint actions of five different strains (a functional complementarity). Four of the strains contribute different enzymes, each of which accomplishes a specific conversion step, while a fifth strain provides protection for all the others from the ever-present danger (to an anaerobic bacterium) of oxygen pollution. It's truly a case of "All for one, one for all," to quote the famous pledge of allegiance in *The Three Musketeers*.

It has also been discovered that bacteria can communicate with one another chemically and that they can coordinate their actions in various ways; freedom of information has very ancient roots. This is especially important in free-ranging bacterial colonies that migrate and forage for food collectively. Perhaps most impressive are the myxobacteria that exhibit two different modes of behavior. They can forage as individuals and then, on signal, join together and become good citizens in a larger community. We now know that this is accomplished by chemical signals and that these signals also have a synergistic property. They become effective only when various individual contributions reach a certain threshold level of concentration. Microbiologist James Shapiro likens bacterial colonies to multicellular organisms, and physicist-cum-microbiologist Eshel Ben-Jacob characterizes them as cybernetic systems – collectively self-aware, adaptive, and even "creative."[26]

Some bacterial strains have even adopted a multicellular division of labor. For instance, colonies of *Myxococcus xanthus* produce a colorful, tree-like structure complete with specialized reproductive pods that are hung from the end of each "branch." And eons before economist Adam Smith (and Plato before him) extolled the benefits of the division of labor in human societies, the ancestors of a single-celled colonial bacterium, *Anabaena*, were already utilizing this principle to solve a difficult dilemma. *Anabaena* is unusual among bacteria in that it engages in both photosynthesis and nitrogen-fixing, a dual capability that gives it a significant functional advantage. However, these two processes are chemically incompatible. The oxygen produced by photosynthesis inactivates the nitrogenase required for nitrogen-fixing. So *Anabaena* has solved this problem by compartmentalizing. When nitrogen is abundantly available in the environment, their cells are uniform in character. However, when ambient nitrogen levels are low, specialized compartments called heterocysts are developed that lack chlorophyll but are able to produce nitrogenase. The heterocysts are then connected to the primary cells by filaments. In effect, a specialization exists which benefits the "whole."[27]

Even more remarkable, and theoretically significant, is the fact that bacteria are profligate, promiscuous gene-sharers – true practitioners of genetic communism. In fact, some microbiologists support Sorin Sonea's idea that bacteria form a single, interconnected world-wide gene pool.[28] Sometimes bacteria engage in "sex," exchanges of genetic materials via direct cellular "bridges." At other times bacteria simply broadcast various gene-bearing plasmids and virus-like fragments as free agents: "To whom it may concern." In either case, the result is an uninhibited flow of genetic information. One consequence of this gene-sharing behavior is much greater collective adaptability. Microbiologist Bill Costerton estimates that the pooling and sharing of genes gives large bacterial communities some 1,500 times greater resistance to antibiotics than is the case with isolated colonies.[29] It's yet another example of a synergy of scale – or perhaps a hybrid, a complementarity of scale.

But more important, it is now recognized that cooperative "horizontal" gene transfers, as opposed to the traditional "vertical inheritance" model, have played a major creative role in evolution. To quote from a recent article on this subject by the distinguished microbiologist Carl Woese:

> Vertically generated and horizontally acquired variation could be viewed as the yin and yang of the evolutionary process.... Horizontal transfer... can call on the diversity of the entire biosphere, molecules and systems that have evolved under all manner of conditions, in a great variety of different cellular environments.... The high, pervasive levels of horizontal gene transfer at early times created an evolutionarily communal state of living systems in the sense that the aboriginal organismal community evolved as a collective whole, not as individual cellular lineages.[30]

Symbiosis as a Form of Synergy

Bacteria also invented positive symbiosis – mutually beneficial partnerships with other types of living organisms. Indeed, symbiosis represents a new "level" of organization in nature. In biologist David Sloan Wilson's characterization, symbiotic organisms have a "shared fate." In effect, two separate sets of genes become functionally interdependent in terms of the workings of natural selection.

Perhaps the earliest instance of symbiosis may have involved motility. Lynn Margulis has proposed that free-swimming, rapidly moving spirochetes (microscopic "snakes" without heads or tails) may have joined forces at an early date with sluggish blue-greens or other energy-producing bacteria to create flagella (Margulis prefers the more cumbersome term "undulipodia"), thus giving both partners new capabilities.[31] Bacteria are also legendary for their many symbiotic associations with plants, animals, and birds. Here are just a few examples:

✳ Virtually all species of ruminant animals – including some 2,000 termites, 10,000 wood-boring beetles, and 200 species of Artiodactyla (deer, camels, antelope, etc.) – are totally dependent on bacteria (and/or fungi or protoctists) to break down the tough cellulose fibers in the plants they consume and convert them to a usable form. In some termite species, in fact, bacteria and other symbionts – totaling perhaps 1 trillion or more – may constitute more than half the total weight of the organism.[32] There is even a tropical, plant-eating bird, the hoatzin (or stinkbird), with a cow-like stomach that can digest plant fibers with incredible efficiency (about 70%) – thanks to its teeming colony of bacterial symbionts.[33]

Without these endosymbionts there would be no ruminant animals, and perhaps, as a consequence, no humans.

* Nitrogen is an essential ingredient for living systems. The atmosphere contains about 78% nitrogen, but plants cannot make direct use of it. They need ions of nitrogen-containing compounds that can be extracted from the soil through the plant roots. But nitrogen is often in scarce supply, and soil depletion is common, for various reasons. Enter *Rhizobia* and other bacteria that invade the roots of legumes (including many vegetables consumed by humans) and stimulate the formation of symbiotic root nodules. *Rhizobia* have mastered the trick of nitrogen-fixing, converting nitrogen from the atmosphere into usable compounds. So the plants supply the *Rhizobia* with energy, and in return the bacteria contribute a dependable supply of an indispensable chemical resource. In his new book, *Tales from the Underground* (2001), David Wolfe notes that: "Essentially all the nitrogen contained within the proteins and genes of plants, animals, and humans has, at one time or another, been funneled through these nitrogen-fixing microbes."[34]

* The African honey guide is an unusual bird with a peculiar taste for beeswax, a substance that is more difficult to digest even than cellulose. In order to obtain beeswax, however, the honey guide must first locate a hive and then attract the attention of a co-conspirator, such as the African badger (or ratel). The reason is that the ratel has the ability to attack and dismember the hive, after which it will reward itself by eating the honey and leaving the wax behind for the birds. However, this unusual example of cooperative predation by two very different species depends on a third co-conspirator. It happens that the honey guides cannot digest beeswax. They are aided by a symbiotic gut bacterium, which produces an enzyme that can break down wax molecules. So this improbable but synergistic feeding relationship is really triangular. And, needless to say, the partnership would not work without the services provided by the bacteria.[35]

* Giant tube worms – with stalks that can reach more than 6 feet in height – thrive on the ocean floors in proximity to hydrothermal vents, fissures that spew forth hot material from deep inside the Earth. The main attraction of these hotspots is not their warmth but their hydrogen sulfide compounds. Hydrogen sulfide can be broken down to extract energy, and the energy can in turn be used to convert the carbon dioxide molecules that are trapped in the surrounding sea water into various organic compounds. However, the tube worms have no mouths or digestive tracts.

They depend entirely on nutritional services provided by a special kind of "chemosynthetic" bacterium.[36]

✳ One of the most extraordinary examples of a symbiotic partnership in nature involves the single-celled protist *Mixotricha paradoxa*. In fact, each *Mixotricha* cell represents an association of at least five different kinds of organisms. In addition to the host cell, there are three external (surface) symbionts, including large spirochetes, small spirochetes, and bacteria. The function of the large spirochetes, if any, is not clear; they may even be parasites. However, the small hair-like spirochetes, which typically number 250,000 or more *per cell*, cover the surface and provide an effective propulsion system for their host through their highly coordinated undulations. Each of these spirochetes, in turn, is associated with a third symbiont, a rod-shaped "anchoring" bacterium. Finally, each host cell contains an internal symbiont – thousands of spherical bacteria that may serve as energy factories (like the mitochondria in other complex cells). But what makes this five-way combination of labor all the more remarkable is that each *Mixotricha* protist is itself an endosymbiont. It populates the intestine of the Australian termite, *Mastotermes darwiniensis*, where it performs the essential service of breaking down the cellulose ingested by its accommodating host.[37]

Eukaryotes Are Us

The emergence of the "eukaryotes" – complex single-celled organisms with a segregated nucleus and an array of internal "organelles" (small, organ-like structures) – is unquestionably one of the major turning points in evolution. What sets the eukaryotes apart is that they have been able to fully exploit the synergistic potential of the division/combination of labor at the cellular level. This, in turn, provided a prerequisite for yet another new level of synergy in evolution – multicellular organisms.

The eukaryote is a microscopic miracle – a highly complex orchestration of activities within and between more than a dozen specialized structures that share in the task of enabling the cell to grow, adapt, repair, and reproduce itself. The cell membrane, the internal cytoskeleton, the chromosomes, nucleolus, ribosomes, Golgi apparatus, endoplasmic reticulum, and all the rest perform various specialized roles and interact (and communicate) with one another in very precise ways. Even the nucleus turns out to be an immensely complex system. But the key to it all, perhaps, is the role played by the mitochondria, ancient symbionts that provide most of the eukaryotes' energy. (In plant cells, the plastids that specialize in photosynthesis are also of prime importance; they capture most of the energy that these cells use, and they too are derived from ancient symbionts.[38])

Prokaryotes (cells without a nucleus), like bacteria, don't have mitochondria. Many rely for their energy needs on a process called glycolysis, which produces molecules of energy-rich ATP (adenosine triphosphate) as their basic energy "currency." Glycolysis is an exceedingly complicated process, utilizing glucose molecules (a six-carbon sugar) as a raw material. In fact, it involves no fewer than nine distinct stages and upwards of 100 biochemical steps. It's yet another example of an orchestrated combination of labor. However, the energy derived from glycolysis is relatively modest. Only two molecules of ATP are produced from each glucose molecule, which then yield only 143 kilocalories of energy per mole. The rest goes to waste in the form of pyruvic acid.

Mitochondria, happily, feed on pyruvic acid (and oxygen). Using an even more complex two-stage process called the "Krebs cycle" and "electron transport," mitochondria are able to extract as many as 36 more molecules of ATP from the pyruvic acid residues, boosting the energy-conversion efficiency of the combined process from less than 5% to more than 40% and the energy yield to as much as 686 kilocalories per mole. It's like adding a turbocharger to your car, or maybe going from a horse and buggy to a 400-horsepower driving machine.

Equally important, the synergy provided by the mitochondria can be multiplied many times over (a synergy of scale). Each eukaryotic cell may have hundreds, or even thousands of these organelles, depending on its particular functional needs. Liver cells, for example, have about 2,500 mitochondria, and muscle cells have several times that number. So the synergistic combination of glycolysis and "respiration" (the technical term for what the mitochondria do) gives the eukaryotes new capabilities for growth (some are 10,000 times larger than bacteria) and various "technological improvements" that would otherwise be impossible. (For instance, their specialized, sequestered chromosomes allow them to carry vastly more genetic information.)

But most significant of all, their larger size and more sophisticated biotechnology allows the eukaryotes to become specialists in an even larger, multicellular combination of labor (namely us), a capability that the prokaryotes lack. The obvious analogy in human societies is the development of energy technologies based on fossil fuels (coal and oil), which have undergirded the Industrial Revolution and all our other technological revolutions since.

On the other hand, this breakthrough in energy production by the eukaryotes, perhaps 1.5 billion years ago, had a collateral consequence – a trade-off that we will meet again and again. The price paid for this new level of synergy in living systems was a greater degree of interdependency; the symbiotic "parts" became dependent on each other, and their fate became tied to the overall performance of the "whole." The once-independent mitochondria provide a case in point. They still reproduce themselves separately from their host cells, utilizing their own DNA. However, they depend entirely on the raw materials (pyruvic acid, oxygen, and other nutrients), as well as the protection supplied by their hosts; they can't exist independently. In fact, most of their DNA has been transferred to the cell nucleus. A similar interdependency is found in many other "obligate" symbiotic partnerships in nature. And, as we shall see, in the human species as well.

"Symbiogenesis"

Synergy of various kinds was almost certainly the key to how the eukaryotes evolved in the first place. One of the leading authorities on the subject, Thomas Cavalier-Smith, argues that the origin of eukaryotes involved "so many drastic transformations of virtually every aspect of the cell that it must be the most complicated and improbable megaevolutionary event in the history of life."[39]

Cavalier-Smith singles out the development of internal membranes and the internal cytoskeleton (again, there is evidence that something like it may have been invented first by bacteria).[40] Not only did the cytoskeleton allow for many internal improvements but, equally important, it allowed for the loss of a rigid outer cell wall in favor of a flexible external membrane. This structural modification in turn led to a vast increase in feeding efficiency. Prokarytic bacteria obtain external nutrients by secreting digestive enzymes into the surrounding environment to break down passing food particles. The products of this process must then be absorbed one molecule at a time through the cell wall. As Maynard Smith and Szathmáry point out, this is a very wasteful method. Eukaryotes, on the other hand, are able to engage in "phagocytosis."[41] They form pockets in their cell membrane that can engulf food particles. These pockets (vacuoles) then fuse with internal organelles (lysosomes) that contain digestive enzymes. Nothing is wasted. Our own digestive systems are the descendants of this breakthrough biotechnology.

Nevertheless, some strategically important microscopic mergers – radical improvements via symbiosis between independent organisms – put the process of eukaryote evolution into overdrive. It was a Russian botanist, Konstantin Mereschkovsky, who originally proposed back in 1905 that the chloroplasts were descended from free-living bacteria and that symbiosis explained their appearance in eukaryotes. Mereschkovsky also coined the term "symbiogenesis" to characterize such evolutionary developments. However, Mereschkovsky's hypothesis (and related work on symbiosis by other Russian scholars) was presented as an alternative to Darwin's theory, so it was rejected and remained largely unknown in the West at that time. But in the 1920s and 1930s, another Russian theorist, B. M. Kozo-Polyansky, recognized that symbiogenesis could also be compatible with Darwinism: "The theory of symbiogenesis is a theory of selection relying on the phenomenon of symbiosis," he wrote.[42] Unfortunately, Kozo-Polyansky's works were published only in Russian and had the misfortune to appear at the height of the Stalinist era; he was virtually unknown in the West until the 1990s.

Symbiogenesis was also proposed independently in the 1920s by an American biologist, Ivan Wallin, who published the idea in a volume called *Symbionticism and the Origin of Species* (1927). Wallin's "rather startling proposal," as he disarmingly characterized it, was that "bacteria, the organisms which are popularly associated with disease, may represent the fundamental causative factor in the origin of species."[43] Claiming that mitochondria could be grown independently of their host cells (a dubious proposition), he was vehemently attacked by his peers. Wallin dropped the subject, and his theory was soon forgotten.

Only in the 1970s did the idea of symbiogenesis begin to gain favor, thanks to new findings and the more solid case developed by Lynn Margulis. At first Margulis was also ridiculed for defending the idea, but she persisted as the evidence in favor of it mounted (especially important was the new technology for analyzing DNA "fingerprints").[44]

Today, it is generally accepted that the mitochondria and the chloroplasts did indeed arise through symbiogenesis, and the Nobel biochemist Christian de Duve, Cavalier-Smith, and others think the peroxisomes did too. (Because the peroxisomes can buffer the cell against oxygen contamination, it is believed that they may have been the first to join the team and played a crucial role in paving the way for later alliances.) One remaining dispute concerns Margulis's proposal (again foreshadowed by a forgotten Russian predecessor) that the centrioles and kinetosomes (basal bodies) – specialized structures with the distinctive $9(3) + 0$ architecture that play a vital role in cell motility – also originated via symbiosis, perhaps among the ancestors of modern spirochetes.[45] Another debate concerns Margulis's thesis that the nucleus itself may be a product of symbiogenesis.[46]

There is no definitive explanation for how, and why, such symbiotic mergers came about in the first place. Several alternative scenarios have been suggested. Many theorists favor a predatory, "infection" model. In the case of the mitochondria, for example, ancestral purple bacteria are presumed to have been invaders that came to prey upon their protist hosts and only later discovered the benefits of mutualism. But other theorists think the key was the development of "phagocytosis" – the loss of a rigid outer cell wall and the ability to "swallow" and digest food internally. In this scenario, the eukaryotes' prospective symbionts were first ingested as prey, not predators.[47]

An intriguing alternative scenario, based on positive symbiosis, was recently proposed by William Martin and Miklós Müller.[48] It is called the "Hydrogen Hypothesis," and it is supported by a variety of genetic and biochemical data. Martin and Müller believe that the process of symbiogenesis (at least in terms of energy production capabilities) was cooperative from the start. In their view, a mutually beneficial association developed between ancient hydrogen-producing archaebacterium and a "methanogen" – a microbe that can utilize hydrogen to extract energy and make sugars, leaving methane as a waste product. The idea came to Martin one day when he was viewing a modern analog, a one-celled eukaryote called *Plagiopyla*. These protists have internal hydrogen-producing organelles called hydrogenosomes, which are surrounded by hydrogen-consuming symbiotic bacteria. Martin and Müller believe that both these organelles and the eurkaryotic mitochondria might trace their ancestry to such a partnership.

Another, purely speculative, mutualistic scenario is based on the functional relationship among photosynthesis, glycolysis, and respiration. It could perhaps be called the "Waste Hypothesis." Early eukaryotes that were able to engage in photosynthesis also generated large quantities of a waste product, oxygen, which is toxic to organic material at high concentrations. At least some of these eukaryotes also utilized glycolysis, which (as mentioned earlier) produces quantities of a second waste product, pyruvic acid. Then along came a purple bacterium that had developed a talent for using waste oxygen and pyruvic acid, via respiration, to meet its energy needs. The partnership was mutualistic from the start, and in time it became mutually obligatory. It is significant that some "archaezoa" (as they are called) that are *not* related to our own direct ancestors apparently engaged in photosynthesis but did not have mitochondria. In other words, the two ancient organisms may have evolved independently, but the symbiosis between the mitochondria and their photosynthetic hosts (a functional complementarity) empowered them to go on to bigger and better synergies.

The Economics of Symbiosis

Whatever the precise sequence of events, the common denominator in each of these proposed scenarios is the "economic" benefits that resulted. Just as many business mergers, joint ventures, and partnerships in human societies occur because various functional complementarities yield greater profits for the participants, even if the benefits may not be equally shared, so "symbiogenesis" in nature has been driven by the synergies – cooperative effects that are not otherwise attainable. The bottom line is the synergy, because the synergy benefits the bottom line – survival and reproduction. As we shall see in Chapter 5, the Synergism Hypothesis holds that functional synergies represent a major source of the variations on which natural selection acts. Let's look at a few contemporary illustrations that involve bacteria as one of the symbionts:

* One striking example involves the African honey guides (and their endosymbiotic bacteria) that were mentioned earlier. It happens that these resourceful birds also form symbiotic partnerships with humans, the nomadic Boran people of northern Kenya. Biologists Hussein Isack and Hans-Ulrich Reyer conducted a systematic study of this behavior pattern some years ago and found that Boran honey-hunting groups were approximately three times as efficient at finding bees' nests when they were guided by the birds. They required an average of 3.2 hours to locate the nest compared with 8.9 hours when they were unassisted. The benefit to the honey guides was even greater. An estimated 96% of the bees' nests that were discovered during the study would not have been accessible to the birds had the humans not used tools to pry them open. Moreover, the Borans' use of smoky fires to suppress the bees' defenses benefited both the humans and their symbionts.[49] (The partnership is also aided by two-way communications – vocalizations that serve as signals.)

* Many species of beetles in the family known as Scolytidae have adopted the strategy of tunneling under the bark of various trees, or into the heartwood, to create protected "galleries" for laying their eggs and hatching their larvae. Normally these invaders would be repelled by the trees' defensive actions, which include filling the cavities with resin. But the beetles are able to overcome these defenses with the help of a symbiotic fungus that kills the wood in the vicinity of their galleries. That's not all, however. It turns out that the beetles also depend on other symbiotic fungi and yeast bacteria to produce a thick lining for the beetles' galleries that serves as a food supply for their larvae. Not only are these creatures interdependent, but the beetles have also evolved a specialized body structure, called a mycangium, for carrying a starter supply of their symbionts with them when they leave their natal nests in search of uninfected trees.[50]

* Biologist Margaret McFall-Ngai has spent many years studying a remarkable example of symbiosis in the waters around the Hawaiian Islands. There is a nocturnal squid, called *Euprymna scolopes*, which hides in the sand in shallow waters during the day and emerges at dusk to forage for prey. What makes this aquatic predator so unusual is that it is aided by a conspicuous light organ that is similar in structure to a vertebrate eye. Even more unusual is the fact that this light organ is powered by a swarm of luminescent bacteria, *Vibrio fischeri*, which combine forces to provide the squid with an ingenious form of camouflage – yes, camouflage. The squid's light organ is not a headlight. It is so designed – complete with a reflector and a transparent lens – that the light emitted by the bacteria is diffused. This produces an effect called "counter-illumination." The light organ serves to eliminate the squid's shadow against the background of dusk-light, moonlight, and starlight coming from above. Without the benefit of this efficient subterfuge, the squid would be far less successful as predators and would themselves be much more vulnerable to larger predators.[51] (Other types of symbiotic "photobacteria" are utilized by various animals for illuminating food, for defense against predators and even for signaling to potential mates.)

Multicellular Synergy

Though bacteria were the first to discover the enormous potential for synergy in a division/combination of labor, multicellular organisms have elevated its use to a whole new level. Some insight into how this occurred is provided by the Volvocales, a primitive order of aquatic green algae that form tight-knit colonies resembling integrated organisms.

Volvocales have been popular with students of evolution ever since the nineteenth century, because they seem to mirror some of the earliest steps toward complex multicellular organization. The smallest of these species (*Gonium*) has only a handful of cells arranged in a disk, while the *Volvox* that give the Volvocales line its name may have some 60,000 cells in the shape of a hollow sphere that is easily visible to the naked eye. Each *Volvox* cell is independent, yet the colony members collaborate closely. For instance, the entire colony is propelled by a fur coat of flagella whose coordinated efforts keep the sphere slowly spinning in the water.

The synergies achieved by *Volvox* were illuminated in a detailed study some years ago by the biologist Graham Bell.[52] Bell noted that the largest of the colonies has a division of labor between a multicellular body and segregated reproductive cells. Bell's analyses suggested some of the benefits. The division of labor and specialization facilitates growth, which results in a much larger overall size. It also results in more efficient reproductive machinery (namely, a larger number of smaller germ cells). The large hollow enclosure in *Volvox* also allows the mother colonies to provide protected envelopes for their numerous daughter colonies; the offspring disperse only when the mother colony finally bursts apart.

But there is another vitally important collective benefit enjoyed by *Volvox*. Bell points out that their larger overall size results in a greater survival rate. It happens that these planktonic algae are subject to predation from filter-feeders like the ubiquitous copepods, but there is an upper limit to the prey size that their predators can consume. Integrated, multicellular colonies are virtually immune from predation by filter-feeders. It's another example of a synergy of scale.

Photosynthesis and Synergy

The awesome power of a combination of labor and a synergy of scale together can best be seen, perhaps, in the progressive evolution of a key biotechnology – photosynthesis.[53] Among the five different types of prokaryotes that engage in photosynthesis, perhaps the best known are the cyanobacteria, which were mistakenly called blue-green algae until it was discovered that they are not eukaryotes after all. Cyanobacteria don't have specialized chloroplasts, like the eukaryotes. They depend for their light-capturing capabilities on a small number of chlorophyll and other "antenna pigments" housed in an internal membrane system that is close to the cell surface. It's a fairly primitive system, but it's sufficient to meet the energy needs of the cyanobacterium; cyanobacteria are too small and too limited in their capabilities to support a larger energy-production system, much less utilize its output.

All this changed with the evolution of eukaryotes. Eukaryotic protists acquired specialized chloroplasts, each of which (at least in modern land plants) contains several thousand photosynthetic "systems" consisting of a "reaction center" and 250–400 chlorophyll and carotenoid molecules – perhaps as many as 1 million antenna pigments altogether. Moreover, each cell may contain 40–50 chloroplasts, or a total of 40 million–50 million antenna pigments.

In other words, specialization gives eukaryotes energy-capturing capabilities that are many orders of magnitude greater than their prokaryote ancestors enjoyed. A crucial corollary, however, is that the productivity increases in chloroplasts are only possible because they are a part of a larger collaborative enterprise; there are no free-living chloroplasts so far as I know. In other words, these functional improvements also created an interdependency (more on this later).

The next major technological improvement in photosynthesis is associated with the evolution of multicellular organisms, which are able to multiply the synergies of the chloroplasts many times over. Now each cell, with its 40 million–50 million antenna pigments, may become part of a vastly larger enterprise in which a great many photosynthetic cells combine forces and develop entire energy-capturing surfaces, each square inch of which might contain 10 million–12 million chloroplasts. And this already huge number (perhaps 50 trillion pigment molecules per square inch) can in turn be multiplied by the light-capturing surface area on a given plant. For a single deciduous tree, the total number of light-capturing pigments might be astronomically large – perhaps 50 million–100 million trillions.

As Adam Smith pointed out in *The Wealth of Nations*, the benefits of an "economy of scale" are limited by the size of the market – the number of consumers who can use what you produce. The same principle applies to photosynthesis. The enormous potential for energy production in the leaves (or needles) of a tree only makes sense if there are a sufficient number of "consumers" available to use it. And of course there are. A giant sequoia tree, for instance, might weigh as much as 4.4 million pounds and consist of 2 million trillion energy-hungry cells. But the record for size is held, very possibly, by a legendary quaking aspen in Utah, which has a single underground root system supporting 47,000 interconnected trunks. This super-tree is estimated to weigh more than 13 million pounds.[54] (By comparison, the largest animal ever, the blue whale, weighs in at a mere 425,000 pounds.) So, if you're not yet ready to go hug a tree, you might at least genuflect to the next one you happen to pass by and pay homage to the synergy it produces.

Many theorists these days eschew the word "progress" with regard to the evolutionary process, because, they claim, it connotes some form of betterment, or puts humankind on some sort of pedestal. Nevertheless, it is still legitimate to speak of "engineering improvements" that gave living systems new capabilities and new powers for growth and diversification. In this light, the energy-capturing technologies developed by humans are not *sui generis* but are part of a broader, "progressive" evolutionary trend that, in turn, has undergirded many other forms of evolutionary "progress." The progressive evolution of complex organisms may or may not increase the long-term odds of survival (look at the dinosaurs), but it is not immaterial to the problem. (We will return to the issue of evolutionary "progress" in Chapter 9.)

Cambrian Synergy

There is much uncertainty about when the first complex multicellular life forms appeared. Estimates range from about 600 million years ago to as much as 1.5 billion years.[55] However, there does seem to be a consensus that the so-called "Cambrian Explosion" – a unique flowering and proliferation of complex organisms around 540 million years ago – was a true "breakthrough" and not simply an artifact of gaps in the fossil record. Great mystery (and much debate) has surrounded this stunning example of nature's creativity. Paleontologist Stephen Jay Gould wrote a best-selling book on the subject, *Wonderful Life: The Burgess Shale and the Nature of History* (1989), in which he drew some provocative (and recently challenged) conclusions about the role of historical contingencies and the historical uniqueness of this breakthrough.[56] In any event, the unresolved question is this: How can we explain what seems to be a contradiction of the traditional assumption – going back to Darwin himself – that evolution proceeds by very gradual, incremental steps?

Darwin was fond of quoting the popular canon of his day, *natura non facit saltum* – nature does not make leaps. But Gould (and others) claimed that nature did make a saltatory leap in the Cambrian period. All of today's animal phyla arose during this relatively small slice of evolutionary time. Echinoderms, mollusks, coelenterates, arthropods, and, of course, our own direct ancestors among the chordates (plus some lesser-known phyla as well) all originated within a period of perhaps 5 million to 10 million years.

The explanation, I believe, will not be found in some obscure historical accident or in a single breakthrough development that remains to be discovered, or literally "unearthed." Rather, it involved another example of the Bingo Effect; it was the "convergent" result of a synergistic combination of many preconditions and much previous "progress" (as we defined it earlier). Perhaps it was also triggered by a threshold event (several have been proposed), but this "event" had the consequences it did only because the time and the context were right – both "external" conditions and prior "internal" (bioengineering) developments converged. If ideas come to the prepared mind, as the saying goes, evolution comes to the prepared organism in a suitable environment. Nature does not make leaps, but it does cross thresholds and achieve potent new forms of synergy.

Certainly there were many external factors – the build-up of oxygen and the protective ozone layer in the atmosphere; a moderating (warming) global climate; a surging proliferation of multicellular life forms, and, very likely, a population explosion during the preceding Ediacaran Period; possibly there was also an overcrowding of available habitats, which may have increased competition, the search for new niches, and the intensity of natural selection.

However, the dynamite was mixed for the Cambrian explosion by an accumulation of functional improvements in living systems over the preceding 2 billion years, culminating in the many soft-bodied multicellular organisms that left few traces behind in the fossil record. The Cambrian breakthrough also depended on many previous inventions – the complex eukaryotic cell, sophisticated sensory abilities, neurological systems, complex reproductive machinery, and, of course, various forms of multicellular organization and a division/combination of labor.

And this says nothing about the many early developments in the molecular biology and biochemistry of life – the invention of DNA and the double helix, the development of proteins and enzymes, glycolysis and the Krebs cycle, and much more. Much has been made recently of the so-called homeobox gene complex, a remarkably conserved gene-set that collectively determines the basic body plans of both *Drosophila* flies and humans, and everything in between. Many theorists also point to the evolution of blastula as an important precursor. Blastulas appear at a very early point in the reproductive cycle for complex organisms, when the developing embyro forms a single layer of undifferentiated cells in the shape of a hollow, fluid-filled sphere. The blastula phase sets the stage for the cellular differentiation/specialization that follows.

From a synergy perspective, the Cambrian breakthrough was not a "revolution." It was a threshold phenomenon resulting from a convergence of many preconditions that together provided the framework for the proliferation of a more complex form of biological organization. Again, it was synergistic – an example of what I have termed the Bingo Effect. (In Chapter 7 we will observe similar threshold effects in the evolution of humankind.) How do we know this was the case? Just apply the synergy test. Take away any of the major contributing factors – the Hox genes, the mitochondria, neuronal networks, and many others. As Margulis and Sagan put it in their recent volume, *What Is Life?*, "the animal explosion had a long microbial fuse." The Cambrian explosion illustrates once again how history and functional synergy have been closely intertwined in the Epic of Evolution. In the final analysis, the Cambrian flowering was not the result of a capricious accident but of an "engineering" breakthrough at the right time and place.[57] An analogous technological breakthrough in human evolution, the automobile "revolution," will be described in Chapter 8.

Social Synergy

Still another level was added to the Magic Castle with the development of social organization among multicellular organisms. In effect, the division/combination of labor and various opportunities for synergies of scale were elevated to a new level; macroscopic synergies became possible that were beyond the capabilities of microscopic organisms, even *en masse*. Some of the many examples of social synergy will be described in Chapter 5 – leaf-cutter ants, honeybees, acorn woodpeckers, meerkats, sea lions, dolphins, communal wasps, prairie dogs, chimpanzees, and others. Here I will use only a single, unique example.

The naked mole-rat is an African rodent species that lives in large underground colonies (usually numbering 75–80 but sometimes more than 200). They subsist by eating plant roots and succulent tubers. Affectionately dubbed "saber-toothed sausages" because they are hairless and have two outsize front teeth for digging, the naked mole-rats represent a particularly significant example of a division/combination of labor in mammals. In fact, these odd-looking animals utilize specialized worker "castes" and a pattern of breeding restrictions that is highly suggestive of the social insects.

Typically (but not always), the breeding is done by a single "queen," with other reproductively suppressed females waiting in the wings. The smallest of the non-breeders, both males and females, engage cooperatively in tunnel-digging, tunnel-cleaning, and nest-making, as well as transporting the colony's pups, foraging for food, and hauling the booty back to strategic locations within the colony's extensive tunnel system. (One investigator, Robert A. Brett, found a mole-rat "city" in Kenya that totaled about 2 miles of underground tunnels and occupied an area equivalent to 20 football fields.) Biologist Paul Sherman, the leader of a group of researchers who have studied these animals extensively, wrote the following description of their tunnel-building activities:

> The animals line up head-to-tail behind an individual who is gnawing [with its outsized, powerful front teeth] on the earth at the end of a developing tunnel. Once a pile of soil has accumulated behind the digger, the next mole-rat in line begins transporting it through the tunnel system, often by sweeping it backward with its hind feet. Colony mates stand on tiptoe and allow the earthmover to pass underneath them; then, in turn, they each take their place at the head of the line. When the earthmover finally arrives at a surface opening, it sweeps its load to a large colony mate that has stationed itself there. This "volcanoer" [so-named because its actions appear to an observer outside to produce miniature volcano eruptions] ejects the dirt in a fine spray with powerful kicks of its hind feet, while the smaller worker rejoins the living conveyor belt.[58]

The vital and dangerous role of defense in a mole-rat colony is also allocated to the largest colony members, who respond to intruders like predatory snakes by trying to kill or bury them or by sealing off the tunnel system to protect the colony. The mole-rat "militia" will also mobilize for defense against intruders from other colonies.

The students of animal behavior find many analogies between the naked mole-rats and eusocial insects like the army ants and honeybees. But in their "politics" and "government," the mole-rats are more convergent with other social mammals, like chimpanzees or humans. As is the case with many other socially organized species, naked mole-rats exhibit a combination of self-organized cooperation ("volunteerism") and social controls that are enforced by various coercive measures ("policing").

The control role of the breeding queen is of central importance. The queen is usually the largest animal in the colony (size usually determines the dominance hierarchy), and she aggressively patrols, prods, shoves, and vocally harangues the other animals to perform their appointed tasks. Indeed, it has been observed that her aggressiveness varies with the relative urgency of the tasks at hand. In addition, the queen acts to suppress breeding and reproduction on the part of the other females, who are always ready to take over that role. (Occasionally other females are allowed to share the breeding function with the queen; why this is so is not known.)

The mole-rat queen also intervenes frequently in the low-level competition that goes on among colony members over such things as nesting sites and the exploitation of food sources. And when the reigning queen dies, there is a sometimes bloody contest among the remaining females to determine her successor.

All of this control activity is facilitated by an elaborate communication system (and information sharing) that includes 17 distinct categories of vocalizations – alarms, recruitment calls, defensive alerts, aggressive threats, breeding signals, etc. In fact, the mole-rats' communication system rivals that of some primate species in its sophistication.

Why do mole-rats utilize this highly cooperative survival strategy? Sherman and his co-workers provide an economic explanation for their behavior:

> We hypothesize that naked mole-rats live in groups because of several ecological factors. The harsh environment, patchy food distribution and the difficulty of burrowing when the soil is dry and hard, as well as intense predation, make dispersal and independent breeding almost impossible. By cooperating to build, maintain and defend a food-rich subterranean fortress, each mole-rat enhances its own survival.[59]

In other words, the mole-rats' strategy is ultimately based on synergy. (Although it is not stressed in the mole-rat research literature, another critically important facilitator is a cooperative relationship between the mole-rats and a bacterial symbiont that can break down the cellulose in succulent tubers.)

Synergy in Ecosystems

Finally, let's look briefly at the role that synergy plays in ecology. The term "ecology" (like the term "economics") is derived from the Greek word

oikos, meaning the "household." Accordingly, ecology refers to the study
of nature's household – the pattern of relationships among the various
organisms in any natural setting, and between these organisms and their
environments.

Darwin himself appreciated the importance of the "economy of nature," a term he frequently used. (The phrase apparently traces to the great
eighteenth-century classifier of species, naturalist Carolus Linnaeus.) In *The
Origin of Species*, Darwin made various references to the "web of complex
relationships" in nature, the "infinitely complex relations" among species,
and the marked "interdependence" of various forms of life. And, in one of
the most widely quoted passages from his masterwork, Darwin wrote:

> It is interesting to contemplate an entangled bank, clothed with many
> plants of many kinds, with birds singing on the bushes, with various
> insects flitting about, and with worms crawling through the damp earth,
> and to reflect that these elaborately constructed forms, so different from
> each other in so complex a manner, have been produced by laws acting
> around us.[60]

"Interdependence," including a subtle interplay of competition and cooperation, is an appropriate way to characterize even a simple ecosystem.
As ecologist Jeff Harvey put it in a recent issue of *Nature*, an ecosystem is
"an integrated system involving the biological, chemical and physical components of a habitat or a region."[61] A classic case in point is the ecosystem of
Lake Superior's Isle Royale, where the efforts of ecologists over the course of
the past century to strike a balance between the local moose population and
predatory wolves have been confounded by various unexpected factors –
climate fluctuations, changes in plant abundance, disease patterns, and even
the effects of inbreeding in the wolf population. In short, ecosystems are
suffused with synergy, both positive and negative. (We'll focus on the latter
in Chapter 4.) Here are just a few brief illustrations:

∗ Fungi are absolutely essential "players" in most terrestrial ecosystems.
  They serve an indispensable role as "decomposers," consuming dead
  matter and returning organic products and nutrients to the soil as natural
  fertilizers. Some biologists refer to fungi as nature's garbage collectors,
  but a better description might be nature's recyclers.
∗ A variation on this theme involves the 14,000 species of dung beetles,
  which have been dubbed nature's pooper-scoopers. In some parts of
  Africa, dung beetles bury and later consume as much as a half ton of

animal detritus per acre per year. In one study in Kenya, 48,000 beetles visited a 6.5-pound pile of elephant dung over two hours and completely demolished it. Without these detrivores, as they are called, the lush African savannas would be wastelands, piled high with droppings from its great herds of herbivores – which would have moved on to greener pastures, as it were. Dung beetle tracks have been found even in fossilized dinosaur feces, going back 75 million years.[62]

* An estimated 80% of all land plants depend on the services provided by symbiotic root fungi, the more than 5,000 species of mycorrhizae that augment the plants' own root systems. The fungi play a key role in the transfer of phosphorus and other minerals to the plant roots, as well as enhancing the plant's water uptake and drought tolerance and its ability to resist pathogens. In fact, seedlings planted in soils where nutrients are abundantly available will still grow poorly or not at all without their symbiotic "fungus roots." The plants return the favor by secreting sugars and amino acids that are absorbed by the fungi. So important is this partnership that biologists Kris Pirozynski and David Malloch have proposed that the initial appearance and spread of terrestrial plant species, perhaps 440 million years ago, was actually the result of a joint venture.[63]

* The flowers that bloom in the spring, to borrow a line from Gilbert and Sullivan's *The Mikado*, have nothing to do with pleasing humans. They are designed to attract pollinators – bees, hummingbirds, butterflies, wasps, beetles, moths, even some birds and mammals. In fact, many plants cannot reproduce without the services of pollinators. For the plants, the sacrifice of some pollen to feed the "pollinivores" – as they are called – is more than repaid by having a vastly more efficient way of transporting their pollen to other flowers, or to other parts of the flower. Rather than having to rely on wind, rain, or gravity, the pollinators can provide targeted delivery services.

* Although most of us are oblivious to the fact, humans are also heavily dependent on "the little things that run the world," as entomologist Stephen Buchmann puts it. Bees alone pollinate an estimated $40 billion a year in agricultural crops in the United States (about one-third of our total), according to the well-known ecologist David Pimentel, including everything from almonds and apples to tomatoes and zucchini. And that excludes pasture crops like alfalfa and clover, plus some 200 million pounds of honey produced at commercial apiaries each year. On a sunny day in June, a single bee might visit 1,000 or more flowers, Pimentel estimates. For the total bee population of New York State alone, that could add up to 1 trillion flower-visits a day.[64]

* The plants' dependence – and ours – on pollinators is the reason why a drastic decline in the number of pollinators in the United States over the past ten years has been termed a "pollination crisis." Honeybees have been especially hard hit, and so have some of our crops, by a combination of problems: parasites (mites), which were first introduced into this country in 1984, plus an invasion from South America of "killer bees" that take over the honeybee hives, exposure of the bees to pesticides, and, perhaps most serious, the effects of commercial and residential development, which disrupts "nectar corridors" and fragments habitats.[65]

* The phenomenon of exponential population growth, mentioned earlier, can profoundly affect the character of an ecosystem over a very short period of time; the mathematics of reproduction offer a tremendous potential for a synergy of scale. In one well-known illustration, ecologist E. J. Kormondy in the 1960s calculated that a single reproductive pair of houseflies could, in theory, produce 5.6 trillion descendants within the course of a single year (seven generations).[66] A real-world approximation happened when rabbits were introduced into Australia by an English gentleman in 1859, the same year that *The Origin of Species* was published. The original dozen began to breed, well, like rabbits. Six years later the Englishman had killed 20,000 on his own estate alone. By 1887 some 20 million had been killed just in the province of New South Wales. By 1950, much of Australia was being denuded of its vegetation. A virus was then introduced to control the rabbit population. But coevolution over time has led to a stable host–parasite relationship. So the war continues.[67] Of course, few if any organisms ever achieve their maximum potential growth rate for very long. They are constrained by a web of ecological dependencies. But in the short run (our lifetime), exponential growth can destabilize an ecosystem and wreak havoc in the natural environment.

An Oasis of Synergy

Perhaps the most stunning example of ecological interdependency can be found in reef corals. These complex, highly videogenic communities (they are popular with TV nature series) have been called "ecological oases," and rightly so.[68] They are found in relatively barren tropical areas where the water temperatures are higher than 18° Celsius (64.4° Fahrenheit), yet they are by far the most diversified of all marine ecosystems. Some of these "coral cities" are huge. In fact, many pacific islands are superstructures built on a scaffolding of coral, and the Great Barrier Reef off the northeast coast of Australia is more than 1,200 miles long and provides food and shelter for untold millions of organisms – reef fish, sponges, sea urchins, marine worms, crustaceans, and more.

The key to a coral reef community is an indispensable symbiotic partnership between the polyps that are responsible for building the coral and various green algae and dinoflagellates that engage in photosynthesis. The polyps have a tubular structure, with a mouth and a ring of tentacles at the top and a base that attaches to the scaffolding of calcium carbonate (limestone) skeletons left behind as a biological inheritance from their innumerable ancestors. The polyps take many of the raw materials they need, particularly calcium and oxygen, from the surrounding waters. They also feed on the organisms that pass by in the water, especially the larvae produced by the reef's other inhabitants. However, they depend for their energy needs, as well as certain minerals, on the green algae and dinoflagellates. In return, the photosynthesizers acquire a stationary platform and various chemicals, including urea and certain needed amino acids. It is a synergistic relationship; neither the coral polyps nor their partners could survive independently.

The coral reef in turn provides food and shelter for a tremendous variety of other creatures that earn their livings in a complex web of predatory, competitive, and cooperative relationships (niches). There are diurnal grazers and nocturnal grazers; there are diurnal predators and nocturnal predators; and there are numerous symbiotic relationships. For instance, certain shrimps benefit from the keen eyesight of bottom-dwelling gobies; there are pearl fish that hide in the mantles of tropical oysters; clownfish often find a safe haven in the stinging tentacles of giant sea anemones; and more than 50 species of small "cleaner-fish" sustain themselves by safely removing parasites, food scraps, loose scales, and other detritus from larger, predatory fish, sometimes at designated "cleaning stations" where the cleaners' appreciative clients may wait patiently in line.[69]

The most colorful example of symbiosis in the coral community, perhaps, is the well-documented partnership between sea anemones and hermit crabs. The anemones provide camouflage and protection for the crabs with their formidable stinging tentacles, while the crabs provide large abandoned snail shells (which they appropriate for hiding places) as attachment structures for the anemones. But more important, as the crabs haul their shells around they also provide transportation services for their sea anemone partners. It's a mutually beneficial arrangement.

Synergy in the Biosphere

The synergies found in local ecosystems can also be seen, in a more diffuse but equally important way, in the global ecosystem – the biosphere. The concept of the Earth as a single interdependent system, including both the physical environment and all living organisms, was first popularized by the Russian geologist Vladimir Vernadsky in his 1926 book, *The Biosphere*. Vernadsky recognized that the boundary line between living matter and the non-living environment is fuzzy and highly permeable. In fact, we now know that living organisms have had a major role in shaping the character of our physical environment and that the biosphere is as much a product of evolution as is, say, our bipedalism and our exalted brain. Evolution has been a unified, multileveled process.

The most dramatic example is, of course, the gradual transformation of the Earth's atmosphere over perhaps 1.7 billion years from being virtually oxygen-free to an oxygen-rich environment that is capable of supporting aerobic (oxygen-breathing) life forms. This profoundly important change was collectively accomplished by the contributions of countless photosynthetic bacteria, green algae, and land plants, and it provides a cardinal illustration of a cumulative synergy of scale. Our uniquely oxygenated atmosphere is a combined effect produced by countless individuals over an incomprehensibly long period of time. Equally important, it is another example of a synergy that was time-dependent – an irreducible product of a historical process.

There are many less-well-known examples of such biologically induced synergies at the global level: the moderation of surface and air temperatures, the salinity and pH of the oceans, chemical balances in the oceans and terrestrial soils, the concentration of various inorganic minerals in underground veins, the vast deposits of limestone, the accumulation of fertile humus soils, the mix of gases in the atmosphere, the ozone layer that partially protects us from UV radiation, and more.

Equally significant is the collective (synergistic) impact of living organisms on the dynamics of global weather patterns. Not only do the vast carpets of plankton in the oceans consume much of the Earth's carbon dioxide and produce much of its oxygen but they also seed the clouds with sulfur compounds. For instance, there is a huge green belt of algae (*Emiliania huxleyi*), affectionately known as "Emily," lying off the coast of Scotland that is some 30 miles wide and 120 miles long. Not only does Emily pump sulfur oxides into the atmosphere along with oxygen, but over time it has also deposited a thick layer of its ornate calcium carbonate shells on the ocean floor.[70]

Good Gaia!

Many environmentalists, and some scientists, have endorsed geochemist James Lovelock's highly contentious and much-debated Gaia Hypothesis – the notion that the Earth and its biota are in fact a single, self-regulating, "homeostatic" organism – literally. "The Earth is alive ... the Earth is a superorganism," claimed Lovelock early on.[71] For instance, Lovelock argues that certain global processes are feedback-driven; oxygen and carbon dioxide levels in the atmosphere are a case in point.

As a metaphor for the global impact of biological activity (including our own), the Gaia Hypothesis may be useful, but I will side with those who believe the literal version is mistaken. Lovelock equates interactive/interdependent forms of unintentional causation – what he calls "automatic feedback" – with purposeful behavior. But "feedback" control is technically not possible without a cybernetic control mechanism – a thermostat or a governor that regulates the behavior of the system. Many processes in the natural world mimic feedback-driven behavior, but they are in fact fortuitous (convergent) effects; they are susceptible to radical change over time precisely because they are not controlled by an independent feedback mechanism. (We will consider this widespread misunderstanding further in the next chapter.)

We can use a human example to illustrate. The daily dance of the New York or London Stock Exchange depends on a balance between buyers and sellers; prices go up or down when transactions occur. And, in every transaction, there must be a seller and a buyer. Yet this perfect balance of buyers and sellers seems to recur day after day. Miraculous? Of course not. An economic version of Gaia? Hardly. It is a self-organized process that follows a complex set of trading procedures and "rules." Sometimes, in fact, the daily balance is only achieved through the actions of the floor traders (the "specialists"), who are charged with maintaining "an orderly market" in a given stock and who may have to buy or sell from their own personal portfolios to keep the process under control.

So there is an element of cybernetic control related to the market in any given stock, but the prices, and the course of the market as a whole, are not controlled. Its collective behavior from day to day is a synergistic effect of many individual actions. In fact, trading activity for a given stock, or the market as a whole, does break down from time to time, for various reasons. No investor who lived through the one-day 500-point meltdown (20%) on the New York Stock Exchange in October 1987 will ever forget it. I would argue that it is the same with oxygen levels and the like in the atmosphere. If there is currently an orderly market in nature, it is historically contingent – not cybernetic. To claim otherwise is to sneak purposiveness into nature in an inappropriate way.[72]

The Whole Shebang

To sum up our brief survey, the Big Whoosh was not some unimaginably remote and obscure event that merely started up the cosmic clock. It initiated an ongoing historical process – only partly shaped by physical "laws" that may well be historical artifacts – a process proceeding with all deliberate speed on a time-scale that we cannot even comprehend. At its heart, the Epic of Evolution has been a richly creative enterprise, and we are an integral part of this "Whole Shebang" (to reiterate Timothy Ferris's title), however small our own personal part may be.

Every organ, every cell, every molecule, every atom, and every quark in our bodies traces back to the primordial stuff. (If we are indeed composed of some 10 trillion of those complex eukaryotic cells, the number of quarks must be astronomical.) We have a direct historical linkage to the Great Beginning, and to all the smaller beginnings ever since, from the formation of the Earth to the origin of life, the invention of the genetic code, the emergence of the eukaryotes, the very first dawn for our primate ancestors, and the very moment of our own conception with the joining of a single egg and sperm.

Most of us roam the parapets of this magic castle in self-absorption. We seem to be unaware of the intricate, many-layered structure below us and all around us; we remain oblivious to our vital connection with its intricate architecture. Worse yet, many of us are ignorant of how completely we are supported and sustained by the myriad of synergies that produced this grand edifice. How many of us are even aware of the strong nuclear force, the covalent bonds, the cell membranes and cytoskeletons that literally hold us together? Some among us believe we can discount the details and reduce it all to a few fundamental laws or a grand "final theory." The problem with this ambitious goal is that it leaves out much of what is important about the Epic of Evolution. (There is more to be said on this subject in Chapter 9.)

We are certainly no more myopic in our self-induced ignorance than any other living species, but we alone have the ability to do better. We have the intellectual capacity to understand the cosmos and our place in it. We are in fact the legatees of an awe-inspiring history, and the golden thread of that history has been synergy. Synergy sustains us, but it also presents us with a plethora of opportunities – and problems – which we will explore further in due course. To borrow a line from the distinguished paleontologist George Gaylord Simpson (the title of his most famous book), in the truest sense synergy is "the meaning of evolution."

4   "Black Magic"

That old black magic has me in its spell,
That old black magic that you weave so well,
Those icy fingers up and down my spine,
The same old witchcraft when your eyes meet mine . . .

Johnny Mercer/Harold Arlen

From nature's chain whatever link you strike,
tenth or ten thousandth, breaks the chain alike.

Alexander Pope

I try hard to be punctual for appointments. It's a discipline that is ingrained from my successive careers as a naval aviator, a science journalist (*Newsweek*), a college professor (in Human Biology at Stanford University), and a private-sector consultant. Hence, I was acutely embarrassed recently when I was more than a half-hour late for a meeting with a visiting colleague at a hotel located in an unfamiliar nearby town. As I pulled into the hotel parking lot, it dawned on me that I had been a victim of negative synergy. No single thing was responsible; there had been a "convergence" of problems. First, I had to make an unforeseen stop for gas. Then there was an accident on the highway; traffic had slowed to a crawl. This cost me several more minutes. My friend's directions were a bit ambiguous, so I got off the highway at the wrong exit. It took me a few minutes to identify the problem, check the map, and make my way back to another highway on-ramp. When at last I arrived at the correct exit, the street signs were so poor and dimly lit in the gathering darkness that I made two wrong turns and had to get the map out again. To top it off, when I finally arrived at the hotel parking lot, it was full. I had to drive up the street to hunt for a parking space. Fortunately, my friend was a connoisseur of synergy and understood my plight. But just in case, I paid for his dinner.

Negative Synergy Is Not an Oxymoron

Negative synergy – sometimes called "dysergy" – strikes many people as a contradiction, an oxymoron. How can synergy be negative? By definition (supposedly) synergy is a good thing; the whole is greater than the sum of its parts, and all that.

In fact, not at all. Synergy is neither intrinsically good nor bad. To borrow a phrase, synergy happens, and our attitude toward it depends on our values, and our interests. A snowstorm in the mountains might be a costly nightmare for a long-distance trucking firm, but for a ski resort operator it could be an unalloyed delight. When a flu virus multiplies and spreads rapidly through the population, the silver lining (so to speak) could be a boost in sales and profits for the drug manufacturers. If the boos coming from the audience at a political rally drown out the candidate's speech, the hecklers will be happy but the speaker will no doubt feel differently. A slave system may be very profitable for the owners but the slaves will probably not appreciate the benefits. A plunging stock market may be bad news for everyone – except for the investors who sold short. A mafia "family" may pride itself on its authoritarian efficiency, but the police and the public may feel menaced. And the synergy of scale that occurs when the voters go to the polls will have diametrically different consequences for the winners and the losers. Whether synergy is a good thing or not depends on your point of view. If a synergistic effect thwarts your purpose, whatever it may be, you are likely to view it negatively, and vice versa.

A classic illustration, the causes of which may never be fully known, involved the "meltdown" in California's energy markets and a devastating spike in energy prices in late 2000 and early 2001. While energy trading companies like Enron and Dynegy profited handsomely from this highly suspicious short-term crisis (prices have since declined and stabilized), the cost to Californians in terms of excess electricity charges alone may exceed $30 billion. And this says nothing of the lost revenues to many California companies and lost wages to California workers from rolling blackouts and other constraints on energy usage. For the energy trading companies, however, it was a case of rampant positive synergy.

Whatever the full explanation may turn out to be, for the residents of California the energy crisis was a model of negative synergy. A major underlying cause was that, despite a booming economy, the state of California and its utilities had underinvested in power plants during the 1990s, and energy deficits were unavoidable. A state plan to deregulate its energy market was also deeply flawed. It freed up prices but capped what could be charged by the utilities while, at the same time, barring them from signing long-term contracts. In other words, the utilities were at the mercy of the spot market, which was highly vulnerable to manipulation. Add to that a drought in the Pacific Northwest, which reduced regional hydroelectric power output, plus a suspicious rise in natural gas prices, and the remarkable coincidence that a record one-third of California's power generators went off-line for "repairs" during the most critical period.[1]

Of course, many forms of synergy are more or less universally viewed as unwelcome and undesirable, and we probably experience these negative synergies just about as often as we do positive synergy. In fact, there is a mirror image on the dark side for every one of the different kinds of synergy identified in Chapter 2. It could be called the antimatter of the synergy principle. (To balance the ledger a bit, we will rely here primarily on illustrations from human societies.) Thus, if the U.S. Postal Service is a good example of positive synergy, the highly efficient organization that ran the concentration camps for the Nazis is an example of the evil that organized social synergy can sometimes produce. Likewise, in the classic business merger situation, the whole is sometimes much less than the sum of its parts. For various reasons, the newly merged partners may do much worse than either of them did before the marriage, and so may the market price of their stock. One critic, business professor Mark Sirower, detailed this phenomenon in a book called *The Synergy Trap: How Companies Lose the Acquisition Game* (1998).[2]

There are innumerable negative synergies of scale in our everyday lives – overloaded telephone circuits that produce busy signals; the long lines at the checkout counter; the acoustics in a crowded restaurant, where everyone must shout to be heard and everyone goes home a little hoarse; rush-hour highway congestion (in California alone, the cost is estimated to be about 1.5 million person-hours each day); the combination of electrical appliances that collectively blow the fuse; or the accumulation of dirt particles in a water filter that clogs the cold water pipe. And when 4 inches of rain fell on New York City within a 3-hour period back in 1999, flooding brought the city to a standstill.

Many synergies of scale produce an admixture of good and bad effects. Take nuclear power. The dense concentration of radioactive materials in the core of a nuclear reactor provides a low-cost source of electricity (a positive synergy), but in the longer term it also produces a dangerous negative synergy – a lethal mass of radioactive wastes. Indeed, weapons-grade plutonium will remain hazardous for at least 250,000 years. Likewise, a "glut" of some commodity, say oil, may drive down prices (good for the consumers), as well as corporate revenues (bad for the producers). Lately, of course, the reverse dynamic has been at work; producers have been holding down the supply and consumers have been feeling the pain.

A variation on this theme is what businessmen euphemistically refer to as "overcapacity." If the manufacturers of automobiles, for example, produce many more vehicles than there are potential buyers, this can become a collective problem. Price competition may increase sharply, factories may operate at far below their designed capacity, and profit margins may decline across the board. Indeed, the hallowed economic "law" of supply and demand is really a law about synergies of scale and, especially, threshold phenomena.

"Dysergy" in Cyberspace

The Internet provides daily examples of both positive and negative synergies of scale. A single message posted on a website can inform millions of people around the world, or raise millions of dollars for a political campaign. Yet even a single virus can do untold mischief to millions of computers and require costly anti-virus measures. Economies of truly enormous scale allow companies like Amazon.com and many other e-commerce pioneers to sell books, and a growing mountain of other things, more efficiently (at lower cost) than ever before. But there can also be diseconomies of scale. To cite just one recent example, a scheme developed by a computer engineer in Las Vegas and posted at an on-line gathering place allowed computer nerds around the country to subvert the business plan of a Texas company, Netpliance. The company was marketing a dedicated Internet terminal at below cost ($99), with the intention of making up the difference with Internet access fees. The plan began to unravel when the computer-techies started buying up the terminals, converting them into low-cost personal computers, and reselling them at bargain prices without the Internet tie-in.[3]

The double-edged nature of the Internet is in fact a paradigm for the double-edged nature of synergy. Though it is a dream come true for the scholars who benefit from remote access to the Library of Congress, at the same time it is a nightmare for the parents who are concerned about the intrusive seductions of on-line pornography. Law enforcement officials and even cops on the beat now have instant access to criminal identification databases from around the world, but would-be bomb makers also have access to a veritable *Joy of Cooking* for do-it-yourself explosive devices. And if the Internet has opened up vast opportunities for e-commerce, it has also created a major threat to many traditional businesses, from publishers to travel agents.

Threshold Effects

Threshold effects with negative consequences (synergy plus or minus one) are also a commonplace in human societies. When airline "load factors" – the percentage of seats that are occupied on a given flight – drop below a certain level (the percentage was once in the mid-sixties but it's now said to vary considerably), the airline will lose money. When the sale price of a product falls below the "fixed cost" for making it, the producer will soon be out of business. When a river reaches the flood stage after a heavy rainfall, it may spill over the levee and inundate surrounding communities. When too many high-rise buildings have been added to a low-rise town, the small-town feeling may be lost. And a single factory may contribute only a small amount of toxic waste to a once-pristine river, but when there are several factories doing so, the fish may die in droves and swimming may be banned.

There are also many cases of threshold effects where too much of a good thing becomes a bad. Adding a little salt to your meal may enhance the flavor, but there's obviously a limit. A mass-transit system benefits from having more riders to share the costs – up to a point. Adding more boats or larger nets to a fishing fleet may increase the catch – and profits – over the short term. But, as we have learned, over-fishing is likely to result in a catastrophic long-term decline in fish stocks. When the share of TV time devoted to commercials in the United States rose to a record 16 minutes and 43 seconds per hour at the end of 1999, and with it an increase in complaints from viewers, some advertising agencies and their clients began to voice serious concern that this relentless sales bombardment was beginning to diminish the effectiveness for all the advertisers.[4] Book publishers had a similar revelation a few years ago when the number of titles published annually hit 58,000. With 159 new books appearing each day, even the speediest Renaissance reader could not keep up.

Negative synergy can also be caused by functional complementarities of various kinds. Drug interactions are an obvious example. Perhaps the most highly publicized case in recent times was the tragic death of Karen Ann Quinlin in the mid-1970s from the combination of Valium (a tranquilizer), Darvon (a painkiller), and gin and tonic, which caused massive brain damage and several years of lingering life-support. Heavy consumption of Acetaminophen (the active ingredient in Tylenol and other drugs) together with alcohol can cause liver damage. Alcohol is also known to interact negatively with various arthritis drugs, anti-anxiety drugs, diabetes drugs, blood pressure drugs, and ulcer drugs. Even grapefruit juice and various drugs, like Plendil and Procardia, may have harmful effects when they're consumed together (headaches, irregular heartbeat, nausea, etc.). The well-known drug researchers Joe and Teresa Graedon, who have specialized in ferreting out harmful drug combinations, find that these side effects are often underreported and underrated by consumers. The title of their recent book, *Dangerous Drug Interactions*, says it all.[5]

Many diseases have multiple contributing causes, a negative complementarity. The risk of heart disease is affected by various factors – a genetic predisposition, stress, lack of exercise, a weight problem, and diet. Diabetes, likewise, involves a synergistic effect. In the so-called Type I version of the disease, which is more severe and usually strikes in childhood, a number of contributing genes seem to be involved. But equally important are various environmental "triggers," like viral diseases (mumps, etc.) and possibly even the consumption of cow's milk. It is suspected that these environmental influences may kill off the insulin-producing beta cells in the pancreas. However, this occurs only in biologically vulnerable bodies. Type II diabetes, which is far more common, strikes later in life and is strongly influenced by obesity. An overweight person who is diabetes-prone may develop a resistance to the sugar-regulating effects of insulin.[6]

Environmental problems like air pollution are often products of negative synergy. Mexico City, with 22 million inhabitants, is perhaps the most notorious example. The World Health Organization ranks it among the most contaminated cities in the world, with more than 300 bad air days a year. As a result, the WHO estimates that more than 1 million residents suffer permanent breathing difficulties, and many thousands die each year from respiratory diseases that are exacerbated by air pollution. The causes of Mexico City's smog are many: geography and weather patterns; pollution by the heavy concentration of industries whose pollution restrictions are inadequately policed by corrupt inspectors; smoke-belching buses that are exempted from pollution-control requirements; rapid growth in the number of automobiles, whose frequent auto-exhaust inspections are widely circumvented with bribes; and more.[7] Like many other such synergistic problems, there are no simple answers.

Negative Social Synergy

Functional complementarities may also play a negative role in shaping our social behavior. The famous experiments by psychologist Stanley Milgram in the 1960s on the so-called "aggressive triad" provide a classic illustration. Milgram showed that the odds of violent acts toward another person are greatly enhanced when one person in a position of authority can make the decision without actually having to carry it out and gives instructions to a subordinate who bears no responsibility for the action and merely follows orders.[8] It has been said, with good reason, that there would be fewer wars if our political leaders also had to do the fighting.

There are many other kinds of negative social synergy. For instance, the management guru James A. Stoner discovered that groups are much more likely than individuals to take risks under exactly the same circumstances. He called it the "risky shift."[9] Psychologist Robert Zajonc also showed that humans and many other animals may perform various tasks better, or worse, in the presence of their fellows. Zajonc called it "social facilitation."[10] Of course, the best-known form of socially induced synergies include what Gustav Le Bon in his classic work *The Crowd* (1896) called the "collective mind," though it is more familiarly known today as "mob psychology." People in crowds will often behave in ways they never would if they were not swayed by what others were doing – cheering, booing, brawling, looting, or lynching.[11]

The division or combination of labor in human (and animal) societies can also produce negative synergy, but often indirectly because the participants are much less likely to continue working together if it is punishing to them to do so. One exception, of course, is slave systems, where the cooperation is coerced and the benefits and costs are highly skewed. Another place to look for negative synergies is in competitive situations. Thus, if a pack of wolves is more successful at capturing black-tailed deer than a lone wolf, then it is clearly a case of negative synergy from the point of view of their prey. Likewise, if a coalition of male lions can prevent a lone male from gaining mating privileges with the females, the loser will be a victim of negative synergy. If a large, efficient manufacturer of computer software can drive smaller, less efficient firms out of business, in the mythology of capitalism that's progress. And if a large, well-organized army overwhelms a smaller foe, the positive synergy for the winners will be offset by the negative synergy for their defeated enemies.

The *Titanic* Effect

As my own personal experience with a late dinner date suggested, convergent, fortuitous synergies can be "recipes" for calamities. Human history is replete with examples of seemingly unrelated factors that combined to produce unintended and unpredictable effects. When the outcome is a major disaster, it could be called the "*Titanic* effect," after the horrific tragedy that many millions of movie-goers recently relived vicariously with the Academy Award–winning movie. What if the ship that was promoted as "unsinkable" had been provided with enough life boats? What if the shipbuilder had used a higher grade, less brittle steel, or better rivets? What if the President of the White Star Line (Bruce Ismay) had not been aboard the ship and pressuring the captain for a speed record? What if any of the numerous telegraph warnings about icebergs had been heeded? What if the lookouts had been provided with binoculars? What if the duty officer had not mistakenly reversed the engines before attempting an emergency turn? What if the nearby merchantman, probably the Californian, had not ignored the many distress signals? What if...? The tragedy of the *Titanic* was in fact a horrendous synergistic effect. If just one of those "what ifs" had been different, the disaster might not have occurred.[12]

Many convergent negative synergies are routine, everyday occurrences. For a motorist, the combination of high speed, wet pavement, a sharp curve, oncoming headlights, alcohol, and failure to wear a seat belt is the classic "recipe" for a fatal accident. Again, if any one of those contributing factors is changed, the odds might be greatly reduced. In a similar way, when a well-known local restaurant succumbed to bankruptcy recently, there was no single cause. It involved a combination of factors – a downtown location where parking had become a major headache, some fashionable new competitors in town, a dated menu with an uninspired chef, an aging owner who left the business largely in the hands of a salaried manager, indifferent service (low pay and low tips), and, at the end, a nearly deserted dining area with a depressing atmosphere.

Synergy Minus One

A special case of negative synergy involves what I call "synergy minus one," a phenomenon that was introduced earlier as a way of testing for synergy. The term is borrowed from those play-along audio tapes for musicians that were popular a few years ago called "Music Minus One." In my usage, it refers to a loss of synergy, say when some part is removed from the whole. The examples are almost endless. Here are just a few:

* What happens, or rather fails to happen, when you misdial a single digit in a telephone number, or when there is a single error in the name you enter for a computer file, or a typo in the e-mail address you are using.
* The effect on your travel plans of having just one problem with your car – a flat tire, a dead battery, misplaced car keys, an empty fuel tank, the loss of a timing chain, or even a personal injury like a broken leg in a skiing accident.
* The many recipes that have to be abandoned when a single ingredient is missing – say, if there is no sugar in the house, or butter, or eggs.
* A baseball team without a pitcher, a football team without a quarterback, a varsity eight without a coxswain, or an America's Cup yacht without a helmsman.
* The many words that lose their meaning when one letter is removed. Remember the example in Chapter 1 – rt, ct, and bt? For that matter, try ra, ca and ba.
* The failure of even a single link in an anchor chain (in Pliny's poetic metaphor).

The essence of the synergy-minus-one idea was captured in the famous children's tale about how, for want of a missing horseshoe nail, the entire war was lost. Synergy minus one has also provided a plot device for innumerable Hollywood movies. Do you remember what happened when the power failed at Jurassic Park? A menagerie of fearsome dinosaurs ran amok. There are also the thousand Grade-B "potboilers," going back even to the silent movie era, where the plot hinges on a cotter pin that was removed from a wagon wheel, or a section of track that was removed from in front of an on-rushing train, or a signal light that was shot out by a gunman, or a rifle that jams just before the big showdown. In the current generation of action movies, only the technologies have improved, not the plots.

Of course, there are many notorious real-world cases of synergy minus one. The rash of long-distance telephone system outages around the country in 1990 was finally traced to an error that appeared in three lines of computer code, out of 2 million altogether. Likewise, the failure of the innovative new automated baggage-handling system delayed the grand opening of the multibillion dollar new Denver Airport for several months. More recently, the tardy completion of a high-speed rail link to the new Hong Kong Airport had the same effect. Likewise, the Alaska Airlines flight that plunged into the Pacific Ocean in 1998 was doomed by the failure of a single part, a threaded screw mechanism that controlled the horizontal stabilizer. And, for the *Apollo 13* spacecraft, a single defective part out of literally millions (a damaged coil inside an oxygen tank) sparked an explosion that forced the cancellation of a planned lunar landing and produced a harrowing near-tragedy for the crew.

Or consider this lead for a *Wall Street Journal* story a few years back: "It was the smallest of mistakes. A detail in a joint, incorrectly designed, weakened a walkway suspended over a lobby of a Hyatt Regency Hotel in Kansas City. The walkway collapsed during a crowded dance party last July, killing 113 people and injuring 186 more."

One of the most bizarre recent examples of synergy minus one involved, of all things, dust. On October 15, 1997, air traffic around the country became snarled and 150 flights were canceled altogether at Newark Airport when airborne dust from a ceiling renovation project at the Long Island control center forced controllers to abandon their work stations and operate with a rotating skeleton crew for several hours. We do not ordinarily think of clean air in the control center as a part of our air traffic system, but it is. And so, of course, is electrical power. Despite the tens of millions of dollars that have been spent by the Federal Aviation Administration to provide backup power systems at these vitally important facilities in recent years, there have been a number of power failures. The predictable result has been disrupted air traffic around the country, sometimes for hours.[13] More recently, we have witnessed similar system-wide disruptions as a result of security breaches at major airports (most notably, Atlanta and San Francisco).

Synergy Goes to War

Synergy minus one also has a distinguished war record – a bloodstained history of dubious achievements. Among the many examples that were described in *The Guinness Book of Naval Blunders* and its companion volume, *The Guinness Book of Military Blunders*, one of the most notorious involved a series of failures by German submarines during the early days of World War Two. Although the U-boats were ultimately destined to play a great part and do much damage to Allied merchant shipping, at the outset the vaunted new magnetic detonators on the German torpedoes were thoroughly unreliable, and the German Admiralty was scandalously slow to acknowledge and fix the problem.

The most famous incident occurred when Lieutenant Wilhelm Zahn and his crew launched a daring attack against one of the most important targets of the war. As it happened, on the morning of October 30, 1939, a high-level conference was taking place aboard the British flagship *Nelson* in the North Sea. The meeting included the then First Lord of the Admiralty, Winston Churchill, as well as the First Sea Lord, the Commander-in-Chief of the Home Fleet, and several other high-ranking naval brass. It would have been quite a haul for any U-boat skipper, though Zahn was unaware of it at the time. At great risk, Zahn and his crew closed on the *Nelson,* fired three torpedoes, and listened in with passive sonar as two of the torpedoes, struck the ship with a metallic "thunk" and then sank to the bottom without exploding. The third torpedo missed altogether. The U-boat managed to escape, but Zahn and his crew were so demoralized by the episode that they were later removed from submarine duty.[14]

Sometimes, of course, it takes more than one defect to destroy the synergy. It may require a negative "convergence."(A colleague, the biologist/anthropologist Pete Richerson, helpfully suggested the term "synergy minus *n*.") The loss of a few leaves will not perceptively reduce the shade under a deciduous tree, but by the late fall you may get a sunburn sitting under it. Or when a high-flying Internet stock comes crashing to the ground, it may truly be a case of death by a thousand cuts (or many thousands). In some cases, the removal of a part does not destroy the synergy but merely alters it. Take the word "where." Remove the "w" and it becomes "here." Remove the "e" and it becomes "her." Remove the "r" and it becomes "he." But that's the end of it. There are many similar examples in chemistry and genetics.

A classic war-related example of a negative convergence (synergy minus *n*) was the doomed Iranian hostage rescue mission in 1980. This bold military gamble failed because of the "coincidence" of three small but critical failures in three of the eight RH-53 helicopters that were assigned to the mission. A clogged cooling vent that had inadvertently been covered led to the failure of two key navigation instruments on one of the choppers. A cracked nut led to a hydraulic leak and consequent failure of the hydraulic pump on the second craft. And a dangerous hairline crack in one of the rotor blades apparently downed the third. With three helicopters out of action, the mission had to be aborted.[15]

Y2K: A Close Call

Perhaps the most important recent example of the synergy-minus-one phenomenon was the much-ballyhooed Y2K computer problem. To be precise, it was actually a case of synergy minus two – two digits that had been left out of the designation for the current year in all but the most recent computer programs. In the wake of a mostly smooth transition to the year 2000, many people assumed that the problem was nonexistent, or greatly overrated. This is exactly wrong. In point of fact, a serious global disaster was averted only because of a heroic behind-the-scenes remedial effort that, added together, dwarfs any other collective, international undertaking since World War Two. It was also by far the biggest technological mobilization in history. All together, some $500 billion was spent to fix the problem, world-wide, including $100 billion in the United States alone.[16]

Though they were mostly unpublicized, many potential Y2K disasters were prevented by the development of ingenious "fixes" that were widely shared, mainly through the Internet. When the big night came, there were actually a great many computer failures, power outages, and the like, but most had insignificant consequences precisely because a vast army of technicians and managers were ready to deploy quick fixes and "workarounds." In the news media, unfortunately, these achievements were largely ignored. But this is nothing new. The media tend as a rule to invert the old adage.

Failure attracts a horde of reporters but success is an orphan (except in competitive sports and politics).

Synergy plus-or-minus one has often played a decisive role in politics. Some of the examples are legendary:

* When the British Parliament failed by one vote to enact major reforms in 1653, Cromwell abolished Parliament and terminated the Commonwealth.
* France made the transition from monarchy to republic in 1875 by one vote.
* Rutherford B. Hayes won the Presidency of the United States in 1876 by one electoral vote.
* It was one vote that gave Adolph Hitler the leadership of the Nazi party in Germany in 1923.
* And, not least, George Bush "won" the American Presidency in 2000 by one Supreme Court vote.[17]

For the losing side in each of these historic votes, it was a case of synergy minus one. But win or lose, any system of decision making based on casting votes is grounded in synergy; it is the combined effect of the individual choices that determines the outcome. And voting systems are not different in principle from any system where winning (or losing) depends on some combination – like getting three cherries or three oranges in a slot machine, or a scratcher, or the correct sequence of numbers on a lottery ticket or a Bingo card.

The Paradox of Dependency

Synergy minus one highlights another aspect of the natural world and human societies that was alluded to earlier. The more valuable some resource may be – the greater its "utility" to use the economists' jargon – the greater may be the cost of losing it. In other words, our dependency increases in direct proportion to the value of some resource or technology. An obvious example is a major power outage in midwinter. When the electrical power grid for a substantial part of Eastern Canada and New England was knocked out for several weeks during the great ice storm of 1998, many thousands of people who depended on electrical heating, electrical stoves, etc., were so severely affected that they had to move to emergency shelters. *The New York Times*, in an editorial, pondered the deeper meaning:

> A storm like this reveals the shallowness of technological civilization – how swiftly the grid collapses. But it also reveals its depth – into how many reaches of ordinary life electricity has penetrated and how high above the fundamental concerns of existence it allows us to float. This storm demonstrates the slow, brutal strength of the cold. The wonder is not that the cold is so powerful, but that we are so seldom aware of its power, of its capacity to fracture a community....[18]

The paradox of dependency provides us with regular reminders of its destructive potential – computer systems that greatly facilitate commerce but bring it to a complete halt when the system goes down; or the airline flight that is canceled due to a jammed cargo door or a malfunctioning fuel transfer system (though some skeptics suspect these malfunctions are more likely to occur when the load factor is low and the passengers can be shifted to a later flight); or the panic buying when supplies of some vital commodity, like food or fuel, are threatened.

Though we are loathe to acknowledge it, the paradox of dependency is one of the central features of any modern society. Like Gulliver with the Lilliputians, we are bound by many thousands of invisible threads into an immense network of organized human activities, a vast combination of labor that provides for our needs. The classical economist Adam Smith, the founding father of modern capitalism, penned one of the classic appreciations of this interdependency in *The Wealth of Nations* (1776):

> Observe the accommodation of the most common artificer or day-labourer in a civilized and thriving country, and you will perceive that the number of people whose industry a part, though but a small part, has been employed in procuring him this accommodation, exceeds all computation. The woollen coat, for example, which covers the day-labourer, as course and rough as it may appear, is the product of the joint labour of a great multitude of workmen. The shepherd, the sorter of the wool, the wool-comber or carder, the dyer, the scribbler, the spinner, the weaver, the fuller, the dresser, with many others, must all join their different arts in order to complete even this homely production. How many merchants and carriers, besides, must have been employed in transporting the materials from some of those workmen to others who often live in a very distant part of the country! How much commerce and navigation in particular, how many ship-builders, sailors, sail-makers, rope-makers, must have been employed in order to bring together the different [dyes]

made use of by the dyer, which often come from the remotest corners of the world! What a variety of labour too is necessary in order to produce the tools of the meanest of these workmen! To say nothing of such complicated machines as the ship of the sailor, the mill of the fuller, or even the loom of the weaver.... If we examine, I say, all of these things, and consider what a variety of labour is employed about each of them, we shall be sensible that without the assistance and cooperation of many thousands, the very meanest person in a civilized country could not be ... accommodated.[19]

We are, most of us, totally dependent on the complex economic systems we have created, though we often take them for granted. We only become aware of our dependency when something goes wrong – synergy minus one. U.S. hospitals have recently provided us with some alarming examples. The discovery that hospital blood supplies are frequently contaminated with HIV and other blood-borne pathogens has exposed some tragic consequences. But even more serious was the finding in a systematic study that tens of thousands of patients have died in hospitals from such human lapses as being given the wrong drug or the wrong dosage.

One of many lessons from the recent terrorist attacks on the World Trade Center and the Pentagon is that our large, complex societies are at once interdependent and ultimately very vulnerable. The potential terrorist threat to our water supplies, power plants, transportation systems, even public health (from anthrax and the like) are both real and disturbing. And these threats pale compared to the destruction and loss of life that can be caused even by small "weapons of mass destruction." Whether or not the broader implications – our global interdependency and the contingent nature of our economic and political order – will come to be more broadly appreciated and acted upon remains to be seen (more on this in the last chapter).

"Devolution" as a Loss of Synergy

One of the most important forms of synergy minus one (or more) historically, has been associated with political "devolution."[20] The term has become a buzzword lately as empires, nations, bureaucracies, and even business firms collapse, divide, downsize, outsource, and in various ways become less than they once were. On the one hand, devolution has been associated with the current trend in Western countries toward reducing or relinquishing the central government's role (power and resources) in relation to various social programs and services – welfare, aid to education, health care, railroads, public utilities, and the like. States and provinces, and even the private sector, are being granted greater responsibility for these functions.

However, the term "devolution" is also widely used in connection with a broader trend that involves the breaking up of entire political systems – empires and nation-states. Devolution in this sense often involves the redrawing of political boundaries. Whole populations may be divided into new political units. Thus the British Commonwealth today exists largely on paper; the Soviet Union is long gone; the pieces of the former Yugoslavia maintain an uneasy peace; the United Kingdom is in the process of devolving as we speak; and there was recently a near miss in Canada when the issue was put to a vote in Quebec. In all such cases, what the changes reflect is that the old "system" is no longer capable of acting as a unified, coordinated whole; the potential for collective synergies is gone.

The term devolution also evokes a more significant historical pattern – the many societies and political systems over the millennia that have disaggregated or disappeared altogether. The list of case studies includes, among many others, the Mayans, the Incas, the Aztecs, the Olmec, Teotihuacan, the Anastazi, the Hohokam, the Sumerians, the Babylonians, the Akkadians, the Hittites, the Minoans, Mohenjo-Daro, the Easter Islanders, the Moriori, the Tasmanians, the Maasai, the Hawaiian and Zulu kingdoms, Han China, Persia, the Mongol Empire, the Ottoman Empire, Byzantium, Carthage, and, of course, Rome.

There is a long tradition of scholarship on political devolution, needless to say, from Edward Gibbon's *Decline and Fall of the Roman Empire* to the writings of Oswald Spengler, Arnold Toynbee, Herbert Simon, David Kennedy, various systems and chaos theorists, and, of course, many modern-day environmentalists (the Club of Rome and the "Limits to Growth" theorists come to mind). Many case studies of devolution can also be found in the research literature in anthropology and archeology. Norman Yoffee and George Cowgill, for instance, have collected 11 examples in their edited volume, *The Collapse of Ancient States and Civilizations* (1988). Their most important overall conclusion was that every case was different. There was no consistent pattern and no "prime mover."[21]

Sometimes a single missing "ingredient" (synergy minus one) seems to have been enough – the depletion of natural resources for the Easter Islanders, a drastic change of climate for the Akkadians and other ancient

Middle Eastern civilizations[22] (and perhaps the Mayans and Teotihuacan as well), or an imported disease epidemic for the Native American civilization of the Mississippi Valley.[23] But perhaps the most frequent cause of political devolution has been military conquest. Sometimes a society may simply continue under new management, but in many other cases – the Trojans, the Carthaginians, the Moriori, the Iroquois Confederacy, to name a few – military defeat was followed by destruction and even virtual annihilation.

A Biblical Example

Synergy, or rather the lack of it, has contributed to many famous military defeats. The failure of the Greek city-states to unite against Philip of Macedon doomed them all to conquest and domination. Had the Gallic and British tribes been able to achieve a united front, they might have avoided being incorporated into the Roman Empire. In nineteenth-century Africa, the Nuer tribes, acting together, were able to expand at the expense of the disunited Dinka. And the British employed a calculated divide-and-conquer policy to add the jewel of India to their imperial crown.

"Divide and conquer" is, of course, one of the oldest gambits in the playbook of military strategy, but sometimes the fatal wound is self-inflicted. Perhaps the earliest recorded example is described in the Old Testament. In approximately 930 BC, King Solomon – ruler of the twelve independent "tribes" that collectively comprised ancient Israel – died suddenly and was succeeded by his 41-year-old son, Rehoboam. King Solomon had generated great unrest and near-revolts among the twelve tribes due to a heavy burden of taxation in the form of forced labor. The new king, well aware of the unrest, agreed to meet with a delegation of representatives from each of the tribes. They presented Rehoboam with a demand that he lighten up on taxes, in return for which they would serve him loyally. But Rehoboam, who should have known better, was persuaded by his circle of hot-headed young counselors not to make any concessions. In a classic act of political folly, Rehoboam chose to threaten the tribal leaders with even harsher measures: "Whereas my father laid upon you a heavy yoke, I will add to your yoke. Whereas my father chastised you with whips, I shall chastise you with scorpions." Confronted with such a bald threat, ten of the twelve tribes seceded. A protracted civil war soon erupted, and the bitterly divided nation was never reunited. Two hundred years later, a fractionated and weakened Israel was conquered by the Assyrians. The twelve tribes were driven from their lands and forcibly dispersed.[24]

The Rise and Fall of Rome

Many cases of political devolution are more complicated than this, of course. They involve a synergistic nexus – a "downward spiral" of negative influences that interact with one another ("synergy minus n"). Rome seems to be everybody's favorite example, and recent scholarship has shed new light on some of the underlying factors that shaped both the rise and the decline of the Roman empire. There are many lessons for our time.[25]

The rise of Rome to preeminence in the ancient Mediterranean world was a synergistic process that combined military successes, important administrative and legal innovations that enabled the Roman government to exert political control over a vast domain, and (not least) a dynamic process of economic growth that was highly symbiotic. Rome in the early days was essentially a yeoman economy consisting mainly of small farmers and urban craftsmen. The economy enjoyed agricultural self-sufficiency, a vigorous and technically superior crafts industry, a growing infrastructure of roads, harbors, aqueducts, etc., and an expanding network of mutually advantageous foreign trade relationships. These trade patterns were based in part on a region-wide distribution of resources (metals, wood, grain, hemp, etc.), together with locally based technologies for processing them. But the advantage of relatively cheap waterborne transportation was equally important. Though the Roman system of roads is legendary, its commercial value was secondary; in those days, land transport doubled the value added to a manufactured item about every 100 miles. So Rome's seaport location on the Tiber River (and the protections afforded by its navy) gave it a major competitive edge in overseas trade. Rome ultimately created a huge fleet of merchant vessels that served a vast common market from Britain to Egypt.

As the Roman economy grew, so did its population. In 400 BC, there were only about 150,000 Roman citizens on the entire Italian peninsula. By 70 BC, the number in metropolitan Rome alone had mushroomed to 500,000, according to the census that year. (And this didn't count women, children, slaves, and freed men.) Forty years later, the total number of citizens had jumped to 4 million, though many of them now lived in various Roman provinces.

Of course, Rome's military prowess was also crucially important. Its highly trained legions, with advanced military technology (iron swords, shields, helmets, body armor, massive siege weapons, and mounted cavalry) and superior tactics (such as the famed phalanx), allowed Rome to win a succession of wars. Meanwhile, its expanding navy, with hundreds of formidable triremes and quinqueremes, eventually gained control of the entire Mediterranean. This, in turn, provided secure trade routes for its merchants. *Mare Nostrum* (our sea) it came to be called.

The Roman Empire was built by winning battles, on land and sea. Yet it followed no preexisting plan or agenda. Some accounts imply that Rome's affluence derived from the booty and tribute that flowed in from its many conquests. This certainly helped. Rome's military triumphs provided a growing supply of slave labor and enough captured wealth to support the Roman treasury (and the army) without having to levy taxes. But plunder was not the engine of the Roman economy. Economic synergy – a vast combination of labor – was the key. The Roman war machine was undergirded by Rome's economic strength.

The growing number of Roman provinces may not have been thrilled to be under Roman rule, needless to say, but there were valued trade-offs. In addition to extensive trade relationships within a unified, protected, and legally regulated economic system (with a common currency and a common commercial language), the provinces also benefited from internal security and Roman protection against external enemies, especially the "barbarians" living beyond the borders of the Empire. For many Roman subjects, the *Pax Romana* had very real benefits, at least at first.

Pox Romana

The unraveling of this impressive system over several centuries provides a model of dysergy. As Rome's population exploded, the agricultural base of the Italian peninsula underwent a transformation. In some areas, the soil became badly deteriorated and its productivity declined. Meanwhile, key domestic resources, such as timber, were seriously overexploited. Rome eventually became more dependent on foreign imports. The agricultural system also underwent a radical change, from self-sufficient small farms (predominantly) to vast land-holdings called *latifundia* that were worked inefficiently by slave labor, principally for profitable cash crops such as wine and olive oil. As a result, many small farmers were displaced from their land over time and had to migrate to the cities to find work. Rome was thus compelled to import increasing quantities of grain from its provinces in Sicily, Egypt, and elsewhere to feed its rapidly growing urban population.

Meanwhile, the economic foundation of metropolitan Rome had begun to decay. The once flourishing crafts industries and ancillary commercial activities were gradually undercut by the rise of competing industries out at the periphery of the empire through a process known as "import replacement." Even though the tools, pottery, terracotta lamps, and amphora (the ubiquitous two-handled earthenware storage vessels) produced in the provinces may have been inferior to those that were made in Rome, they were much cheaper, and in time this cost advantage undermined Rome's export markets. For instance, what is now the city of Bordeaux in southern France (then Gaul) emerged as the center of European pottery-making by the third century AD. To make matters worse, the wealth of Rome's upper classes began to flow out of Italy and into investments abroad, while rents and taxes extracted from the Roman peasantry frequently reduced them to extreme poverty and drove less productive lands out of cultivation, further deepening Rome's dependence on imports.

In effect, Rome was becoming a parasite on its provinces. By Julius Caesar's time, some 350,000 people in the city of Rome were on the dole, perhaps 20% of the total population. The treasury was being drained by heavy foreign grain purchases to feed a growing number of indigents (thus the famous "bread and circuses" policy). By then, food prices had multiplied as the currency was progressively debased and severe inflation took hold, which further increased the demands on the treasury. Out of necessity, the Roman government began to impose increasingly burdensome taxes on the provinces, partly to feed its domestic population but also to support the increased military expenditures necessitated by the growing popular resentment against Roman taxes. Rome was caught in what could be called a vicious (or may be "viscous") circle, a nexus of dysergy.

To make matters worse, Rome's wealth was increasingly concentrated in the hands of a politically dominant rentier class (a growing proportion of which included the hierarchy of the early Christian church), while an increasing number of fees, commissions, and kickbacks were levied to support an entrenched and growing bureaucracy. The result was a vast parasitical superstructure on top of an impoverished economic base. The stage had been set for Caesar's seizure of dictatorial power in 49 BC, and for an increasingly strife-torn *Imperium* marked by frequent revolts and civil wars, administrative and military decay and corruption and, ultimately, a split between the eastern and western parts of the empire.

Some data collected by the well-known sociologist Pitirim Sorokin illuminates this negative trend. During the period from 425 to 201 BC, there were only four major incidents of internal conflict reported. Between 200 BC and AD 25, the number jumped to 41. From AD 25 to 250, there were 49 more and from AD 250 to 475, the number increased further to 61. The first century AD was the most turbulent of all.[26] In other words, for an increasing number of Rome's subjects, the perceived costs of the empire came to outweigh the benefits. In the end, few were left who were willing to defend it. The *Pax Romana* had become the *Pox Romana,* and in AD 410, the Visigoths under Alaric administered a military *coup de grâce* and then proceeded to sack, plunder, and pillage the city.[27]

The path downhill was not a straight line, however. There were periods of reform and resurgence. After two disastrous plagues and a decline in the number of new slaves, there were actually labor shortages, especially on the great estates. Indeed, some parts of the empire flourished – Britain, North Africa, and especially the eastern half of the empire, where the Emperor Constantine built the "New Rome" (Constantinople) – even as the Italian peninsula languished. But after AD 250, the decline became more pronounced, and there was a succession of political crises. As Ramsay MacMullen put it in his authoritative study on *Corruption and the Decline of Rome* (1988), the empire began to lose "bits and pieces of itself" to the "barbarians."

Toward the end there were many signs of decay: declining trade, especially where pirates or bandits controlled the trade routes; also the abandonment of many older villas and little or no new construction; towns that began to fortify themselves against raiders in a kind of reversion to the pre-Roman "autarky," with every town looking out for itself; plus a heavy burden of taxes, widespread corruption, and a decline in public service. But perhaps the most telling indicator was what happened to the Roman army, according to the evidence compiled by Giovanni Forni.[28] Forni estimated that, in AD 41, two-thirds of the army recruits came from Italy and adjacent areas, with a total force numbering about 300,000. By AD 300, well into the process of decline, the Roman army numbered more than 1 million men, and three-quarters of the new recruits came from the Rhine and Danube provinces while Italy provided fewer than 5%. In the empire's last full century, Rome was defended primarily by Romanized "barbarians." And as barbarian settlements, barbarian culture, and barbarian political control spread through Europe, Rome's influence progressively waned. Finally, the barbarian "defenders" turned on Rome itself. In the end it was not so much an invasion as a revolution from within against a corrupt, parasitical, and increasingly ineffectual regime.[29]

The Lessons of History

The epitaph penned by the distinguished ancient historian A. H. M. Jones, almost a half century ago, still seems valid:

> I have outlined what seem to me to be the principal causes of the empire's decline, and of its collapse in the West. It would be difficult, and probably profitless, to attempt to weigh their relative importance, for they inter-acted upon one another so as to form a single complex .... The decline and fall of the Roman empire was the result of a complex of interacting causes which the historian disentangles at his peril.[30]

Is it possible that this classic example of political devolution could have been prevented? Could the decline of the Roman empire have been re-versed? The answer is yes in theory but no in fact. If a concerted set of demographic, economic, and political changes had occurred, a revitaliza-tion of Rome is at least conceivable. There was nothing inevitable about it; it followed no law of history. But this alternative scenario would have required a level of understanding about the basic causes and ultimately fa-tal consequences that did not then exist. It would also have required new political institutions and practices that were not invented until many cen-turies later, beginning with the *Magna Carta*. And it would have required a new set of economic, social, and political relationships that were very unlikely at that juncture; some historical convergences cannot plausibly be rewritten. However, it may be possible for us to learn from this classic ex-ample of political devolution. We'll talk about some of these lessons in the final chapter.

High and Low Synergy

There is one other categorical distinction regarding synergy that should be noted briefly. Anthropologist Ruth Benedict back in the 1950s was the first to draw a contrast between what she referred to as "high" and "low" synergy. The trouble is, her use of the term "synergy" was heavily laden with moral and ideological overtones. "I spoke of societies with high social synergy where their institutions insure mutual advantage from their undertakings, and societies with low social synergy where the advantage of one individual becomes a victory over another, and the majority who are not victorious must shift as they can."[31] To Benedict, synergy and distributive equity were lumped together; it had to be a win–win relationship to be synergistic.

Nevertheless, Benedict's distinction is a useful one if it's confined strictly to functional differences. A high-synergy system is one in which all of the parts work together effectively – when the potential for various combined effects is fully realized. (How the resulting benefits may be distributed is another matter.) An Indianapolis 500 race car, tuned to perfection, may roar out in front of its competitors. A clunker with worn spark plugs and timing that is badly out of adjustment may have trouble pulling away from a stoplight. A stew is a stew, but some recipes produce a tasteful, nutritious meal while others may end up in the garbage can. Some businesses have got it all together, and the results show up in the black ink on the bottom line. Others seem to be the victims of just one damn thing after another and end up in Chapter 11, or worse.

A detailed study that compared two western U.S. hospitals some years ago provides a well-documented illustration.[32] (Their names will not be used to protect client confidentiality.) In many ways, these two institutions were so much alike that they provided about as close to a controlled social experiment as one can get in the real world. Both hospitals were privately owned, non-profit institutions operated by different religious orders. They were located in the same state and were therefore subject to the same regulatory environment. Both served urban, residential, low-income minority areas. They were of comparable size. And they had about the same socioeconomic mix of patients in terms of the reimbursements received for services.

Nevertheless, Hospital A (as we will call it) had a consistent 6–8% net income each year, while Hospital B had been averaging less than 1% and had lost money in one recent year. Why? There were no obvious differences in any of the "external" factors. If anything, Hospital A had a slight disadvantage with an older facility, 50 fewer beds, a heavier load of charity and low-income patients, and more double-occupancy rooms (which are harder to keep filled). The most striking difference between the two hospitals, it turned out, was in how they were managed.

The chief executive officer (CEO) of Hospital A was open, supportive, and sensitive to his staff and the needs of the institution. He was also decisive, well organized, and committed to the development of his hospital. This hospital also had a formal planning system that ran throughout the organization, so that planning was an integrated, system-wide process in which each unit participated. The management also tried to promote a sense of everyone participating in a common effort. Staff morale was high, and everyone tended to line up behind organizational objectives; nobody was sabotaging the plans. In addition, this hospital had made extensive efforts to build a good image in the community, with a board of trustees drawn from the community rather than from the religious order. There were many efforts to involve the community in its planning process. As a result, Hospital A ran a lean, efficient operation with tight cost controls and an incredibly high occupancy rate of 95%.

In contrast, the CEO of Hospital B was authoritarian and closed. He handed down orders rather than consulting with his staff. Because he was ambitious and wanted to move up in the corporate hierarchy, he calculated every move. The hospital had no formal planning process at all. When planning was done, it was ad hoc and often at cross-purposes with other parts of the organization. Furthermore, every one of the hospital's trustees was a member of the corporate religious order and came from out of town. As a result, Hospital B ran a very lax operation. Staff productivity was low. There was a chitchat organizational culture and many operating inefficiencies. Patient volume was only 75 to 80% of the maximum, despite the hospital's supposed advantages. Overall operating expenses at this hospital were about 10% higher than for Hospital A.

The conclusion seems well founded. Given similar resources and circumstances, good management can make the difference between a high- and low-synergy organization.

Negative Synergy in Nature

The natural world, too, has been shaped by the double-edged character of synergy and dysergy. For instance, destructive population explosions are a common occurrence in nature and may even destabilize an ecosystem. (Good for the winners, but not for everyone else.) Likewise, the awesome synergies of scale in an army ant raiding party makes them the scourge of the many tropical forests where they are found. (Good for them; bad for their prey.) When a tight-knit colony of *Volvox* cells frustrates the filter-feeders, it's positive synergy for the algae and bad for their predators. And if the symbiotic partnership between honey guides and humans benefits both the birds and the Boran tribesmen, it is definitely not welcomed by their honeybee victims. In fact, nature often resembles a zero-sum game, with synergy as the trump card. A hungry predator may be defeated by a mob of defenders. A large coalition of males with designs on a group of females may be able to drive away a smaller coalition of suitors. Or a large group of animals may be able to defend a food patch against all comers.

Synergy minus one is also frequently found in genetics. More than 300 human disorders and disease syndromes are associated with a single defective gene, or a deleterious mutation.[33] Similarly, the ubiquitous "transposons" (or transposable elements), often erroneously called "junk DNA," in fact may play a very destructive role, interrupting the normal functioning of the genome in various ways. Among other things, transposons are associated with diseases like hemophilia, leukemia, and even breast cancer.[34] Certain behaviors, or other morphological (bodily) traits, may also be distorted or eliminated by mutational changes, as shown by experiments in various laboratory species – inbred mice, *Drosophila* flies, and the nematode worm *Caenorhabditis elegans*, among others. A dramatic example in mice is known as "small eye." If this gene mutates, the embryonic mouse will develop without eyes.[35]

On the other hand, the relationship between living organisms and parasites of various kinds might be considered examples of "synergy plus one," a destructive addition to a synergistic system. In fact, it is estimated that every living organism carries at least one and sometimes many parasites, external, internal, or both. For instance, a rich variety of ugly-looking endoparasites (as viewed by scanning electron microscopes) infect humans world-wide. Trypanosome, the protozoan responsible for sleeping sickness, infects many thousands of Africans each year. Some 200 million people, all told, are infected by the blood fluke *Schistosoma*. And hookworms are even more pervasive. These free-loading critters suck blood from the intestines of some 1.3 billion people, world-wide.[36]

"Ecological parasitism," a sustained relationship between two or more species, is equally pervasive, and it is often difficult to determine whether it is in fact symbiotic, parasitic, or sometimes both, depending on the context.[37] To illustrate, the so-called VAM (vesicular–arbuscular mycorrhizal) fungi are generally considered to be models of mutualism with many species of plants. Careful studies have shown that VAM fungi do in fact enhance plant growth in low-phosphorous soils, but in high-phosphorous soils, or in low sunlight conditions (when photosynthetic activity is reduced), they may actually become parasitic and reduce plant growth.[38] In a similar vein, avian brood parasitism, such as the cowbirds' practice of infiltrating their eggs into other species' nests, is ordinarily harmful to the hosts, who end up nurturing somebody else's nestlings. But cowbird chicks also eat botfly larvae, which can infect a host's nestlings and lower their fitness. So, whenever there are heavy botfly infestations, the parasitic cowbirds may actually enhance their hosts' reproductive success.[39]

In sum, positive and negative synergy are everywhere deeply intertwined, and both have played a major role in shaping the course of evolution. As we shall see in the next chapter, synergy is more than simply an interesting or commonplace effect. It is also an immensely important source of evolutionary novelties and a wellspring of creativity in evolution. Natural selection is often portrayed by evolutionists as the chief "mechanism" of innovation in nature. But this is inaccurate. Natural selection is more like an editor than a sculptor; it "chooses" among the options that are generated for "testing" in a given environment. And synergy is the source of many of these options; it's an engine of evolutionary innovation. The time has come to discuss the Synergism Hypothesis in detail.

5   The Synergism Hypothesis

The real act of discovery consists not in finding new lands
but in seeing with new eyes.

Marcel Proust

Often the most important contribution a scientist can make
is to discover a new way of seeing old theories or facts.

Richard Dawkins

Why has there been so little appreciation – if not reverence – for synergy over the years? If human societies in fact depend on synergy, why is the word not on everyone's lips?

Why Not Synergy?

One reason is that there is a common tendency to define the term narrowly. Many people equate synergy with the cliche (a paraphrase of Aristotle) that "the whole is greater than the sum of its parts." They assume that synergy is limited to organized "systems," or to phenomena with quantitative properties ($2 + 2 = 5$), like the GE ad suggested. Others identify the term only with some specific synergistic effect – business deals, drug interactions, group "syntality," technological innovations, the division of labor, or the like. They don't recognize that "synergy" is an umbrella term for irreducible cooperative effects of all kinds.

Another reason for underrating synergy has to do with the way our minds seem to work. As a rule, we tend to focus either on the whole or some part, but not on how the parts interact to produce wholes. Most of us have trouble visualizing complexity, especially when we are unable even to see or directly experience the parts and the relationships between them. The "microcosmos" is incomprehensibly small. Atoms, molecules, and even cells are only visible to us in huge numbers, only when they produce wholes of various kinds. By the same token, even when we do bother to look under the hood of our car to check the oil, most of us are unable to appreciate what is going on inside; the parts that are visible to us do not reveal their secrets.

Another problem is our tendency to economize mentally. The military has a word for it: the "need to know." As a rule, we pay only selective attention to the enormous quantity of information that bombards us every day. We learn only what suits our purposes, or what is unavoidable. We may learn how to drive a car, but we may be only mildly curious about how it works. We may learn how to operate a computer but may not have a clue about what goes on inside of it, or how the software is written. And the same goes for cell phones, CD players, television sets, and many other miracles of modern technology. In fact, most of us have only a general idea about how our own bodies are constructed and how the various parts work together. We tend to pay attention only when something goes wrong with it. For instance, how many of us know the functions of the pituitary gland?

Scientists and engineers study these complex systems more closely, of course, but they also tend to view the world through the lenses of their own specialized concepts, paradigms, and theories. To many scientists, in fact, synergy is an unfamiliar term. This is not because there is any dearth of synergistic phenomena in their field but because the synergies travel under various aliases – cooperativity, emergence, mutualism, co-determination, networking, symbiosis, interdependent effects, coevolution, dynamical attractors, heterosis, epistasis, linkage effects, catenated (chain-linked) processes, even some of the terms used in Chapter 2: threshold effects, phase transitions, and a division of labor. It's rather like the parable of the blind men and the elephant. Each blind man, being able to touch only a part of the elephant, variously describes it as being like a wall, a tree, a rope, a snake, and a fan.

There are a few notable exceptions, though. Certain "hard science" disciplines, like biochemistry, molecular biology, pharmacology, and neurobiology make liberal use of the term. In large measure this is due to the fact that these disciplines are deeply involved in studying cooperative phenomena and their functional effects. However, the research in these fields is highly technical; to most of us, it is arcane and obscure. Here are just a few "synergy" titles that were picked at random from the weekly sweep by my on-line search service: "Sites of Interaction of Streptogramin A and B Antibiotics in the Peptidyl Transferase Loop of 23S RNA and the Synergism of their Inhibitory Mechanisms" (*Journal of Molecular Biology*); "Synergism of Nitric Oxide and Iron in Killing the Transformed Murine Oligodendrocyte Cell Line N20.1" (*Journal of Neurochemistry*); "Synergy between an Antiangiogenic Integrin Alpha Antagonist and an Antibody–Cytokine Fusion Protein Eradicates Spontaneous Tumor Metastases" (*Proceedings of the National Academy of Sciences*); "Folic Acid-enhanced Synergy for the Combination of Trimetrexate Plus the Glycinamide Ribonucleotide Formyltransferase Inhibitor" (*Biochemical Pharmacology*).

Unfortunately, the synergies that are routinely discovered in these disciplines have not for the most part inspired other biologists or social scientists, much less the general public, to go looking for synergy. One indicator is a computer search of a major biological sciences data base for the five years from 1991 through 1995. A total of 3,379 "synerg" references were found, but 95% of these – based on a random sample – were confined to various "hard science" disciplines. Only a handful were related to disciplines that are concerned with social behavior, like ethology, behavioral ecology, primatology, sociobiology, or even "symbiology" (the study of symbiosis). But times are changing.

The Vicissitudes of Synergy

Evolutionary biology – the prime arena for evolutionary theories – has always had a special problem with the concept of synergy. Ever since Darwin, competition has been viewed by mainstream biologists as the dominant influence in evolution.[1] Darwin himself had a deep and subtle understanding of the complexities and paradoxes that abound in the natural world, and he fully appreciated the interplay in nature between competition, cooperation, and ecological interdependence. Nonetheless, the concept of "natural selection" implies a competitive "struggle for existence," as Darwin put it. The full title, seldom-quoted, of Darwin's masterwork is: *The Origin of Species, Or the Preservation of Favoured Races in the Struggle for Life.*

Not surprisingly, the concept of synergy, which focuses on cooperative relationships of various kinds, has been of relatively little interest to mainstream evolutionary biologists, until recently. Many evolutionists over the years have equated cooperation with "altruism," or self-sacrifice – a very problematic phenomenon in a process presumably governed by "selfish genes" (in Richard Dawkins' famous metaphor). In fact, Edward O. Wilson launched his discipline-defining volume, *Sociobiology: The New Synthesis*, in 1975 with the surprising assertion that altruism was "the central theoretical problem of sociobiology."[2]

Even symbiosis, which involves "mutualistic" relationships between members of two or more different species, has long been treated by evolutionary biologists as a subsidiary, even marginal phenomenon, while cooperation among members of the same species was viewed as an exception that required special circumstances, in order to overcome the ever-present tendency to cheat. This is what the so-called "game theory" models of cooperation were developed to illuminate. Indeed, when a mathematical formalization of "tit-for-tat" – a game-theory version of "you scratch my back and I'll scratch yours" – was jointly proposed by Robert Axelrod and William Hamilton in 1981, it was hailed as a great theoretical breakthrough (see below).

Another obstacle for the synergy concept has been the gene-centered orientation in evolutionary theory. Although there was a flurry of interest in "emergent evolution" and "holism" in the 1920s, the so-called "Modern Synthesis" of the 1930s and 1940s all but banished such diversionary themes. For many years, especially in the heyday of the population genetics pioneers – Haldane, Fisher, Wright, Morgan, Dobzhansky, and a few others, as well as some non-geneticists like Julian Huxley, Mayr, and Simpson – evolution was conceived as, quintessentially, an incremental process of differential selection among randomly occurring gene "mutations" in hypothetical "gene pools" (though there have always been minority views as well).[3]

Within the rigorously quantitative paradigm of population genetics, the magic key to evolution was deemed to be the "selection coefficient," the rate at which one variant of a gene (an "allele") might replace various alternatives over successive generations. Accordingly, evolution was defined as being equivalent to a change of gene frequencies over time in a population, or "deme." This had the effect of transmuting evolution into a statistical measure – a bookkeeping exercise. To be sure, it was recognized that there are various biases in the selection process, linkage effects (epistasis), partial penetrance, heterosis, the population structure, random drift, and the like. Nevertheless, as biologist George C. Williams put it in his influential book, *Adaptation and Natural Selection* (1966), selection at any higher levels of organization in nature than individual genes is largely "impotent . . . not an appreciable factor in evolution."[4] (Williams has since retreated from this salient.) Ernst Mayr, one of the current deans among evolutionary biologists and a frequent critic of such monolithic reductionism, has called it "bean-bag genetics."

Perhaps the ultimate reductionist conceit was Richard Dawkins' image, in *The Selfish Gene* (1976), of an organism as a "robot vehicle" or a "survival machine" manipulated by its genes.[5] If the chicken is merely the egg's way of making another egg, as the saying goes, the organism is only the gene's way of making another gene, so it was said.

The problem with this formulation is that genes don't go about their selfish business in splendid isolation for the simple reason that they can't survive and reproduce alone; they are dependent upon the combined effects (the synergies) they produce in cooperation with other, functionally different genes. An organism is at bottom a more or less complex combination of labor. Thus, among the many important findings of the human genome sequencing project is the discovery that there are 1,195 genes specifically involved in producing our hearts, 2,164 genes associated with our white blood cells, and 3,195 genes that shape the human brain.[6] As Dawkins himself observed in a later popularization, *The Blind Watchmaker* (1986), "In a sense, the whole process of embryonic development can be looked upon as a cooperative venture, jointly run by thousands of genes together. Embryos are put together by all the working genes in the developing organism, in collaboration with one another . . . We have a picture of teams of genes all evolving toward cooperative solutions to problems . . . It is the 'team' that evolves."[7] (More on this below.)

The Ideological Debate about Evolution

Another significant obstacle for the concept of synergy in evolution has been its (presumed) political implications. Unfortunately, evolutionary theory has been an ideological battleground over the years, as warring factions of liberals, conservatives, socialists, and apologists for *laissez-faire* capitalism have variously used (or misused) the theory of biological evolution to advance their economic, social, and political agendas, or to attack their opponents.

At one extreme, perhaps, was the Russian anarchist and naturalist, Prince Pëtr Kropotkin, whose turn-of-the-century polemic, *Mutual Aid: A*

Factor of Evolution (1902), was written to refute the "nature, red in tooth and claw" school led by "Darwin's bulldog," Thomas Henry Huxley. Huxley had characterized nature as a "gladiator's show." In rebuttal, Kropotkin, a Russian émigré, argued that the abundant evidence of co-operation in nature falsified Huxley's one-sided interpretation of Darwin's theory. "During the journeys which I made in my youth in Eastern Siberia and Northern Manchuria ... I failed to find – although I was eagerly look-ing for it – that bitter struggle for the means of subsistence, *among animals belonging to the same species* [his emphasis], which was considered by most Darwinists (though not always by Darwin himself) as the dominant charac-teristic of the struggle for life ..."[8] Kropotkin claimed that cooperation is more important than competition in nature and is the key to "progressive" evolution.

At the opposite pole were the Social Darwinists like William Graham Sumner, E.B. Tylor, Albert Keller, Gustav Ratzenhoffer, and others, who took their inspiration from Herbert Spencer's writings. Social progress, they argued, was fueled by competition and the "survival of the fittest," in Spencer's provocative term. Fairly typical was this pronunciamento by Tylor: "The institutions which can best hold their own in the world gradu-ally supersede the less fit ones, and ... this incessant conflict determines the general resultant course of culture."[9] Business magnate John D. Rockefeller, in a Sunday school address, assured his audience that "The growth of large business is merely a survival of the fittest ... This is not an evil tendency in business. It is merely the working out of a law of nature and a law of God."[10] However, it was the steel baron and philanthropist, Andrew Carnegie – never a man to mince words – who penned the most inflamma-tory expression of the Social Darwinist credo, in an 1889 essay known as "The Gospel of Wealth":

> While the law [of competition] may be sometimes hard for the individual, it is best for the race, because it ensures the survival of the fittest in every department. We accept and welcome, therefore ... great inequality of environment, the concentration of business, industrial and commercial, in the hands of the few, and the law of competition between these, as being not only beneficial, but essential for the future progress of the race.[11]

The political/ideological debate over the course of the past century has ranged somewhere in the middle between these two extremes. Political liberalism (in the American sense) predominated in the immediate post–World War Two era, and so did the view that human societies are *sui generis*.

Many social theorists of that era adopted the naive (and self-serving) view that biological evolution was something that had happened in the past, and that biology was largely irrelevant to an understanding of human behavior. Unlike other animals, it was said, humans are infinitely malleable products of their different cultures; human nature is a *tabula rasa*. Ashley Montagu, a well-known and widely published social scientist of that era, was famous for his oft-repeated statement that "all specifically human behavior humans have had to learn from other humans." Meanwhile, the pioneering sociology of Herbert Spencer was displaced by the work of the much-younger but "politically correct" French sociologist Émile Durkheim. Durkheim was shamelessly hailed as the "founding father" of sociology and was lionized for such assertions as: "Every time that a social phenomenon is directly explained by a psychological phenomenon, we may be sure that the explanation is false."[12]

New voices were already challenging this extreme point of view when Edward O. Wilson launched the science of sociobiology with the publication of his massive 1975 synthesis. Wilson unleashed a fury of attacks from the left, especially from the self-styled "Sociobiology Study Group," by claiming (in his famous last chapter on humankind) that human behavior is governed by invisible "epigenetic rules." Wilson also declared that one of the objectives of sociobiology was "to reformulate the foundations of the social sciences." He suggested that the humanities and social sciences should be reconceived as "specialized branches of biology."[13] It's not surprising, in hindsight, that many mainstream social scientists and humanists felt deeply threatened by sociobiology – whatever their politics. On the other side of the ledger, the new science of sociobiology also catalyzed a wealth of new research and theory on the biological basis of behavior. Wilson's book was a landmark event in the biological sciences.

The new trend in academia toward the biologizing of human societies was soon reinforced by events in the marketplace and the political arena – especially the competitive challenge of an emerging global economy, along with the revitalization of political conservatism, the election of Ronald Reagan as President of the United States of America and the collapse of the Soviet Union. These and other influences conspired to push the political pendulum in the other direction. Just as "Neo-Darwinism" (and "inclusive fitness theory") was emerging in the 1980s as the reigning paradigm in evolutionary biology, so the proponents of a resurgent free-market capitalism – like their Social Darwinist forebears – began to clothe their capitalist arguments in the rhetoric of no-holds-barred "Darwinian competition" and the "survival of the fittest." Between the Neo-Darwinists and Neo-Conservatives, it was hard to tell which was the dog and which the tail; perhaps they both wagged each other.

Under these circumstances, it's not surprising that a theory, first proposed in 1983, about the causal role of cooperative phenomena in evolution – especially the assertion that synergy played a key role in the evolution of complexity at all levels of living systems – was not widely noticed. But that was then. Times – and attitudes – have changed.[14]

A Favorable Tide?

There is a metaphor in Shakespeare's *Hamlet* that has been borrowed by many modern authors, perhaps because it seems to capture an eternal truth: "There is a tide in the affairs of men, which, taken at the flood, leads on to fortune; omitted, all the voyage of their life is bound in shallows and in miseries." Thus, in the 1930s the historian Arthur Schlesinger (senior) used Shakespeare's famous image in a widely acclaimed article called "The Tides of American Politics" (1939). In the 1960s, the French historian Jacques Pirenne wrote a magisterial volume that was translated and published in English as *The Tides of History* (1962). Political scientist Karl Deutsch also used the metaphor in the title of his classic text on the *Tides among Nations* (1979).

More recently, an on-line search of the Internet bookseller Amazon.com produced a total of 274 current titles that include the word "tides." There are books on corporate tides, the tides of power, tides of migration, tides of change, the tides of reform, China against the tides, the tides of war, the tides of love, political tides in the Arab world, and, of course, many volumes related to ocean tides.

Although we like to think that science is free from "extraneous" social and political influences, of course this is not so. Thomas Kuhn, in his celebrated volume on *The Structure of Scientific Revolutions* (1962), argued that science is very much influenced by the tidal effects associated with different "paradigms." Ideas and theories that fit within or support the currently dominant framework of basic assumptions and theories in a given discipline are more likely to be favorably received. On the other hand, conflicting work, especially if it challenges the reigning paradigm, is often ignored or rejected. Kuhn's specific scenario for scientific revolutions has been much debated. Nevertheless, there seems to be widespread agreement that Kuhn's core idea is valid, even if the dynamics may be somewhat different from his original formulation.

A classic case in point is the Nobel Prize–winning biologist Barbara McClintock's work on the so-called "jumping genes" – genetic rearrangements during the development of an organism via "transposons" (transposable elements) that can produce variations in the fully developed "phenotype" of an organism (such as the different color patterns in maize). This phenomenon, painstakingly documented by McClintock over many years, remained in the shadows until late in her life. The main reason was that it contradicted the then reigning "central dogma" of molecular biology – namely, that the genes are expressed during development in a linear, deterministic fashion (DNA to RNA to proteins). Now, of course, it is recognized that development is a much more complex process and that a variety of non-linear, self-organizing, feedback-dependent influences may affect the outcome.[15]

A similar tide-change seems to be taking place currently in evolutionary biology, and perhaps also in the political arena, with respect to the role of cooperation, symbiosis, and even synergy. One early sign was the adoption of the synergy concept in the 1980s by the eminent biologist John Maynard Smith, who developed a "synergistic selection" model to characterize the interdependent functional effects that can arise from altruistic cooperation. (Maynard Smith later broadened the concept to accord with a strictly functional interpretation of cooperation, whether altruistic or not.) Also important was the growing body of work in game theory (so-called) on the evolution of cooperation, using the methodology pioneered by Maynard Smith.[16]

Another significant contribution was made by biologist Leo Buss in his 1987 book on the evolution of higher levels of organization. Buss invoked the concept of synergy, though in a narrow sense and without much elaboration. His focus was on how potential conflicts between the prospective "parts" of a more inclusive system could be resolved.[17] The biologically oriented psychologist David Smillie also utilized the concept of synergy in connection with his study of social interactions in nature, but again without elaboration.[18]

Biologist David Sloan Wilson and various colleagues have also played an important role with their dogged efforts over the past 20 years to put the much-criticized concept of "group selection" on a new footing. Although Wilson's approach remains gene-centered, he stresses the role of what he calls a "shared fate" among individual cooperators, which implies a functional interdependency.[19] (Group selection remains controversial, but at least it is no longer treated as a heresy.[20]) Another significant contribution includes the experimental work of biologist Lee Dugatkin on cooperation, along with his recent books on the varieties of cooperation in nature, *Cooperation among Animals* (1997) and *Cheating Monkeys and Citizen Bees* (1999).[21]

Especially important, however, is the work of biologist Lynn Margulis on the role of "symbiogenesis" in evolution, particularly in relation to the origin of complex, eukaryotic cells. Now recognized as a major theoretical contribution, this concept – which (as noted earlier) traces to a group of Russian botanists at the turn of the last century – has focused our attention on a domain in which synergistic functional effects have been of decisive importance.[22] (Indeed, the relatively new discipline of endocytobiology – centered in Europe – is concerned especially with investigating symbiotic and synergistic phenomena of various kinds at the cellular level.)

But perhaps the most significant sign that a favorable tide now exists for the synergy concept are the two books coauthored by John Maynard Smith and Eörs Szathmáry on the evolution of complexity, *The Major Transitions in Evolution* (1995) and *The Origins of Life* (1999), which (independently) identify and feature the causal role of synergy at various levels of biological organization.[23] Maynard Smith now recognizes the "universal" importance of functional synergy (in a personal communication), as does Ernst Mayr (also in a personal communication). Indeed, in the past few years new research and theoretical work on mutualism – both within and between species – has reached the flood stage. To quote ecologist Judith Bronstein: "Mutualisms . . . have finally come to be recognized as critical components of ecological and evolutionary processes . . . Every organism on earth is probably involved in at least one and usually several mutualisms during its lifetime." Indeed, cooperation seems to be an evolutionary modality whose time has come. There was even a recent report about cooperative (altruistic) sperms.[24]

Arthur Koestler, the well-known novelist and polymath (he was the one who coined the terms "holon" and "holarchy" to characterize multilevel systems), got it right 30 years ago. In the landmark book co-edited with James R. Smythies entitled *Beyond Reductionism* (1969), Koestler made the remark (quoted in the Prologue) that "true novelty occurs when things are put together for the first time that had been separate." Water molecules, lichen, Velcro, and a myriad of other combinations of parts – from neutrons to nation-states – can produce unique collective effects that the parts cannot produce alone. Nor, most likely, can these synergies be produced by a different combination of parts, or even the same parts in a different environment. Synergy is, in fact, a bottomless well of creativity in evolution. More important, synergy has also played a major *causal* role in evolution, and especially the evolution of complexity.

This may sound like a contradiction of Darwin's theory and a cavalier rejection of 150 years (and more) of evolutionary biology. But, in fact, the opposite is true. This theory involves only a different way of viewing the same phenomena – a shift of focus (and emphasis) to a different aspect of the evolutionary process. It is entirely consistent with Darwin's theory. Call it an economic approach – or perhaps "bioeconomics." The Synergism Hypothesis views evolution as an ecological and economic process – a survival enterprise – in which living systems and their genes are embedded. Let me briefly summarize this theory.

(For the record, a number of other theorists have recently drawn attention to the economic aspect of evolution, including paleontologist Niles Eldredge in his 1995 book, *Reinventing Darwin*, and ecologist Steven A. Frank in his *Foundations of Social Evolution*, in 1998. Frank begins his book thus: "The theory of natural selection has always had an affinity with economic principles." The Synergism Hypothesis is also consistent with the call for "pluralism" by Elliott Sober and David Sloan Wilson in their 1998 book on altruism and multilevel selection theory, *Unto Others*.[25])

Where Is Natural Selection?

In the conventional evolutionary paradigm, natural selection is viewed as the primary "mechanism" of evolutionary change. Evolution is characterized as a process of "blind variation and selective retention," in psychologist Donald Campbell's popular slogan. Accordingly, paleontologist George Gaylord Simpson asserted that "The mechanism of adaptation is natural selection . . . [It] usually operates in favor of maintained or increased adaptation to a given way of life." Similarly, Ernst Mayr informs us that "Natural selection does its best to favor the production of programs guaranteeing behavior that increases fitness." And Edward O. Wilson, in *Sociobiology*, assured us that "natural selection is the agent that molds virtually all of the characters of species."[26]

The problem is that natural selection is not a "mechanism." Natural selection does not *do* anything; nothing is ever actively "selected" (although sexual selection and predator–prey interactions are special cases). Nor can the sources of causation be localized either within an organism or externally in the environment. In fact, the term natural selection is a metaphor for an important aspect, or property of the ongoing evolutionary process. (Darwin's inspiration for his metaphor was the "artificial selection" practiced by animal breeders.) Natural selection is really an "umbrella concept" that refers to whatever functionally significant factors (as opposed to historical contingencies, fortuitous effects, or physical laws) are responsible in a given context for causing differential survival and reproduction. Properly conceptualized, these causal factors are always relational; they are defined both by organism(s) and their environment(s), and by the interactions between them.

This crucially important point can be illustrated with a textbook example of evolutionary change – "industrial melanism." Until the Industrial Revolution, a "cryptic" (light-colored) strain of peppered moths called *Biston betularia*, predominated in the English countryside over a darker "melanic" form. The wing coloration of the cryptic strain provided camouflage from avian predators (like thrushes) as the moths rested on the mottled trunks of lichen-encrusted trees. This gave them an advantage over the darker form (*carbonaria*) that stood out. As a result, the melanic form was relatively rare. But as soot progressively blackened the tree trunks in areas near England's growing industrial cities, the relative frequency of the two forms was eventually reversed; the birds began to prey more heavily on the light, cryptic strain while the darker, melanic strain became less visible.[27]

The question is, where in this example was natural selection "located"? What was the "mechanism"? The short answer is that natural selection included the entire configuration of factors that combined to influence differential survival and reproduction. In this case, an alteration in the relationship between the coloration of the trees and the wing pigmentation of the moths, as a result of industrial pollution, was an important proximate factor. But this factor was important only because of the inflexible resting behavior of the moths and the feeding habits and perceptual abilities of the birds. If the moths had been subject only to insect-eating bats that use "sonar" rather than a visual detection system to catch insects on the wing, the change in background coloration would not have made any difference. Nor would it have mattered if there were not genetically based differences in wing coloration that allowed for "selection" between two alternative forms.

Hence, one cannot (technically) speak of "mechanisms" or fix on a particular "selection pressure" in explaining the workings of natural selection; these are only shorthand expressions. One must focus on the interactions that occur within an organism and between the organism and its environment(s), inclusive of other organisms. Natural selection as a *causal agency* refers to the functional consequences produced by adaptively significant changes in a given organism–environment relationship. In other words, natural selection is a consequence of the bioeconomic "payoffs."

The "Causes" of Evolution

What are the factors that are responsible for initiating changes in these relationships? In other words, what are the sources of the "variations" that affect natural selection? The answer, of course, is many things. It could be a functionally significant mutation, a chromosomal transposition, a change in the physical environment, a change in one species that affects another species, a developmental change, or it could be a change in behavior that results in a new organism–environment relationship. (More on behavior as a cause of evolutionary change in Chapter 6.)

In fact, a whole sequence of changes may ripple through a complex network of relationships. For instance, a climate change might alter the ecology, which might induce a behavioral shift to a new habitat, which might encourage an alteration in nutritional habits, which might precipitate changes in the interactions among different species, resulting ultimately in the differential survival and reproduction of alternative physical traits and the genes that support them. However, it is the functional consequences of these changes that *cause* natural selection.

Another way of putting it is that natural selection does not "select" genes; it differentially rewards, or disfavors, the functional effects produced by genes in a given context (the phenotype). As biologists Russell Lande and Stevan J. Arnold put it in a classic, though underappreciated, article in the journal *Evolution* back in 1983: "Natural selection acts on phenotypes, regardless of their genetic basis, and produces immediate phenotypic effects within a generation that can be measured without recourse to principles of heredity or evolution." Biologist Alan Grafen calls it the "phenotypic gambit."[28]

Just to underscore this more subtle conception of natural selection, let's look at one more example. Certain English land snails (*Cepaea nemoralis*) are subject to predation from thrushes, which have developed the clever habit of capturing the snails and then breaking open their shells with stones. In other words, a remarkable behavioral innovation (tool use) in one species became a cause of natural selection in another species. (We'll highlight the role of "tools" in evolution in Chapters 6 and 7.)

However, the impact of natural selection in the snails was also shaped by two additional factors, one genetic and the other ecological. It happens that *C. nemoralis* exhibit genetically determined variations in shell banding patterns, which in turn provides varying degrees of camouflage. The result is that the more "cryptic" genotypes have been less intensively preyed upon than those that are more visible to predators. However, the pattern of predation by thrushes (and the frequencies of the different snail genotypes) also varies greatly from one location to the next. The reason is that the thrush populations, being subject themselves to predators (like hawks), display a strong preference for well-sheltered localities. Paradoxically, the snails are much less subject to predation in more open areas.[29]

In both of the examples above, the "causes" of natural selection were the functional effects of various organism–environment changes, insofar as they impacted on differential survival and reproduction. Another way of putting it is that causation in evolution is iterative; it also runs backwards from our conventional view of things. In evolutionary change, effects are also causes. To use biologist Ernst Mayr's well-known distinction, it is the "proximate" functional effects arising from any change in the organism–environment relationship that are the causes of the "ultimate" (transgenerational) changes in the genes, and the gene pool, of a species.

A stunning illustration of this causal dynamic can be found in the long-range research program among "Darwin's finches" in the Galápagos Islands, led by zoologist Peter Grant and his wife, Rosemary. Over the years, the Grants have documented many evolutionary changes in these closely related bird species, particularly in the mix of beak sizes and shapes, in response to pronounced environmental fluctuations. During drought periods, for instance, the larger ground finches with bigger beaks survive better than their smaller cousins. Small seeds become scarce during the lean years, so the only alternative food source for a seed-eater is much larger, tougher seeds that must be cracked open to get at their kernels. Birds with bigger, stronger beaks have an obvious functional advantage, and this is the proximate cause of their differential survival.[30]

The Synergism Hypothesis

The Synergism Hypothesis represents an extension of this line of reasoning. I call it "Holistic Darwinism," because the focus is on the selection of wholes, and the combinations of genes that produce those wholes. Simply stated, cooperative interactions of various kinds, however they may occur, can produce novel combined effects – synergies – that in turn become the causes of differential selection. The "parts" that are responsible for producing the synergies (and their genes) then become interdependent "units" of evolutionary change. In other words, it is the "payoffs" associated with various synergistic effects in a given context that constitute the underlying cause of cooperative relationships – and complex organization – in nature. The synergy produced by the "whole" provides the functional benefits that may differentially favor the survival and reproduction of the "parts." Although it may seem like backwards logic, the thesis is that functional synergy is the underlying cause of cooperation (and organization) in living systems, not the other way around. To repeat, the Synergism Hypothesis is really, at heart, an "economic" theory of complexity in evolution.[31]

Because this may be an alien idea, let me restate it in a slightly different way. The functional effects produced by cooperation (and organization) are the very cause of complexity in evolution. The "mechanism" (as it were) underlying the evolution of complex systems is none other than the combined functional effects that these systems produce. It is the synergies that are the proximate causes of natural selection, or "synergistic selection." Synergistic effects represent an independent source of the "variations" which may be "acted upon" by natural selection.

In fact, this paradigm is very similar to the way economists tell us that markets work in human societies. When a new "widget" is developed, its ultimate fate – its survival and reproductive success, so to speak – is ultimately determined by how well it succeeds in the marketplace. If the widget sells well, the "supply" is likely to increase, or so economic theory tells us. If not, the widget will soon go extinct. (There is a cultural analogue of natural selection in human societies that we'll talk further about in the next chapter.) Many factors – internal and external – may contribute to these synergies. Moreover, the synergies are always historically contingent and situation-specific. They are not the predictable product of a prime mover, or the inexorable outcome of any self-organizing fractal dynamic, much less the working out of some deterministic "law" of evolution. History matters – a lot.

The evolution of birds, one of the major turning points (and major puzzles) in evolution, provides a good illustration. We have known since the discovery of *Archaeopteryx* fossils many years ago that the oldest birds go back perhaps 150 million years, but we have not known until recently how this remarkable adaptation occurred, given the paucity of fossils and other evidence. Now things have changed. The picture that has emerged shows that flight was not the result of a single breakthrough – a megamutation, or a "punctuated equilibrium." It involved a suite of synergistic functional changes that occurred progressively over many millions of years, even to some degree independently of each other and for very different purposes.

In accordance with the theory first proposed by John Ostrom in the 1970s, it seems likely that birds evolved from a line of small, two-legged, meat-eating dinosaurs called theropods. It was the theropods' adaptation to the fast pursuit of prey on the ground that led to many of the bioengineering changes that much later on facilitated flight. (I will mention only a few of the more important developments here.) Bipedalism, which pre-dated even the theropods, was a key precursor. It induced many progressive changes in the theropod hind limbs, including a reduction in the number of digits, rotation of one of the toes (perhaps for better balance), and various changes in the ankle joints and foot bones for increased mobility and speed on the ground. But more important, bipedalism freed up the forelimbs to develop in other ways, initially for grasping prey and only later for gliding and then full flight. As theropod evolution progressed, there were other significant skeletal changes in the shoulder girdle and the forelimbs. The sternum also became much larger and the clavicles fused to form the wishbone found in modern birds. These latter two changes created strong anchoring points for the muscles that most likely developed initially in relation to ground pursuit and, later on, for gliding, soaring, and flight. Likewise, the tail gradually became shorter and stiffer and the tail vertebrae became compressed into what is called the "pygostyle," which controls the tail feathers in modern birds.

As two of the experts on bird evolution, Kevin Padian and Luis M. Chiappe, observed in a recent review article on the subject: "In summary, a great many skeletal features that were once thought of as uniquely avian innovations ... were already present in theropods before the evolution of birds. Those features generally served different uses than they did in birds and were only later co-opted for flight and other characteristically avian functions, eventually including life in the trees."[32] Even feathers, traditionally the defining characteristic of birds, may well have evolved for a different purpose (as evidenced by some recent fossil finds in China) – perhaps to serve as a lightweight form of insulation or even for display (or both).[33] By the same token, egg-laying and nesting in primitive theropods as opposed to internal, placental reproduction was an adaptation that evolved in a different context but later facilitated the transition to flight.

So how did birds evolve? In sum, it was the result of a synergistic combination of "pre-adaptations" (or "exaptations" if you prefer the Gould and Vrba neologism) and progressive functional changes that together produced significant advantages. And these synergistic effects in turn encouraged other functional changes – development of the flight stroke, the evolution of the all-important "thumb wing" that allows for flight control at low speeds, and other refinements "in the same direction." How do we know that this process was synergistic? We need only to take away one of the major "parts" – say feathers, or bipedalism, or egg-laying, or even the relatively small size of the theropods. (There has never been a Boeing 747 equivalent in birds, and for good reason. Some flightless birds, like the now-extinct moas of New Zealand, weighed as much as 1,000 pounds, but the largest known bird of prey was a menacing 30-pound giant eagle, also now extinct.) In fact, the evolution of birds provides another example of the Bingo Effect – a suite of adaptive changes that ultimately crossed a threshold to produce a synergistic new "package."

Synergy and the Evolution of Complexity

Synergy has played a key role in the progressive evolution of complex systems in nature. However, complexity is not an end in itself; it's a consequence of innovations that produce more potent forms of synergy. Synergy is the "driver"(metaphorically speaking).

Sponges provide a handy illustration.[34] Although sponges come in many different sizes and shapes, the "model" sponge looks more like an urn or a vase than your typical kitchen sponge. Sponges are also the most primitive of all animals in terms of complexity. Indeed, they are often confused with plants because they are immobile and have no internal organs, no mouth, no gut, no sensory apparatus or even a nervous system. They are more like a colony of cooperating independent cells. Sponges even have their own separate classification (Porifera, or "pore-bearers"), and they may have evolved separately from other animals.

Sponges also earn their living in one of the simplest possible ways, as filter-feeders. They pull water into an internal cavity through large pores in their "skin," which consists of an outer layer of epithelial cells and a gelatinous inner layer with a skeleton of thin, bony "spicules." The sponge's internal cavity is in turn lined with a layer of specialized "collar cells" (choanocytes) that are equipped with a whip-like flagellum and numerous filaments. These collar cells combine forces to move the water through the sponge and then push it out through a large opening at the top called the osculum. As the water passes through, the collar cell filaments extract oxygen and food particles (microbes and organic debris of various kinds). These vital nutrients are then distributed to the non-feeding cells via another specialized set of mobile transporter cells called amoebocytes. The amoebocytes are also responsible for carrying wastes and for manufacturing and distributing various kinds of skeletal materials – calcium carbonate, silica, spongin (a tough protein-like substance), or some combination of these, depending upon the type of sponge.

Reproduction in sponges is also (typically) a cooperative effort. Although the freshwater forms frequently reproduce asexually (often by casting off "gemmules" that are somewhat like seed pods), most sponges are hermaphrodites, meaning that they produce both sperm cells and eggs. The sperm cells are launched into the sponge's cavity and are ejected through the osculum in the hope that they will find their way to another sponge's cavity. When a sperm is lucky enough to enter a recipient sponge, it may be captured by one of the collar cells and then transferred to an amoebocyte, which in turn carries it to an awaiting egg. Eventually, the fertilized egg will become a free-swimming larva and will venture out on its own to find an appropriate site for developing into a new adult. It is really a unique reproductive system.

That's about all there is to the sponge story, except for the chemicals they produce to repel potential predators. The combination of labor in sponges involves fewer cell types (six) than the number of workers in Adam Smith's pin factory – namely, epithelial cells, pore cells, collar cells, amoebocytes and two kinds of sex cells. (Some larger sponges also have specialized cells that aid in opening and closing their oscula.)

The point of this story is that even the minimal level of complexity found in sponges is tied directly to the functional effects that the parts produce together – the synergies. Each part is specialized for the role it plays in the "system." Each part is also completely dependent upon the other parts; no part could exist without the services of the others, and only together can they survive and reproduce successfully. Furthermore, the properties and capabilities of each part cannot be understood without reference to its role in the operation of the system as a whole. Nor can we understand the whole without an appreciation of how the parts work together.

In fact, sponges display several different forms of synergy – functional complementarities, a combination (division) of labor, synergies of scale, and even structural (Gestalt) synergies. For instance, the shape of the (classic) sponge, with its exit opening located at the top, utilizes physics to help pull water through its cavity, rather like the updraft in a chimney. As a result, a sponge can typically process a quantity of water equal to its own volume in less than ten seconds. Likewise, in the larger sponges – some taller than a human – the internal walls may be elaborately folded. This has the effect of greatly increasing the surface area available for filtering and feeding, a necessity for meeting the increased nutritional needs of a larger organism. (There are an estimated 10,000 species of sponges all told, and they are found throughout the world, from the Bahamas to the Antarctic, and from intertidal zones to the deep sea floor.)

The Selfish Genome

How do we know this is a synergistic system? Just take away a major part – say the amoebocytes, or the collar cells, or the epithelial cells, or the skeletal spicules. Sponges would not exist without the synergy that their parts produce together. By the same token, imagine what would happen if one were to change its accustomed environment, say by putting a sponge into a nutrient-free swimming pool, or into an ice pack. Any theory of complexity based on the operation of deterministic laws cannot deal with the effects of different contexts, or a contingent process; but a functional (economic) theory focused on synergistic relationships can. The Synergism Hypothesis asserts that it was the functional synergies (the economic benefits, broadly defined) that were responsible for the evolution of sponges, not some hidden law of complexity. Indeed, there is some evidence that sponges originated as a symbiotic union between a primitive host and the once-independent ancestors of the choanocytes. About 150 species of very similar one-celled "choanoflagellates" – some free-swimming and others that attach themselves to a substrate – still exist today. There is even circumstantial evidence that the choanoflagellates are themselves a product of a symbiotic union between an ancient protozoan and free-swimming spirochetes.[35]

By the same token, a gene-centered explanation cannot account for sponges without reference to Maynard Smith's concept of "synergistic selection" – the proposition that the genes for sponges were selected as an interdependent "team" and that the selfish interests of the genes were subordinated to the functional needs of the genome as a whole. Call it the "selfish genome" paradigm (a term that seems increasingly apropos in light of recent discoveries about how the genome operates).[36] In fact, biologist Mark Ridley, in an important new book with a title that is similar to a 1996 article of mine, *The Cooperative Gene* (2001), argues that biological complexity presents a major obstacle at the genetic level, due to the vastly increased risk of a "mutational meltdown." Given the natural error rate of perhaps 1 for every 100 nucleotide letters, any complex species with a large genome and many multiplications during ontogeny would very soon become extinct in the absence of various means (everything from proofreading and repair enzymes to sexual reproduction) for reducing and/or correcting this mutational burden.[37] In other words, selfish genes must be ruthlessly suppressed.

Seeing with New Eyes

It is a well-worn cliche that you are more likely to find what you are looking for and, conversely, that you will *fail* to see what you are not looking for, no matter how important it is. Contrary to the narrow, Neo-Darwinian dogma, evolution is not just about competition between genes. Or even about helping your relatives. It's also about win–win cooperation, and about competition via cooperation. Most important of all, though, it's about the costs and benefits of competition and/or cooperation – and the costs and benefits of complexity. In light of the many developments described earlier, a more balanced perspective is coming into view. Nowadays the term synergy is in the air, and the research literature is burgeoning with discoveries of cooperation and synergy in nature. Here are just a few illustrations:

* Long before Walt Disney discovered meerkats (the Afrikaner name for mongoose) and gave them a featured role in *The Lion King*, these highly gregarious small mammals were using the synergy principle in various ways to cope with the challenge of survival in a tough environment, the Kalihari Desert and other marginal areas in southern Africa. Renowned for their ability to stand tall on their hind legs to scan the horizon while

using their long tails to form a tripod for balance, meerkats live in elaborate underground burrows with multiple-family groups of up to 30 or more animals.

Among other things, meerkats benefit from huddling closely together for warmth during the cold desert nights (environmental conditioning); they also hunt collectively and jointly defend their burrows with noisy displays and threatening charges (synergies of scale); they take turns standing sentry duty to watch out for predators, such as hyenas, jackals, and eagles (cost-sharing); and they use various signals to communicate with their companions, including sharp warning cries when danger appears (information-sharing). There is even a rough division of labor in meerkats. The males are primarily responsible for defending the burrow and its dozen or more entrances, while the females and immature males share in nurturing the infants. In addition, meerkats economize on building and maintaining their burrows by sharing their quarters with non-competitive solitary yellow mongooses and social ground squirrels.[38]

* The construction of communal nests by the social wasp *Polybia occidentalis* is a complex activity requiring the coordination of various tasks (a combination of labor). To study the economics, biologist Robert Jeanne conducted a comparative study of small versus large colonies, as well as the nest-construction technique used by social wasps compared to the less efficient method of solitary wasps. Jeanne found that small colonies required almost twice as many worker–minutes to complete the same amount of construction (due mainly to materials handling inefficiencies that larger colonies could minimize). In addition, he was able to determine that social wasps could collect and process a given amount of nest material with 2.6 times fewer foraging trips than solitary wasps required, thanks to information-sharing. (An added advantage was that the social foragers drastically reduced the duration of their exposure to predators while they were out in the field.) In other words, the synergies in communal nest construction were measurable.[39]

* In a comparative study of reproduction among southern sea lions during a single breeding season at Punta Norte, Argentina, Claudio Campagna and his co-workers found that only 1 of 143 pups born to group-living females died before the end of the season, as compared to 60% mortality among the 57 pups born to solitary mating pairs (a synergy of scale). Pups in colonies were better protected from harassment and infanticide caused by roving subordinate males and were far less likely to become

separated from their groups and die of starvation. This is another case where the synergies were measurable.[40]

* By any definition, the leaf-cutter ants of the genus *Atta* are among the most spectacular examples in nature of a "superorganism."[41] Edward O. Wilson, who has studied them extensively, reports that the leaf-cutters are true agriculturalists. They grow fungi ("mushrooms") in extensive underground chambers. The fungi are provisioned with great masses of fresh vegetation that is harvested by hordes of workers and brought back to the nest in a complex division of labor that involves several categories (and sizes) of workers and a highly orchestrated combination of labor. Each new colony is founded by a new "queen" that digs a nest and personally nurtures the first few workers. Eventually, however, the queen retires to a lifetime of producing many millions of daughters, most of whom become workers and soldiers. As Bert Hölldobler and Wilson, in their delightful 1994 book, *Journey to the Ants*, conclude, the leaf-cutter ants have validated "the idea of the ant colony as a tightly regulated unit, a whole that indeed transcends the parts."[42] In fact, it has recently been learned that the whole is even more elaborate than we supposed. Cameron Currie and his colleagues have shown that the leaf-cutters also depend upon a symbiotic relationship with a *Streptomyces* bacterium that produces an antibiotic to fight a parasitic mold. As a bonus, the bacteria also stimulate the growth of the fungus gardens.[43]

* Acorn woodpeckers are unique among cooperatively breeding birds in that they also collaborate in filling one or more communal "granaries" with acorns.[44] Woodpeckers eat many different kinds of fruits, nuts, seeds, even insects and small invertebrate animals. But in certain parts of the western United States, Central America, and the Northern Andes in Colombia, where the climate is moderate and oak trees are plentiful, acorns provide a staple food in such abundance that acorn woodpeckers can survive through the winter months without having to migrate. Indeed, an extended family might occupy and defend the same territory over many generations.

The key to this sedentary lifestyle, however, is the woodpeckers' collective ability to store and protect a very large cache of acorns for winter use. Their granaries are the trunks and branches of large trees, where acorn storage holes have been laboriously drilled over many years and are replenished with fresh acorns from year to year. In fact, some large granaries – trees that are so pock-marked they appear to be afflicted with arboreal measles – may have as many as 50,000 acorn holes. Based on estimates of the time required to drill a single hole (30–60 minutes) and the average amount of time devoted to this activity by the woodpeckers (perhaps 5%), a single large granary could represent the work of several decades!

There are a number of important advantages in using trees to sequester acorns. One is that the granaries can more easily be defended by collective action (mobbing) against other birds and squirrels. Another is that the trees provide a favorable storage site that allows the acorns to dry out and remain free of mold and rot through the wet winter months; there are almost no losses (no inefficiencies) compared to storage in the ground. But, most important, the "capital investment" associated with building the granary only needs to be done once in the lifetime of a given tree. In other words, the granaries represent a cumulative "synergy of scale" with a major historical element. No acorn woodpecker could develop or stock a granary alone, much less defend it. This is clearly a case of a "strength in numbers" that benefits each individual bird. In effect, acorn woodpeckers inherit not only their genes but their granaries.

Behavioral biologists Walter Koenig and Ronald Mumme, who have studied the acorn woodpeckers extensively, draw this conclusion: "The cumulative investment in time and energy devoted to creating granaries is obviously extraordinary ... There is, however, a big payoff for groups whose granaries are sufficiently large to supply stored acorns throughout the subsequent spring: not only do such birds have increased over-winter survivorship, but their reproductive success is greatly enhanced."[45]

These examples are not isolated exceptions. They are representative of a growing research literature that, I believe, supports the Synergism Hypothesis.

Growing Support for This Theory

The Synergism Hypothesis, to repeat, makes the claim that the evolution of cooperation (and complex organization) in nature is directly tied to synergistic functional effects of various kinds; "synergistic selection" is the agency. Thus hunting or foraging collaboratively – a behavior found in many insects, birds, fish, and mammals – may increase the size of the prey that can be pursued, the likelihood of success in capturing prey or the collective probability of finding a food "patch." Joint action against potential

predators – alarm calling, herding, communal nesting, synchronized reproduction, coordinated defensive measures, and more – may greatly reduce
the individual's risk of becoming somebody else's meal.

Likewise, shared defense of food resources – a practice common to social insects and social carnivores alike – may provide greater food security
for all. Cooperation in nest-building, and in the nurturing and protection
of the young, may significantly improve the collective odds of reproductive success. Coordinated movement and migration, including the use of
formations to increase aerodynamic or hydrodynamic efficiency, may reduce individual energy expenditures and/or facilitate navigation. Forming
a coalition against competitors may improve the chances of acquiring a
mate, or a nest-site, or access to needed resources (such as a water-hole, a
food patch, or potential prey). Indeed, every one of the different kinds of
synergy identified in Chapter 2 can be found in nature, and most are found
in every kingdom, phylum, class, and order, if not in every family, genus,
and species.[46]

These synergies are often quantifiable, and in nature they are related
more or less directly to the "bottom line" – survival and reproduction. A
school (or "shoal" in the biologists' lingo) of dwarf herring using coordinated evasive maneuvers can greatly reduce their joint risk of being eaten
by a predatory barracuda.[47] Groups of immature ravens, acting in concert,
are often able to overcome the resistance of a mature adult and gain access
to feed at an animal carcass.[48] In prairie dog colonies, the overall level of
anti-predator protection increases with group size, yet the average amount
of time each individual member needs to spend looking out for predators decreases.[49] When lions hunt large game animals, capture efficiency
(captures per chase) and the number of multiple kills obtained increases
significantly with the size of the lion group (up to a point).[50]

One advantage of sociality in the paper wasp *Polistes bellicosus* is that
larger groups can quickly rebuild their nests after being damaged by a
predator, which in turn improves their chances of reproductive success.[51]
A solitary hyena is unlikely to be able to separate a wildebeest mother from
its calf, but when hyenas hunt in pairs they coordinate their efforts. One
hyena distracts the mother while the other catches the calf.[52] A number
of ant species establish colonies with multiple queens. This seems paradoxical, since ant colonies often engage in what has been characterized
as all-out warfare. A possible explanation was found in a study of the
desert seed-harvester ant (*Messor pergandei*) by Steven Rissing and Gregory
Pollack. Colonies with multiple queens gain a significant advantage because

they can mount much larger raiding parties against other colonies founded by a single queen.[53] Similarly, Sara Cahan and Glennis Julian found in a laboratory experiment that multiple-founder colonies of leaf-cutter ants (*Acromyrmex versicolor*) were significantly more successful in establishing fungus gardens.[54]

In the case of the fire ant (*Solenopsis invicta*), the benefits of forming co-operative colonies are more a matter of cost- and risk-reduction. Walter Tschinkel, in a recent experimental study, found that pregnant fire ant queens had a strong affinity for using preformed holes rather than investing in the excavation of new breeding chambers, even though they might have to share their ready-made quarters with other queens. Not only did this strategy save a lot of work but it had the great advantage of allowing the queens to minimize their exposure to environmental desiccation and predators.[55]

None of these findings would have surprised Darwin, although it might surprise many of us to learn that Darwin himself mustered an array of evidence regarding the benefits of sociality, in *The Descent of Man* (1871). The field research conducted today is far more sophisticated and systematic, of course, but Darwin's "anecdotal evidence" is consistent and is still useful. Here is an excerpt:

> Animals of many kinds are social; we find even distinct species living together; for example, some American monkeys; and united flocks of rooks, jackdaws and starlings.... We will confine our attention to the higher social animals, and pass over the insects, although some of these are social, and aid one another in many important ways. The most common mutual service in the higher animals is to warn one another of danger by means of the united senses of all.... Wild horses and cattle do not, I believe, make any danger-signal; but the attitude of any one of them who first discovers an enemy, warns the others. Rabbits stamp loudly on the ground with their hind feet as a signal; sheep and chamois do the same with their fore feet, uttering likewise a whistle. Many birds and some mammals, post sentinels, which in the case of seals are said generally to be the females. The leader of a troop of monkeys acts as the sentinel, and utters cries expressive both of danger and of safety. Social animals perform many little services for one another; horses nibble, and cows lick each other for external parasites; and Brehm states that after a troop of *Cercopithecus griseoriridis* has rushed through a thorny brake, each monkey stretches itself on a branch, and another monkey sitting by, "conscientiously" examines its fur, and extracts every thorn or burr.

Animals also render more important services to one another; thus wolves and some other beasts of prey hunt in packs, and aid one another in attacking their victims. Pelicans fish in concert.... Social animals [also] mutually defend each other. Bull bisons in North America, when there is danger, drive the cows and calves into the middle of the herd, while they defend the outside.[56]

Later on in this same passage, Darwin also expressed a view of the causes of social evolution that, I believe, can fairly be interpreted as supportive of the Synergism Hypothesis. Darwin argued that it was the functional benefits produced by sociality that drove the associated biological changes:

It is often assumed that animals were in the first place rendered social, and that they feel as a consequence uncomfortable when separated from each other, and comfortable while together; but it is a more probable view that these sensations were first developed in order that those animals that would profit by living in society should be induced to live together, in the same manner as the sense of hunger and the pleasure of eating were, no doubt, first acquired in order to induce animals to eat.

The Economics of Synergy

If synergy can be found literally everywhere in nature, it is also highly contingent. The reason, in a nutshell, is that it is always subject to economic criteria, namely, the costs and benefits in a given context and how these are allocated among the "parts." This fundamental economic consideration has important implications.[57]

In the first place, there is no such thing in biological evolution as "order for free," in biophysicist Stuart Kauffman's term (see Chapter 9). Any form of order, whether self-organized or directed by some organizing mechanism, has costs associated with producing it (energetic, informational, and otherwise). Moreover, biological systems have operating costs as well as capital costs. In a physical system it may be possible to create a stable form of order that is more or less permanently fixed. Not so in living systems.

Accordingly, natural selection functions in a very real sense like a business entrepreneur or a venture capitalist. If the benefits of any "adaptation" do not, on balance, outweigh the costs (if it is not "profitable" in terms of its impact on the survival and reproduction), the system that is responsible for producing the adaptation is very likely to be declared bankrupt and go out of business. Thus it may not make economic sense to form a herd, or a school, or a communal nest if there are no predators about, especially if proximity encourages the spread of parasites or concentrates the competition for locally scarce resources. Nor does it make sense to huddle together for warmth at high noon during the summer months, or to huddle against dehydration during the rainy season. And group hunting is not advantageous if the prey are small and easily caught by an individual hunter without assistance. Deterministic approaches to complexity are blind to such functional contingencies, while the Synergism Hypothesis predicts that cooperation – and complexity – are ultimately dependent upon these "economic" criteria; such contingencies are the rule.

The research literature in behavioral biology contains many case studies of the contingent relationship between cooperation and synergy. For instance, Craig Packer and Lore Ruttan conducted a systematic reanalysis of the data on hunting behaviors from 28 different studies encompassing 60 species. These researchers concluded that cooperative hunting was only likely to occur when there was a potential for synergy – per capita food acquisition that was greater than the amount each participant could capture on its own. Moreover, the critical factor was not the degree of relatedness among the cooperators but the size, abundance, and character of the prey and the degree of preexisting gregariousness among the hunters, which might have been due to other synergistic benefits (like mutual protection against predation).[58]

Recall also the examples cited in Chapter 4. It is now well established that the mycorrhizal fungi that form symbiotic associations with many modern land plants have variable effects. They enhance plant growth in low phosphorus soils, but in high phosphorus soils or low sunlight conditions (when photosynthetic activity is reduced) mycorrhiza may actually become parasitic and reduce plant growth.[59] Likewise, the cowbirds' practice of infiltrating their eggs into other species' nests is ordinarily harmful to the "hosts"; they end up nurturing somebody else's nestlings. But whenever there are heavy botfly infestations, the parasitic cowbirds, which eat botfly larva, may actually enhance their hosts' fitness.[60]

Because many forms of cooperation involve trade-offs – costs as well as benefits – the synergies that occur in nature are also highly sensitive to threshold effects. For instance, biologists William Shields and Janice Crook found that breeding success in groups of barn swallows (*Hirundo rustica*) was better in small groups than in larger groups; they observed that larger groups were much more heavily infested with parasites.[61]

Similarly, biologist Robin Dunbar re-examined the data for 18 baboon troops throughout sub-Saharan Africa and concluded that very large groups experienced various behavioral symptoms that seemed to be stress-related (and very human-like) – less time spent resting, less social activity, a faster travel-pace between various sites, more group fragmentation.[62]

A more complex cost–benefit relationship was observed among social spiders (*Anelosimus eximius*) in the rainforest of Ecuador by biologists Leticia Avilés and Paul Tufiño. These researchers discovered that colony size had very different effects on three different fitness "components" – the likelihood of reproduction by the females, average clutch size, and survival of the offspring to maturity. The net result was that colonies of moderate size had much higher lifetime reproductive success than did either very small or very large colonies.[63]

Cooperation Versus Competition

Despite the traditional assumption that cooperation and competition in nature are "dichotomous" (either–or) choices and are at odds with each other, this is not so; competition and cooperation are commonly intertwined, or juxtaposed. In fact, many forms of cooperation in nature are related to improving one's competitive ability. Thus coalitions of males, both in primates and social carnivores, are commonly formed in order to compete with other males for mating privileges, or for access to a nesting site, or other resources. In bonobos, coalitions of females can effectively blunt male bullying behaviors. Likewise, temporary coalitions of immature animals and birds may unite to gain access to a contested food source controlled by an adult. Many other coalitions arise for self-defense against potential predators. (We will revisit this idea in Chapter 7, when we consider the role of synergy in the evolution of humankind.)

Equally important, many of the same animals may cooperate for certain purposes and compete at other times. For instance, prairie dog colonies benefit greatly from joint protection against various predators, but the trade-off is a high level of interpersonal aggression over food resources and mating privileges.[64] Similarly, groups of male lions may cooperate in taking "possession" of a group of females but will then quarrel among themselves over individual mating rights.[65] This sequence is reversed in male elephant seals. As described earlier, when they are finished with their often prolonged and sometimes bloody battles for dominance during the winter–spring breeding season, everyone but the new alpha male and perhaps a few male "helpers" will huddle together in tightly packed "rookeries" to share heat.[66]

Testing for Synergy

How can the Synergism Hypothesis be tested? One way was suggested earlier – experiments or "thought experiments" in which a major part is removed from the "whole" (synergy minus one). In many cases, a single deletion, subtraction, or omission will be sufficient to eliminate the synergy. Take away the heme group from a hemoglobin molecule, or the energy-producing mitochondria from a complex eukaryotic cell, or a major gene from the so-called "homeobox" gene complex (which establishes the basic body plans for complex organisms, from *Drosophila* flies to humans), or, for that matter, remove one of the two hyena hunting partners.

Of course, many synergies-minus-one do not destroy the system but will only attenuate or diminish its effects, as we noted in Chapter 4. The removal of one emperor penguin from the huddle, or one acorn woodpecker from its communal nest, or one meerkat from its burrow, or one dwarf herring from a shoal might have little or no effect on the rest of the group, though it might be fatal for the individual (which provides indirect support for the Synergism Hypothesis). Sometimes it requires multiple wounds to kill a system. In other cases, the consequences may depend on the functional importance of the missing part. The removal of a tooth from the human body will (most likely) have only minor consequences, but the removal of a major organ can be fatal. By the same token, the operation of a car will not be affected by the loss of a decorative chrome strip (although it may adversely affect the resale price). The loss of a spark plug will diminish the car's performance but may not stop the engine. But the loss of a tire will render the car virtually inoperable – unless the tire is replaced by a spare. (Many evolved systems have redundancies to insure against consequences of losing a vital part.) In any case, complex systems are dependent upon their parts and, very often, vice versa. And, if we wish to understand why the wholes exist and how they came to be, we can do so by observing the combined effects they produce in a given context and the consequences of introducing some dysfunction.

A second way of testing the Synergism Hypothesis involves the use of a standard research methodology in the life sciences and behavioral sciences alike – comparative studies. Often a comparative study will allow for the precise measurement of a synergistic effect. Some examples have already been mentioned:

* Planaria (flatworms) that can collectively detoxify a silver colloid solution.
* Emperor penguins that can drastically reduce their energy expenditures.
* The internal temperature of a beehive or an army ant nest.
* Wasp colonies with multiple queens that can outcompete colonies with single queens.
* Nest-construction efficiencies achieved by social wasps compared to individuals.
* Lower predation rates in larger meerkat groups with more sentinels.
* Higher pup survival rates in social groups of sea lions versus isolated mating pairs.
* The hunting success of cooperating hyenas in contrast with those that fail to cooperate.

A comparison of the choanocytes in sponges with the very similar free-swimming choanoflagellates provide an especially apt example. The choanoflagellate protists are able to capture only enough nutrients to meet their own individual needs. By contrast, the choanocytes that line the inner walls of sponges are able, through their collective efforts, to process much greater quantities of water per cell and can generate enough surplus nutrients to provide for the needs of a vastly larger organism.

Another ready-made example, from the research literature on symbiosis, is lichen. Many lichen partnerships are facultative; in some environments, the fungi and green algae (or cyanobacteria) also live independently. In a comparative study, symbiologist John Raven found that overall nutrient and energy uptake was significantly better in the lichen partnerships than in their non-symbiotic cousins.[67]

Finally, a classic experiment in ecology provides a textbook illustration of how to measure synergy. The experiment was designed to study the effects of sunlight and two different fertilizers (nitrate and phosphorus) on the growth of a small woodlands flower (*Impatiens parviflora*). One significant finding was that varying amounts of increased sunlight made little difference during the 5-week test period without the addition of fertilizers. Furthermore, the use of either nitrate or phosphorous (essential ingredients for proteins and amino acids) alone made only an incremental difference. But when the plants were treated with the two fertilizers together, they weighed 50% more at the end of the test period than either of the two single-fertilizer groups and almost twice as much as the non-fertilized "controls."[68] The results were clear-cut. The interaction between sunlight, nitrogen, and phosphorus in plant growth is synergistic, and the consequences are measurable – as any skilled gardener already knows.

Neo-Darwinism Compared

The ultimate test of this theory, or any other scientific theory, involves the "so-what" question. What does this theory explain, and is it better or more satisfactory than other alternative explanations? Can it deal with unexplained puzzles, or paradoxes that other theories cannot? And can it make predictions that don't necessarily follow logically from other existing theories? I believe the answers in all cases are yes.

First, a little background is in order. The cardinal assumption of Neo-Darwinism is that the genes are free agents that pursue their self-interests in relentless competition with other genes. This being the case, cooperation and social organization in nature are a theoretical conundrum, especially if one assumes that cooperation depends upon altruism, or self-sacrifice. Early on, this is exactly what was assumed to be the case, and the proposed solution to this apparent problem is what has come to be known as "inclusive fitness theory" (or "kin selection," in Maynard Smith's term). Actually, the idea was first proposed by Darwin himself. In *The Descent of Man*, Darwin coined the term "family selection" to suggest that sociality might first arise within families, where self-sacrifices would be offset by benefits to close relatives.

When the foundations of modern population genetics were being laid in the 1920s and 1930s, the problem of sociality was raised once again, and again the solution that later came to be called inclusive fitness was proposed by the pioneer geneticists. "Insofar as it makes for the survival of one's descendants and near relations," wrote the formidable J.B.S. Haldane in his seminal book *The Causes of Evolution* (1932), "altruistic behavior is a kind of Darwinian fitness, and may be expected to spread as a result of natural selection."[69] However, it was William D. Hamilton's pair of classic papers on "The Genetical Evolution of Social Behavior" (1964) that formalized the idea and gave it wings.[70] In the process, Hamilton reinforced the tendency to equate sociality with altruism. He identified only three kinds of social behavior in his original papers – altruism (self-sacrifice), selfishness (gaining at another's expense), and spite. Only later on did he add "reciprocity" to his list. Hamilton's model suggested that sociality (read altruism) could be accounted for in terms of its benefits to close kin.

The key to inclusive fitness theory is the coefficient of relationship (r) between two individuals. According to the conventional wisdom of population genetics, two siblings have one-half their genes in common. So their coefficient of relationship is $r = 1/2$. Likewise, two first cousins have one-eighth of their genes in common, or $r = 1/8$. Assuming for the sake of argument that there is such a thing as an altruist gene, it could spread in a population via natural selection if the increases in the summed fitness of close relatives due to self-sacrificing behavior by an altruistic relative were sufficient to offset that relative's loss of fitness. In formal terms, the gain–loss ratio (k) must exceed the reciprocal of the average coefficient of relationship, or $k > 1/r$. Legend has it that Haldane anticipated Hamilton's model with some bar-room bravado: "I would gladly give up my life for two brothers or eight cousins." (Maynard Smith, a student of Haldane, says the number was later increased to ten; Haldane wanted to show a profit.)

Inclusive fitness theory has proven to be an important predictor of social behavior in nature, in the limited sense that sociality is frequently correlated with close kinship. Many sociobiologists and behavioral ecologists assume that the relationship is also causal, and all-important. However, there are several problems with this presumption. One is that the sociality–kinship correlation is far from perfect. There are also many examples in nature, and more are being found all the time, where sociality is not tied to kinship or is only loosely associated, most notably in groups that include a mixture of close kin and unrelated cooperators. In other words, kinship is not a necessary precondition for cooperation. This would seem to contradict the basic assumption of inclusive fitness theory. Another way of putting it is that if $r = 0$, inclusive fitness theory would logically seem to predict that sociality will not occur. Yet it does, nonetheless.

To cite some examples: In birds, many species have been found where immature "helpers-at-the-nest" will aid not only their younger siblings but also unrelated nestlings.[71] There are also studies of food-sharing in birds that is not primarily directed toward relatives.[72] Communal breeding among unrelated birds is also commonplace.[73] Alarm calling is also common and may benefit both related and unrelated birds, and even other species.[74] In insects, the recent discovery that many colonies consist of multiple queens and/or multiple lineages presents a major challenge to inclusive fitness theory. For instance, biologists Joan Strassman, David Queller, and their colleagues, in a study of a multi-queen neotropical wasp (*Paracharter-gus colobopterus*), found that colony workers showed no preferential treatment toward their own kin, and many other studies have confirmed this pattern.[75] Likewise, in mammals there are now many studies – in male and female lions, chimpanzees, capped langurs, Japanese macaques, evening bats, and more – that involve cooperation among unrelated individuals.[76] Then there are "allomothering," "baby sitting," and "adoption" by un-related conspecifics which have been reported in close to 300 species.[77] Something other than kinship must also be involved in social cooperation, as various theorists have come to recognize.[78]

Other problems with inclusive fitness theory relate to some logical com-plications that are not often addressed. If sociality is strongly associated with kinship, what accounts for the fact that a great many species are not at all social? Indeed, Stuart West and his colleagues point out that compe-tition among relatives is common and frequently intense.[79] It seems that kinship is neither necessary nor sufficient for sociality to occur. Also, if kinship is presumed to be causally related to cooperation, how come the most common forms of cooperation in nature occur among altogether dif-ferent species (symbiosis)? In symbiotic relationships, the partners have none of their genes in common (in theory), although this long-held popu-lation genetic dogma is rapidly coming unraveled; about 31% of the genes even in yeast bacteria have homologues in the human genome, and hu-mans share some 99% of their genes with chimpanzees. We don't really give up 50% of our genes when we reproduce sexually but only 50% of the genes that have functionally different effects from those of our partners. Moreover, all genes are not created equal, but that's another story.[80]

Egoistic Cooperation

One answer to these objections has, no doubt, already occurred to readers who are familiar with the recent literature on social evolution and coop-erative behaviors. Sociality and altruism are not necessarily equivalent to one another; it is possible to have cooperation that is not altruistic. I call it "egoistic cooperation," to differentiate it from "altruistic cooperation." (Elliot Sober and David Sloan Wilson also stress the distinction between genetic and psychological altruism.) It may very well be that close kinship is an important "inducement" for the evolution of altruistic cooperation in nature (group selection is another possible path), but there are many other forms of sociality that are mutually beneficial; they may impose short-term costs that are later offset by equivalent or greater benefits, or they may provide immediate net benefits to the cooperators. In short, cooperation can also involve relatively straightforward "economic" calculations of costs and benefits – and the main constraint may be how these costs and benefits are toted up and distributed among the cooperators. However, it took the sociobiologists a while to recognize this simple reality.

The decoupling of kinship, cooperation, and altruism began with the publication of another landmark theoretical paper, Robert Trivers's "The Evolution of Reciprocal Altruism" (1971). Although Trivers's thesis has often been referred to as the Good Samaritan scenario, this is not correct. The biblical Good Samaritan acted without the expectation of repayment; it was an uncompensated act of charity to a stranger of a different "nationality." But under Trivers's scenario (he didn't actually provide a formal model), the term "altruist" is a misnomer. The helper acts with the assurance that a low-cost, low-risk form of assistance performed now will be repaid with interest later on. It's really reciprocity with a delayed repayment schedule. Or, to be precise, it's an investment; the ultimate benefits will greatly outweigh the costs. Reciprocal altruism is really an oxymoron.

In fact, Trivers's paradigm ultimately relied on synergy. Trivers illustrated his concept with a human example that, again, was presaged by J.B.S. Haldane, even to the percentages used. Suppose that two strangers (non-kin) find themselves alone together next to a body of water. One of the two falls in, with a 50% chance of drowning absent a rescuer. If the stranger comes to the victim's aid – with a 5% chance of drowning in the process – but is later repaid when the tables are turned, the aid-giver is actually a net beneficiary. It's really a form of risk-sharing, as described in Chapter 2.

Of course, this was also a highly contrived scenario. It would take a lot of encounters among non-kin at the water's edge, a lot of accidents, and a lot of drownings (because the rescuer gene would be a rare mutation, initially), for such a genetic change to spread through a large population. Indeed, Trivers's model/scenario requires the improbable assumption that "the entire population is sooner or later exposed to the same risk of drowning."[81]

More credible were the three classes of real-world examples that Trivers cited to support his thesis – cleaner-fish, alarm calling in birds, and various types of helping behaviors in humankind. Other examples have been identified in the years since Trivers's paper was published. Perhaps most famous is the blood-sharing among unrelated vampire bats mentioned earlier, although behavioral biologists have also found many examples in chimpanzees and other primates, including coalitions, food-sharing, and mutual grooming behaviors. These are described in some detail in primatologist Frans de Waal's very readable and important book, *Good Natured: The Origins of Right and Wrong in Humans and Other Animals* (1996).

However, all of these real-world examples depend on "personal relationships" (whether kin or not). As Trivers himself put it in his original paper: "It will be argued that under certain conditions natural selection favors these altruistic behaviors because in the long run they benefit the organism performing them." Commenting on various human examples, Trivers noted that "all these forms of behavior often meet the criterion of small cost to the giver and great benefit to the taker." In other words, it is not altruism at all but mutually beneficial, win–win reciprocity – synergy.

Game Theory Revisited

A more decisive decoupling of kinship, altruism, and cooperation occurred when John Maynard Smith introduced game theory (specifically, the so-called "Prisoner's Dilemma") into evolutionary biology in the 1970s and early 1980s.[82] The basic premise of game theory models (so called because they involve calculated moves by each of two independent, self-serving "players"), is that the cooperators are unrelated and that they will only participate if there is an expectation of net benefits. The very purpose is to design the game in such a way that cooperation will be an "evolutionarily stable" strategy and will allow a gene for cooperation to spread in a population, rather than being swamped by "cheaters."

However, synergy was always the secret ingredient – the hidden key in these game theory models. It was simply disguised and hidden in the numbers used to fill in the "cells" of various "payoff" matrices. For instance, in the famous tit-for-tat model presented by political scientist Robert Axelrod and William Hamilton – which should properly be credited to the solution proposed by the legendary systems scientist Anatol Rapoport – the payoffs posited for different player "strategies" were defined in such a way that mutual cooperation would be more rewarding over time than "defection," or cheating. Though for the most part unappreciated, it was the "economic" benefits that undergirded their model.[83]

Game theory has proven to be a fertile and productive theoretical tool, but it has also been limited by its narrow and constricting assumptions about the nature of cooperation. The reason why Axelrod and Hamilton's tit-for-tat model was hailed as a great step forward was that it added the simple assumption that the cooperators would interact more than once. The game would be "iterative," with each player's subsequent "moves" being affected by what had come before; in effect, it added a cumulative form of synergy to the game. (A crucial corollary was that the players should start out by being cooperative and then respond in kind to whatever the other player does.)

A further step toward realism occurred when zoologist Martin Nowak and mathematician Karl Sigmund developed a new kind of game theory model called "Pavlov," which allowed the players to punish cheaters. Called "win–stay, lose–shift," this strategy permitted a player to exclude cheaters from subsequent rounds and, by implication, from future benefits. In other words, the players could manipulate the costs and benefits.[84]

It turns out that Pavlov (and other variants since) conforms well with reality. It is now recognized that "policing" of cooperation and the punishment of cheaters is common in nature, and that cooperation is not so constrained by the threat of cheating as the game theory models implied.[85] As Timothy Clutton-Brock and Gregory Parker point out in a review article on the subject: "Individuals often punish other group members that infringe their interests . . . Punishing strategies are also used to establish and maintain dominance relationships, to discourage parasites and cheats, to discipline offspring or prospective sexual partners and to maintain cooperative behaviour."[86]

A more serious objection is that inclusive fitness theory, reciprocal altruism, tit-for-tat, and other game theory models exclude one of the most important forms of cooperation in nature, interactions that produce combined effects (synergies) that are largely self-policing because they are interdependent. This is frequently the case with symbiotic relationships and in species where there is a socially organized division/combination of labor. Maynard Smith and Szathmáry have suggested a useful metaphor to illustrate this distinction. Suppose that two oarsmen decide to cooperate in rowing a small boat across a river. In one alternative configuration, a "sculling" arrangement, the oarsmen each have two oars and row in tandem. In this situation, it is possible for one oarsman to slack off (to cheat) and let the other one do most of the work. This represents the classical game theory relationship.

Now imagine instead a "rowing" arrangement. In this configuration, each oarsman has only one opposing oar. Now their relationship to the performance of the boat is interdependent. If one of the oarsmen slacks off, the boat will go in circles and will not reach its goal. Interdependence has the effect of making a cooperative relationship self-policing. Maynard Smith and Szathmáry conclude that the rowing model is a better representation of how cooperation (and complexity) evolves in nature. "The intellectual fascination of the Prisoner's Dilemma game may have led us to overestimate its evolutionary importance."[87]

Other Problems

There are other problems with the conventional game theory paradigm. For instance, there are many cases in nature where the alternative to a win–win cooperative effort is not zero (the lowest value possible in a game theory payoff matrix) but death. If you were an animal faced with the prospect of confronting a predator, cooperative defense might be the only logical choice. Cheating would be self-defeating. Another problem is that game theory models have not as a rule dealt with multiple interests, where cooperation in one area – say mutual grooming – may also affect cooperation in other areas, like hunting, meat-sharing, coalition-building, or mutual defense. Nor does game theory capture the sometimes complex interplay between the costs and benefits associated with various choices, or "strategies."

But a more serious problem with the existing game theory models is that they exclude one of the most common forms of cooperation in human societies, "teamwork" that produces what I call "corporate goods." In the corporate goods model (which could include any number of players), the participants may contribute in many different ways to a joint product (say the capture of a large game animal or the manufacture and sale of a "widget"). However, unlike "collective goods" that are indivisible and equally shared (even possibly with non-participants and cheaters), corporate goods can be divided in accordance with various principles, or "rules," or "contracts." Sound familiar?

Neo-Darwinism is particularly insensitive to synergies of scale – the many cases where collective action produces combined effects that would not otherwise be possible. Lee Dugatkin, in a recent book, cites an example (based on some research by Susan Foster) involving the collective behavior of the wrasse, a tropical reef fish that preys on the abundant supply of eggs produced by the much larger sergeant-major damselfish. Because female damselfish aggressively defend their nests, no single wrasse, nor even a small group, can overwhelm the damselfish's defenses. However, very large groups can do so and are rewarded with a gourmet meal of damselfish caviar. Dugatkin is not certain that this is an example of synergy. "Cooperation in group foraging certainly pays off for the wrasse, but whether they do anything as a group that exceeds the sum of their own individual actions is not clear."[88] I disagree. Since success in raiding a damselfish's nest can only be achieved by a large group of wrasse acting in concert, it's an unambiguous example of a synergy of scale.

Dugatkin calls this form of behavior "by-product mutualism," a term coined some years ago by the ornithologist Jerram Brown. Dugatkin explains that by-product mutualism refers to the fact that cooperation "in certain cases" is just a "by-product" of the selfish interests of the participants.[89] From an evolutionary perspective, however, all cooperation is a by-product of the synergies it produces in relation to the ultimately selfish interests of the participants. When cooperation involves altruism and closely related kin are the beneficiaries, it's called inclusive fitness; when there are reciprocal acts of generosity, it's called reciprocal altruism; and when there are mutual benefits, it's called mutualism, or tit-for-tat, or win–win, or a cooperative effect, or whatever. All of these cooperative behaviors are "selfish" in the broad sense of being fitness-related – whatever may be the immediate motivator. And all depend on synergy. Even if the behavior is the "unintentional" aggregate effect of many individual actions, it is still synergistic. And it's the synergies that provide the "rewards" for cooperative behavior.

Actually, Jerram Brown's original usage of the term by-product mutualism was directed to the more limited idea that one animal's independent, uncoordinated pursuit of its own self-interest might "coincidentally" benefit another animal.[90] An example would be a situation where two predators happen to be pursuing the same prey at the same time and "unintentionally" succeed together where neither might have done so alone. Some theorists might claim that this model applies to such things as mobbing behaviors by various birds, fish, and mammals – or perhaps even to the wrasse example. But this begs the question. Would the mobbers also engage in such behaviors by themselves, or does their strength derive in various ways from their numbers? More to the point, would they survive at all if they were to opt for mounting a solo defense (or assault) without regard for what their nest-mates did? If, in fact, mobbers only mob when they can do it collectively and reliably choose "flight" as their best option when they are alone, it is reasonable to assume that collective action is more than merely a "by-product" of individual behaviors. (Sacrificial behaviors by parents in defense of their offspring are, of course, another matter.)

By-product mutualism in Jerram Brown's sense of the term may also provide a model for how "deliberate" cooperation can arise in nature. This model does not require a gene for cooperation. It requires only genes that enable an animal to learn from experience – what the Behaviorist psychologists call "reinforcement" learning, or "operant conditioning." In this model, what starts out as by-product mutualism (inadvertent synergy) may lead over time to an "intentional" pattern of synergies. And if the new behavior in turn leads to subtle genetic changes, this would be an example of "downward causation" in evolution – a concept we'll discuss further in the next chapter.

A documented case – involving the development of cooperative hunting behavior in a troop of olive baboons (known as the "Pumphouse Gang") – provides an important illustration. In the course of studying a group of 49 baboons on a huge ranch near Nairobi, Kenya, over a period of several years, Shirley Strum, Richard Harding, and several other co-workers observed the emergence and spread of a new "cultural" pattern. At first it was confined to a few adult males that opportunistically pursued and captured newborn antelopes or hares. It was a solitary activity and there was no food-sharing. But over the course of time the pattern changed. The amount of predation increased; females and juveniles began to participate; food-sharing became more commonplace; hunting skills and efficiencies improved; most important, the troop began to evolve systematic searches and coordinated attacks.[91] In sum, it was the synergies (the proximate rewards) that drove the behavioral changes, not genetic mutations or natural selection, much less a law of self-organization. We'll come back to this paradigm in the next chapter.

Indirect Reciprocity

One other apparent conundrum that has recently come under greater scrutiny is a class of cooperative behaviors in nature that do not seem to have any relationship at all to reproductive fitness from the standpoint of the "donors." For instance, helping behaviors among unrelated individuals – say the meerkat "baby sitters" or the "helpers at the nest" in various bird species – appear to be an evolutionary puzzle. What's in it for the helpers? Some years ago, biologist Richard Alexander developed the concept of "indirect reciprocity" as a possible explanation. Alexander's argument was that, in a stable, continuing network of cooperators, a donor might ultimately receive a fair return for some helping behavior "indirectly" if it later became the recipient of some other member's generosity. It amounted to a formalization of the old expression "what goes around comes around."

Much more thought and analysis has been devoted to this phenomenon in recent years, and the consensus seems to be that indirect reciprocity may well be a factor in socially organized species, independently of kinship.[92] Significantly, this phenomenon seems to be most likely to occur under the conditions that, most probably, also characterized the evolution of the human species (see Chapter 7). In any event, indirect reciprocity amounts to a subset of the broader category of "reciprocity" (see below), and it is dependent – as always where cooperation is involved – on synergy.

Synergistic Selection

There is one other important approach to understanding cooperation in nature that should be mentioned. The basic idea is that there are cases in which some collective interest with ultimate fitness consequences for all concerned may override the interests of the individuals and constrain natural selection at the individual level in favor of selection for group-serving traits. David Sloan Wilson, the biologist who is most closely associated with the revival of group selection theory in evolutionary biology, calls it "trait-group selection." For obvious reasons, I prefer John Maynard Smith's equivalent term "synergistic selection."

We can illustrate the idea of group selection with a variation on the "sculling" and "rowing" models described earlier. Recall that in the sculling model, one of the two tandem oarsmen could "defect" (cheat) without undermining the attainment of their joint objective. Of course, this was a hypothetical situation. In the real world of small boating, a high wind, or a strong current, or a distant goal might demand the combined efforts of both oarsmen. But now imagine a very different situation, where the boat is in a race against another boat. Now if the two oarsmen want to win the race they will most likely have to make an all-out effort. It has become a group selection game, and the fate of the two oarsmen is totally interdependent, even if they are rowing in tandem.[93]

Darwin himself recognized the multilevel character of natural selection, especially where human evolution is concerned. Modern-day anthropologists will, no doubt, find fault with his specific scenario, but the spirit of the group selection idea was captured in this passage from *The Descent of Man*: "A tribe including many members who, from possessing in a high degree patriotism, fidelity, obedience, courage and sympathy, were always ready to aid one another, and to sacrifice themselves for the common good would be victorious over most other tribes: and this would be natural selection."[94]

A Stalking Horse

Many group-selection advocates, especially in the early part of this century, routinely assumed that the interests of the group (the "public interest") would assure control over individual interests – a proposition that was obviously fraught with ideological and political implications. The issue became contentious only when zoologist V. C. Wynne-Edwards made himself a stalking horse, in Edward O. Wilson's characterization, by propounding a boldly stated version of the group selection hypothesis in his 1962 book, *Animal Dispersion in Relation to Social Behaviour*. Wynne-Edwards asserted that social animals routinely display behaviors that involve curtailment of their own fitness for the good of the group. "The great benefit of sociality," he wrote in a separate article in the British science journal *Nature*, "arises from its capacity to override the advantage of individual members in the interest of the survival of the group as a whole."[95] Notwithstanding the fact that he clearly stated that this could only occur if natural selection were to operate between groups "as evolutionary units," Wynne-Edwards soon became a pariah in evolutionary biology and has been routinely chastised for his heresy ever since.

The reductionist rebuttal, led by George C. Williams in his 1966 book, was that higher-level selection was largely a figment of a "romantic imagination." Williams charged that group-selection theorists ascribe "mystical" properties to emergent phenomena. Emergence can better be explained as a statistical summation of individual behaviors, or fortuitous effects, or a product of the laws of physics. "A wolf can live on elk only when it [coincidentally] attacks its prey in the company of other wolves with similar dietary tendencies. I am not aware, however, of any evidence of functional organization of wolf packs."[96]

David Sloan Wilson's important contribution to the debate (supported by some parallel work by Michael J. Wade and other colleagues) was to dissociate group selection from the problematical effects it might have on an amorphous population of organisms and to focus instead on group selection in functionally interdependent groups – "trait groups."[97] Wilson's argument, in a nutshell, is that you must view a trait group as a distinct unit of selection because the effects, in terms of survival and reproduction, are interdependent and synergistic (although Wilson didn't use that word). Moreover, when a trait group also happens to be socially organized, it is appropriate to use Herbert Spencer's term "superorganism"; the social group, like an organism, can be viewed as an integrated "whole." Wilson together with the well-known philosopher of science, Elliott Sober, have recently published an exhaustive treatment of the subject under the heading of "multilevel selection theory."[98]

A key point that sometimes seems to be overlooked by the opponents of group selection is that it need not involve altruism or be opposed to "selfish" individual selection, as Darwin himself pointed out. When win–win forms of cooperation are involved, and the immediate costs and benefits are more or less equitably distributed, individual and group selection may work hand-in-hand. The reductionist response is that, in such cases, group selection can be collapsed into special cases of individual selection. In ant–aphid symbioses, for instance, ants have genes for cooperating with aphids and aphids have genes for cooperating with ants. That's all you need to know.

However, the reductionist riposte breaks down when the issue of causation is raised. Where did the genes for cooperation between ants and aphids (actually a simplistic formulation) come from? In this well-documented form of symbiosis, certain ant species guard aphids against predators and are rewarded with nutritious "honeydew" excretions from the aphids, which earn their living by "milking" the phloem sap of various plants. Both species have evolved special adaptations (and genetic modifications) to facilitate their cooperative arrangement. But how did these adaptations evolve? We can only guess how this behavior might have arisen in the first place, but it would not have spread and become perfected as a routine pattern of cooperation between the two species without synergy; it had to be interdependent and mutually "reinforcing" (rewarding) at the proximate (behavioral) level. If genetic mutations made it possible, synergy closed the deal.

Turning the Necker Cube

Very often it is possible to view the same phenomenon from more than one perspective. Each viewpoint may illuminate a different aspect of the phenomenon. In a later edition of *The Selfish Gene*, Richard Dawkins used the metaphor of a Necker cube to illustrate this point.[99] A Necker cube is a two-dimensional drawing of a three-dimensional cube that can be perceived in two different ways. (This can easily be demonstrated in a lecture-hall presentation.) However, the Necker cube metaphor can be pushed too far. Reliance on only one perspective may obscure many important properties (and principles) in the phenomenon one is studying. Thus, when a "trait group" is involved – when there are combined, interdependent effects produced by two or more genes together (remember the "rowing" model), the unit of selection is not the individual genes but the "whole." One needs only to apply the synergy test; if the removal of a part (or a gene) diminishes or destroys the synergies that are the proximate cause of natural selection in a given context, then it is a case of synergistic selection; it involves "Holistic Darwinism."

More than that, I would argue that much individual selection can also be viewed as group selection from a functional (synergy) point of view. Even when a single mutant gene is involved, natural selection is almost always focused on the effects that the gene produces in its interactions with other genes and the environment. The mutation may be the proximate cause – the "difference that makes a difference," to use anthropologist Gregory Bateson's well-known formulation. But the effects produced by the mutation depend both on the genetic and the environmental context, which may vary widely.

An illustration can be found in the literature on human diseases. To date, genetic anomalies have been implicated in some 1,200 different disease syndromes – from porphyria to manic depressive psychosis, Huntington's chorea, and Tay–Sachs disease – many of which involve a single mutation. Nevertheless, it is imprecise to say that these mutations "cause" the diseases. As noted earlier, they are examples of synergy minus one. When one of the necessary genes in a complex system is removed or its function is altered by a mutation, it is the disruptive effects on the system that causes the disease (a point well understood by the professionals in the medical field).[100] As Dawkins himself put it in one of his later works:

> In natural selection, genes are always selected for their capacity to flourish in the environment in which they find themselves ... But from each gene's point of view, perhaps the most important part of its environment *is all the other genes that it encounters* [his emphasis] ... Doing well in such environments will turn out to be equivalent to "collaborating" with these other genes."[101]

By the same token, many genetic effects are dependent on their environment. A classic example is favism, an acute hemolytic anemia that is produced by a single mutant gene. However, the incidence of favism varies widely from one country to another, not because the gene is missing from some populations but because the disease is co-determined by the consumption of fava beans (or inhalation of the plant pollen). Because fava beans are grown and consumed primarily in the countries that rim the Mediterranean basin, the disease is concentrated in that part of the world.

Weighing the Synergism Hypothesis

What are the implications of the Synergism Hypothesis? First and foremost, it requires us to break free from a single-minded preoccupation with the role of genes in evolution. It shifts our focus to the functional units and the behaviors that are found in the "economy of nature." Genetic mutations and other molecular-genetic phenomena are only one of the many sources of innovation in the natural world. Furthermore, the fate of a gene is almost always tied to the effects it produces in combination with other genes. It is the functional costs and benefits, and how these are distributed in terms of both proximate survival needs and ultimate reproductive success (the phenotype), that provides the key to explaining how cooperation and complexity have evolved. And the key to the key is synergy.

The Synergism Hypothesis provides a unifying framework for explaining cooperation and complex organization in biological evolution, and, as we shall see, in human evolution as well. Synergy is the common denominator – the cutting edge of Occam's razor. A "bioeconomic" (and phenotypic) focus, which this paradigm requires, frees us from the constraining theoretical boxes associated with a gene-centered approach. Lee Dugatkin, in a recent book, suggests that there are four "paths" to cooperation in nature. However, the path is not the goal, and it is the goal that determines whether or not a given path will be followed. Like the yellow brick road in *The Wizard of Oz*, there must be an Emerald City at the other end. Synergy is the Emerald City. It provides the goals for all four of Dugatkin's paths – and more; I would argue that there are at least five, maybe six distinct paths to cooperation and complexity in evolution.

(1) *Altruism* – including both inclusive fitness and certain forms of group selection.

(2) *Reciprocity* – which covers various kinds of exchanges; this also includes Trivers's reciprocal altruism, but it removes the implication that altruism is in any way involved.

(3) *Functional interdependence* – that is, functional complementarities and the division/combination of labor, including symbioses between species.

(4) *Mutualism* – including all of the many forms of cooperation that are *not* functionally integrated, such as "by-product mutualism" and synergies of scale.

(5) *Parasitism* – a category that some theorists might view as the antithesis of cooperation; some treat parasitism as a form of predation (negative synergy) in slow motion. It implies involuntary cooperation. However, it also represents a form of "operating together," and, as noted above, may shade into mutualism or reciprocity.

A possible sixth category is "slavery" – a very slippery term. Many theorists prefer to put slavery into a separate category. I prefer to include it under the heading of parasitism. It's a form of involuntary cooperation that may or may not have asymmetrical costs and benefits. Indeed, there may be mutual benefits. (The leaf-cutter ants and other species that cultivate "fungus gardens" provide an obvious example.)

In any case, I believe that the presence of functional synergy is a better (and certainly broader-based) predictor of cooperation and complexity in nature than any of the existing models or categories, and, indeed, sheds light on some of the paradoxes and problems described above: why cooperation occurs among both kin and non-kin; the conditional, context-dependent nature of cooperation; the high degree of sensitivity to how the costs and benefits are allocated, including the costs and benefits of *not* cooperating; and the rationale for "policing" cooperation. (Simply put, policing is likely to occur when the synergistic benefits outweigh the policing costs.) The

Synergism Hypothesis also sheds light on such evolutionary puzzles as the multiple independent invention of various forms of symbiosis (say the many different kinds of cleaner-fish or the many different lichen partnerships) and the convergent patterns of social evolution in different parts of the world, especially in insects and birds. In all these cases, it is the bioeconomic benefits – the synergies – that are the "drivers."

Synergy has the power to override kinship. It also has the power to override the "war of every man against every man" – to use the philosopher Thomas Hobbes's famous characterization of "the state of nature." (George Williams may have gone Hobbes one better in an article with the curmudgeonly title, "Mother Nature is a Wicked Old Witch."[102]) However, synergistic selection will occur only when the context and circumstances are right – when there has been a "convergence" of necessary preconditions, facilitators, and incentives. As the economists would say, cooperation and complexity will evolve only when it is "cost-effective." The philosopher John Locke called it a "social contract," but it would be more accurate to call it a bioeconomic contract.[103]

The Synergism Hypothesis also leads to one unambiguous (law-like) prediction that cuts across all of the theoretical schools (shoals) of contemporary Darwinism. Stable cooperation in nature is always associated with synergistic functional effects. Positive synergy is necessary, if not sufficient. When/if the synergy evaporates or disappears, so will the cooperative relationships that sustained it – in due course. (The timing will obviously depend upon the degree of "phenotypic plasticity." A trait that is genetically preprogrammed will be subject to negative selection, but a facultative, learned adaptation will be eliminated much more rapidly.) And no, this statement is not a tautology. Cooperation may or may not produce positive synergy. As we saw in Chapter 4, it may produce negative synergy ("dysergy"), or nothing at all of functional significance. Or the costs may simply outweigh the benefits.

Furthermore, this prediction is not confined to social cooperation among members of the same species. It applies to all forms of biological organization, including human societies. The classical Darwinian paradigm is focused primarily on sexually reproducing organisms of the same species. It is embarrassed by such gaping anomalies as bacteria. Not only are bacteria a far more important part of the natural world than we had previously imagined but (as noted earlier) they are promiscuous gene-sharers; where bacteria are concerned, the selfish gene metaphor is inapposite. Horizontal gene exchange long pre-dated complex multicellular organisms. More important, the gene-centered, competitive model of evolution has little to say about symbiosis between species, a phenomenon that has been gravely underrated in the past and that can only be explained in economic terms – in terms of its functional synergies.

Consider the Alternatives

What Stephen Jay Gould and Niles Eldredge called "ultra Darwinism" does not have a theory of complexity. The most radical of the reductionists would deny that complex organization has any independent reality; complexity doesn't need explaining. When a hard-nosed reductionist uses the term "emergence," he/she means that the whole is an epiphenomenon – an expression of the "interests" of the parts, or the manifestation of underlying laws. Francis Crick, for instance, tells us that, "the properties of a benzene molecule are not in any sense the simple arithmetic sum of the properties of its twelve constituent atoms. Nevertheless, the behavior of benzene . . . can be calculated if we know how these parts react, although we need quantum mechanics to tell us how to do this. It is curious that nobody derives some kind of mystical satisfaction by saying 'The benzene molecule is more than the sum of its parts' . . ." In other words, Crick suggests, complexity is reducible to lower-level causes, whether it be electrons or genes – or the hidden laws of complexity.[104] Similarly, complexity theorist John Holland believes that "laws" of emergence will ultimately be found that will be able to explain complex phenomena like human consciousness.[105]

This is an altogether different usage of the term "emergence" from the so-called emergent evolutionists of the 1920s. These mostly forgotten theorists argued that "wholes" (organisms, groups, species) are significant units of evolutionary change; they have a significance as wholes that can't be derived from an understanding of the parts. One of the leaders of this school, the pioneer experimental psychologist Conwy Lloyd Morgan, saw emergent evolution as a unified creative process and as the central theme in evolution (an echo of Herbert Spencer). Emergent phenomena, from simple organisms to the human mind, were the most important sources of evolutionary "progress," according to Lloyd Morgan.[106] Other theorists, like Jan Christian Smuts – a distinguished scholar, soldier, and two-time prime minister of South Africa – asserted that wholes were the "only" unit of evolutionary change. Citing Darwin's flirtation with Lamarckism as a precedent for his views, Smuts also suggested in his book, *Holistic*

Evolution (1926), that "holistic selection" – the actions and choices of organisms – often precede and precipitate evolutionary changes.[107] This was a prescient idea that is presently "re-emerging." (We'll come back to this in Chapter 6, on "The Sorcerer's Apprentice.")

What are the alternatives to the Synergism Hypothesis?[108] The multi-level selection theory advocated by David Sloan Wilson and others – however important – is an analytical framework, not a theory of complexity. Another proposed alternative marches under the banner of "complexity theory." But this approach amounts to a set of promissory notes based on analogies that are drawn from various kinds of physical and physiological processes (plus some elegant and intriguing computer-modeling related to the dynamics of complex systems – there's no gainsaying that). Complexity theory may yet be able to illuminate additional law-like aspects of the evolutionary process. (Many have been well understood for decades, thanks to the landmark work of D'Arcy Thompson, Bernard Rensch, and others.[109]) Nevertheless, the dream of reducing the evolutionary process to the operation of a few deep "mechanisms" or "laws" that will enable us to understand and predict evolutionary "emergence" from the bottom up, so to speak, is destined to fail.

To be specific, many complexity theorists discount the implicit calculus of bioeconomic costs and benefits that determines many evolutionary "vicissitudes." They are also insensitive to the multileveled character of evolutionary causation. And they are "in denial" about history – about the unavoidably context-dependent nature of evolution. For some, in fact, the very purpose of complexity theory is to cut through the many history-laden particularities to expose the deeper (mathematical) regularities, or the self-organizing dynamic, that they believe will "emerge" from their models. But the intractable problem is that evolution is not ultimately a law-like, deterministic process, although it may well be shaped and constrained in ways that we still do not fully understand. (We will have much more to say on this issue in Chapter 9.)

Another alternative explanation of complexity is Stephen Jay Gould's so-called "drunkard's walk" model. In *Full House* (1996), Gould found "almost chilling" Edward O. Wilson's testimonial about the functional improvements that have been associated with the evolution of complexity in nature. As in some of his other recent writings, Gould wished to deny the relevance of any notion of general "improvement" as a significant aspect of the evolutionary process. He sought to undermine the traditional anthropocentric conceit that there has been "progress" (culminating of course in humankind), or that the so-called "trends" in evolution imply that something has gotten better.

Gould had a point, of course, but objections to the notion of a "driven," law-like trend toward betterment or progress in evolution go back to Darwin himself. What was new – Gould's conceit if you will – was his assertion that the evolutionary process is essentially random in its overall course and that the simplest forms of living systems – bacteria – represent the "modal" trend. For instance, Gould attacked Cope's rule (after the nineteenth-century paleontologist Edward Drinker Cope), which holds that most lineages tend to increase in size over time. This is an artifact, Gould said, of the circumstance that life began at the extreme "left wall" (or tail) in the distribution of biological size and complexity. Therefore, any subsequent directional change could only be toward the "right wall" (greater complexity). Furthermore, complexity in living systems is not a product of natural selection; organisms merely "wander" into greater complexity (like a "drunkard's walk").

There is a kind of "just-so" quality to this notion (to borrow a metaphor from Gould). It implies that systematic size/complexity increases in nature could occur without being "tested" and winnowed by natural selection. In fact, such changes always entail bioeconomic "costs" (energy, for instance) that, at the very least, have to be offset by equivalent "benefits." So intent was Gould on making his anti-progressive case that he even allowed himself to fall into a logical trap – a rare event for such a stellar thinker. In light of some recent, suggestive work by various colleagues, Gould claimed that "a small overall tendency toward decreasing complexity may characterize the history of most lineages."[110] If life began in extreme simplicity, it had to get more complex before it could become less complex. So, if there is any residue of complexity at all remaining in "most lineages," the overall trend, at the margin, had to be in that direction (unless there is some statistical sleight of hand going on).

Indeed, in the process of unpacking his most important (and eccentric) example – the disappearance of .400 hitters in baseball – Gould inadvertently undermined his own argument. Gould's key point was that, despite appearances, baseball hitters have not gotten worse over the years. This trend is a result (in his words) of an "improvement in general play" – i.e., pitching and fielding. What's that? Did he say an "improvement" – a "progressive" advance in functional performance rather than a "drunkard's walk"? So it seems there is a role for functionally based, "progressive" changes after all, at least in the evolution of baseball.

Holistic Darwinism: The New Evolutionary Paradigm

I believe that Holistic Darwinism and the Synergism Hypothesis provide a "viable alternative" to these other claims to a theory of complexity. Once the door has been opened to the reality that nature is organized into hierarchies (or holarchies) of functional units that are not only real but have causal efficacy in their own right, then a multilevel paradigm – as championed by David Sloan Wilson and others – becomes a necessity. And this, in turn, leads to a consideration of the economics of complexity, and the dynamics of Synergistic Selection. Genetic mutations are often characterized as the "engine" of evolution, but the engine is nothing without the car. It is time to focus on the car.

6 "The Sorcerer's Apprentice"

> It is not the organs ... of an animal's body that have given rise
> to its special habits and faculties; but it is, on the contrary,
> its habits, mode of life and environment that have in the course
> of time controlled the shape of its body, the number and state of
> its organs and, lastly, the faculties which it possesses.
>
> Jean Baptiste de Lamarck
Zoological Philosophy

Paul Dukas was a late nineteenth- and early twentieth-century French music critic and composer whose small body of orchestral works might be forgotten today were it not for Walt Disney and the renowned conductor Leopold Stokowski. When Disney teamed up with Stokowski in the late 1930s to produce the then path-breaking cartoon movie *Fantasia* – a melding of classical music and sometimes surreal visual imagery – one of the pieces they chose for the film was Dukas's "The Sorcerer's Apprentice." Thanks to Mickey Mouse, this lively scherzo, and the tale that it accompanied, became the centerpiece of the movie.

Based on a narrative poem by the German literary giant Johann Wolfgang von Goethe, the story concerns a great and powerful sorcerer who leaves home for the day and instructs his apprentice (Mickey) to keep watch over things in his absence. The hard-working apprentice is frustrated, however, by the drudge work he is expected to do (like cleaning the house), and he soon hits on the idea of using some of the sorcerer's magic to conjure a helper. Spying a broom in the corner, Mickey invokes a magic spell and, presto, the broom is transformed into a dutiful drone that goes out to fetch water from a nearby well. So far so good. Then things begin to go awry.

The rest of the story contains an important morality tale for humankind that we will talk about in the last chapter. The main point here is that Disney's movie fantasy also provides a metaphor for an underrated aspect of the evolutionary process. Living systems are at once products of nature's magic and increasingly powerful sorcerer's apprentices; they (we) have developed the ability to make magic of our own.[1] To use the scientific jargon, living organisms are not simply "dependent variables." They are also "independent variables." For better *and* worse, the products of evolution have mimicked Mickey Mouse.

Lamarck's Vision

The notion that living systems have themselves played an important role in shaping the course of life on Earth is hardly new. In fact, the idea goes back to Jean Baptiste de Lamarck, the eighteenth- and early nineteenth-century French naturalist whose *Zoological Philosophy* (1809) was an influential precursor to Darwin's work. Lamarck postulated an inherent "power of life" and a general trend in evolution toward greater complexity that is reminiscent of Aristotle's vision and presaged Herbert Spencer's grandiose "law of evolution" – not to mention the aspirations of some modern-day complexity theorists (see Chapters 8 and 9).

However, Lamarck is best known, even infamous, for his thesis that the course of evolution has been shaped by "habits acquired by conditions" – that is, the direct inheritance of traits developed during the lifetime of an individual organism. (Of course, Lamarck could only guess what the precise mechanism might be; he lived long before the discovery of genes and the genetic code.) The long necks of giraffes served as an illustration. Lamarck proposed that this distinctive animal trait arose because ancestral giraffes had progressively stretched their necks over many generations in order to reach the leaves in the tops of acacia trees, and that the changes were then passed on to their offspring.[2]

Darwin was generally scornful of this idea in his earlier years. "Heaven forfend me from Lamarck's nonsense," he wrote to a friend in 1844. But in *The Origin of Species* Darwin displayed his usual scientific caution by not ruling out the possibility that environmental influences and "the use and disuse of parts" (as he put it) could be one source of biological variation and adaptation in nature. He even cited some possible evidence in its favor. Yet for obvious reasons Darwin considered Lamarckian influences to be a minor subsidiary to natural selection.[3]

The rise of the science of genetics at the turn of the twentieth century put a conclusive end to Lamarck's theory. In a classic (if gruesome) experiment, the pioneer geneticist August Weismann cut off the tails of 20 successive generations of mice without, of course, producing a tailless strain.[4] Other evidence against Lamarck's thesis could be found in the practices of farmers and pet breeders, who routinely dock tails, notch ears, castrate males, spay females, and so on. There are also such human customs as circumcision, pierced ears and noses, shaved heads, and various forms of deliberate mutilation, none of which is heritable.

Unfortunately, in the process of rejecting the Lamarckian view of inheritance, the baby got thrown out with the bath water; the role of behavior as an important factor in evolutionary change was also summarily rejected. Weismann's claim that random mutations are the underlying source of creativity in evolution became one of the cornerstones of the nascent science of genetics and, ultimately, of a gene-centered evolutionary theory. For a time Weismann's "mutation theory" even eclipsed Darwinism, but in the 1930s a new theoretical synthesis was achieved. It was recognized that, if mutations (and the process of "recombination" in sexual reproduction) are responsible for generating biological novelties, they must subsequently be "tested" in the environment by natural selection. This perspective is captured by psychologist Donald Campbell's slogan "blind variation and selective retention." Ernst Mayr has characterized it as a "two-step, tandem process."

Although this mechanistic, gene-centered paradigm appealed to a young science that aspired to mimic the law-governed rigor of the then reigning "queen" of the sciences (physics), it also provided an insufficient account of the dynamics of evolution because it left the organism out of the equation. The "phenotype" was treated as a passive object (a black box) whose fate is determined by the impersonal forces of genes and the environment. But living systems are more than "robot vehicles," as some theorists would have it. They are in fact active participants in the evolutionary process. While Lamarck may have guessed wrong about the machinery of inheritance, he deserves full credit for recognizing the importance of the organism and its behavior as a separate causal agency in evolutionary change.

Lamarck no less than Darwin appreciated that functional adaptation to the environment is a problem for any organism. However, the environment is not fixed, Lamarck observed, and if circumstances (*circonstances*) change, an animal must somehow accommodate itself or it will not survive. Changes in the environment over the course of time can thus be expected to give rise to new needs (*besoins*) that in turn will stimulate the adoption of new "habits." Furthermore, asserted Lamarck, changes in habits come first and structural changes may follow.[5]

Darwin also appreciated the role of behavior in evolutionary change, but his view of its relative importance was more guarded: "It is difficult to tell, and immaterial for us, whether habits generally change first and structure afterwards; or whether slight modifications of structure lead to changed habits; both probably often change almost simultaneously." As always, Darwin provided an appropriate illustration:

> Can a more striking instance of adaptation be given than that of a woodpecker for climbing trees and for seizing insects in the chinks of the bark? Yet in North America there are woodpeckers which feed largely on fruit, and others with elongated wings chase insects on the wing; and on the plains of La Plata, where not a tree grows, there is a woodpecker ... which never climbs a tree![6]

Organic Selection

Many of Darwin's successors were not so ecumenical. However, at the turn of the twentieth century a movement developed concurrently among several British and American scientists who, in effect, Darwinized Lamarckism and assigned to behavior a more prominent role in evolution. Although their perspectives differed somewhat, their views were generally lumped together under psychologist H. Mark Baldwin's term "Organic Selection."

The basic claim of Organic Selection theory was that, in the course of evolution, the first step in producing systematic biological changes might well be changes in behavior, especially among the more "plastic" species. When an animal is in some way able to modify its behavior so that it can "select" a new habitat, or a new mode of adaptation, after a number of generations the change might precipitate congenital changes "in the same direction" that would undergird and perfect the new adaptation. This would occur not because the changes are somehow stamped into the offspring but because the new environment creates a "screen" that would selectively favor individuals with the relevant "somatic variations."[7]

Lamarck's giraffes are a possible case in point. Naturally occurring variations in the neck lengths of ancestral Giraffidae most likely became adaptively significant when these animals acquired, perhaps through trial-and-error, a new "habit" (eating acacia leaves) as a way of surviving in the relatively desiccated environment of the African savanna. We cannot know for certain that this was the case, but some suggestive evidence can be found in a related species of short-necked giraffes called okapi. Significantly, the okapi occupy woodland environments and, as expected, have very different feeding habits.

From a modern perspective, Baldwin's Organic Selection theory was crudely formulated; it was worded ambiguously; it was based on a rudimentary pain–pleasure model of behavior; and it did not lend itself to being "bench-tested" in the way that genetically determined traits can be manipulated in laboratory strains of *Drosophila* (fruit flies). With the emergence of the science of genetics, the Organic Selection idea (and Lamarckism) went into a total eclipse.

However, its theoretical umbra proved to be only temporary. In the early 1950s a modest effort to rehabilitate Organic Selection was undertaken by paleontologist George Gaylord Simpson, who renamed it the "Baldwin effect." However, Simpson treated it as a subsidiary phenomenon. "It does not, however, seem to require any modification of the opinion that the *directive force* [his emphasis] in adaptation, by the Baldwin effect or in any other particular way, is natural selection."[8]

Meanwhile an independent-minded British embryologist and geneticist, Conrad Waddington, challenged the mainstream viewpoint when he produced experimental evidence (published in *Nature* but nonetheless downplayed by other geneticists) for a Darwinized version of Lamarckian inheritance that he called "genetic assimilation." Using the geneticists' favorite experimental species, fruit flies (because they are low-cost, reproduce rapidly, and have many single-gene traits that can be altered readily in selection experiments), Waddington showed that certain developmentally influenced behavioral characters, like sensitivity to various environmental stimuli, could be enhanced through differential selection to the point where the traits would appear "spontaneously," even in the absence of the stimuli.[9]

Waddington also became a vocal critic of the gene-centered view of evolution. As he pointed out, "it is the animal's behavior which to a considerable extent determines the nature of the environment to which it will submit itself and the character of the selective forces with which it will consent to wrestle. This 'feedback' or circularity in a relation between an animal and its environment is rather generally neglected in present-day evolutionary theorizing."[10]

A major reconsideration occurred in the latter 1950s, when the American Psychological Association and the then-new Society for the Study of Evolution jointly organized a set of conferences that resulted in the landmark edited volume called *Behavior and Evolution* (1958).[11] This often-cited work contained a mother-lode of tantalizing ideas: There is the suggestion that adaptive radiations, an important aspect of evolutionary change, might be "fundamentally behavioral in nature" (Simpson), that behavior might often serve as an isolating mechanism in the formation of new species (Spieth), that all organisms, inclusive of their behavior, are "teleonomic" (purposeful) systems (Pittendrigh), and that, in the process of evolutionary change, new behaviors may appear first and genetic changes may follow (Mayr). None of this was associated with the discredited Lamarck, of course.

The "Pacemaker" of Evolution

Mayr elaborated on his views in a major article two years later on "The Emergence of Evolutionary Novelties," in which he characterized behavioral changes as the "pacemaker" of evolution. Although the response to Mayr's argument was underwhelming, to put it mildly, it remains a landmark.[12] Here is just an excerpt:

> A shift into a new niche or adaptive zone requires, almost without exception, a change in behavior. In the days of mutationism ... there was much heated argument over the question whether structure precedes habit or vice versa. The choice was strictly between saltationism and Lamarckism. The entire argument has become meaningless in light of our new genetic insight. It is now quite evident that every habit and behavior has some structural basis but that the evolutionary changes that result from adaptive shifts are often initiated by a change in behavior, to be followed secondarily by a change in structure ... It is very often the new habit which sets up the selection pressure that shifts the mean of the curve of structural variation ...
>
> Darwin was fully aware of this sequence of events. The parasitic wasp *Polynema natans*, in the family Proctotrupidae, lays its eggs mostly in the

eggs of dragonflies. Most of its life cycle, including copulation, takes place under water. [Here Mayr quotes Darwin]: "It often enters the water and dives about by the use not of its legs, but of its wings, and remains as long as four hours beneath the surface; yet it exhibits no modification in structure in accordance with its abnormal habits." ... Other aquatic species of parasitic wasps have since been discovered in the families of Chalcididae, Ichneumonidae, Braconidae, and Agriotypidae. As Darwin stated correctly, none of them has undergone any major structural reorganization following the shift to a new adaptive zone.

The shift from water to land, as mentioned above, was likewise made possible by a prior shift in habits, in this case, in locomotor habits. There is agreement about this between the students of vertebrates ... and of arthropods ... *The study of behavior differences among related species and genera is apt to throw much light on the sequence of events that trigger the emergence of evolutionary novelties* [emphasis added] ...

The tentative answer to our question "What controls the emergence of evolutionary novelties" can be stated as follows: Changes of evolutionary significance are rarely, except on the cellular level, the direct results of mutation pressure. Exceptions are purely ecotypic adaptations, such as cryptic coloration. The emergence of new structures is normally due to the acquisition of a new function by an existing structure. In both cases the resulting "new" structure is merely a modification of a preceding structure. The selection pressure in favor of the structural modification is greatly increased by a shift to a new ecological niche, by the acquisition of a new habit, or by both.[13]

Legendary Learners

The chorus of voices proclaiming the creative role of behavior in evolution rose to a crescendo of sorts in the late 1950s and 1960s, with the publication of such important books as W.H. Thorpe's *Learning and Instinct in Animals* (1963), C.H. Waddington's *The Strategy of the Genes* (1957) and *The Nature of Life* (1961), Alistair Hardy's *The Living Stream* (1965), Lancelot Law Whyte's *Internal Factors in Evolution* (1965), Robert A. Hinde's *Animal Behaviour: A Synthesis of Ethology and Comparative Psychology* (1966), and Arthur Koestler's *The Ghost in the Machine* (1967). Many of the examples of novel behaviors that were described in these works have become legendary:

✳ One of the most frequently cited examples concerns the discovery in the late 1940s that British blue tits had developed the clever habit of prying open the foil caps from the milk bottles that were delivered directly to peoples' front stoops. Like those bobbing toy birds that perch on the edge of a water glass, the blue tits then proceeded to dunk their heads and drink the cream. (It is said that the practice spread rapidly and eventually crossed the English Channel.)[14]

* Also legendary is psychologist Wolfgang Köhler's famous experiments with chimpanzees in the 1920s and 1930s. In one case, captive chimpanzees were able to solve the problem of how to reach bunches of bananas that were suspended high overhead by stacking wooden boxes on top of one another to create a makeshift ladder. There was also the feat of a chimpanzee named "Sultan," who learned to join two sticks together to reach through the bars of his cage and rake in food items that were out of reach.[15]

* Primatologist Jane Goodall, a legend herself, has observed many examples of novel behaviors among the chimpanzees she studied at Gombe Stream in Tanzania. One classic incident involved a low-ranking male in the community (Mike), who discovered that he could terrorize the other males by banging loudly on an empty steel drum, rolling the drum down hill, and otherwise using it for threat displays. Mike exploited the intimidating effects of his new "weapon" to rise in the hierarchy and become the dominant male – until the novelty wore off.[16]

* Even more famous, perhaps, are the classic experiments by the Nobel Prize–winning entomologist Karl von Frisch on learning in honeybees. One way of testing for creative problem-solving behavior, von Frisch reasoned, is to present an animal with a problem that evolution could not have anticipated and then observe the response. So von Frisch contrived an experiment in which *Apis mellifera* foragers were confronted with a unique situation where their artificial food sources were systematically moved further and further away from the hive. In an eerie resemblance to true insight, the bees learned to anticipate the moves and began to wait for the food at the presumptive new locations.[17]

* Honeybees are also good problem-solvers. Experienced honeybees normally avoid alfalfa, whose flowers possess spring-loaded anthers (the sacks at the top of the pollen-bearing stamens) that deliver a sharp blow to any bee that attempts to enter. But modern, large-scale agricultural practices sometimes leave the honeybee with the choice of alfalfa or starvation. In these situations, the bees have learned to avoid being clubbed by foraging only among alfalfa flowers whose anthers have already been

tripped, or else eating a hole in the back of the flower to reach the nectar.[18]

* The case of Imo, the young female "genius" in a Japanese macaque colony, is especially compelling because this inventive young monkey devised two novel food-processing techniques that subsequently spread to other members of her troop. Shortly after primatologists began providing sweet potatoes to a free-ranging group on Koshima Island off the coast of Japan in the early 1950s, they observed Imo taking soiled potatoes to a nearby stream, and later to the ocean, to wash them off before eating them. Other members of Imo's troop observed this behavior and began to emulate it. Later on the researchers decided to provision the animals with wheat, which they scattered on the beach. Again, it was Imo that first began taking handfuls of grain mixed with sand into the water, where she could wash them off and use the water as a natural separator (because the sand will sink and the grain will float). And again, Imo's innovation soon spread to others. Today, potato washing and grain separation are established cultural patterns in that troop, though Imo herself is long gone.[19]

During the past 30 years, the research literature on learning and innovation in living organisms – from "smart bacteria" to human-tutored apes and playful dolphins – has grown to cataract proportions.[20] (Indeed, there is now so much of it that some excellent earlier work is being overlooked and forgotten.) The examples are almost endless: worms, fruit flies, honeybees, guppies, stickleback fish, ravens, various songbirds, hens, rats, gorillas, chimpanzees, elephants, dolphins, whales, and many others. (In the index to their new book on *Animal Traditions*, Eytan Avital and Eva Jablonka list well over 200 different species.[21])

Especially enlightening are the numerous studies of species that have adapted in one way or another to the spread of humankind and human settlements. Domestic rats, cockroaches, spiders, snakes, gophers, and other uninvited guests often make themselves at home in our homes. There are the raccoons that haunt suburban garbage cans; the hummingbirds that use our artificial feeders; and the birds that nest in chimneys and under the eaves of our houses. Deer pose a special problem. Once terrified of humans and for good reason, they now casually dine on golf course fairways or drop by for a snack in suburban backyards.

Some forms of animal learning have created serious problems for humans. A notorious case involves the black bears at Yosemite National

Park that have become proficient at foraging for food in garbage bins, visitors' campsites, and even in parked vehicles. (It is reported that they prefer Hondas.) Then there are the wily, omnivorous coyotes that seem to thrive almost everywhere in North America and think nothing of lunching on our garden vegetables or devouring our pet cats. Especially troublesome are the alligators that regularly menace Floridians. Bill Adler's *Outwitting Critters,* along with his two other books on how to cope with "crafty wildlife" and "backyard bandits," have been popular sellers in the "how to" and "home care" sections of local bookstores in recent years. Adler's general conclusion: Our opponents are intelligent, resourceful, and devious.[22] Or, as the well-known animal psychologist Donald R. Griffin put it in the title of a recent commentary: "Animals know more than we used to think."[23]

The "Sentient Symphony"

Lynn Margulis and Dorion Sagan, in their 1995 book, *What Is Life?*, called it the "sentient symphony." What is most significant about the behavior of living organisms, they claimed, is their ability to make choices. (For the record, Waddington mounted a very similar argument back in the 1950s.) Margulis and Sagan, allowing themselves a bit of poetic license, wrote: "At even the most primordial level living seems to entail sensation, choosing, mind."[24] They cite compass-carrying bacteria, harboring fragments of magnetite, that can orient themselves to the north or south pole. Other bacteria can sense and respond to light, heat, sugars, and acids, among other substances. Bacteria and other mobile microbes can also make "selections" among various food choices. *Amoeba proteus* will prey on *Tetrahymena* but will avoid *Copromonas.* And paramecia prefer to feast on small ciliates and other protists but will reluctantly feed on bacteria if their favored foods are unavailable. Likewise, the shell-forming foraminifera are highly selective about the raw materials they use for building their shells, when choices are available.

We also know that primitive *Escherichia coli* bacteria, *Drosophila* flies, ants, bees, flatworms, laboratory mice, pigeons, guppies, cuttlefish, octopuses, dolphins, gorillas, and chimpanzees, among many other species, can learn novel responses to novel conditions, via "classical" and "operant" conditioning. (One cynic has pointed out that behavioral scientists have only confirmed what pet lovers and circus animal trainers have known for centuries.) More important, our respect for the "cognitive" abilities of various animals continues to grow. Innumerable studies have documented that many species are capable of sophisticated cost–benefit calculations, sometimes involving several variables, including the perceived risks, energetic costs, time expenditures, nutrient quality, resource alternatives, relative abundance, and more. Animals are constantly required to make "decisions" about habitats, foraging, food options, travel routes, nest-sites, even mates. Many of these decisions are under tight genetic control, with "preprogrammed" selection criteria. But many more are also, at least in part, the products of past experience, trial-and-error learning, observation, and even, perhaps, some insight learning.[25] As biologist Robert Foley has pointed out, a woodpecker that could not tell the difference between a telephone pole and a living tree would not survive for very long.[26]

Indeed, biologists Simon Gilroy and Anthony Trewavas find that even plants make "decisions." In the marine alga *Fucus*, for example, at least 17 environmental conditions can be "sensed," and the information that it collects is then either summed or integrated synergistically as appropriate. Gilroy and Trewavas conclude: "What is required of plant-cell signal-transduction studies . . . is to account for 'intelligent' decision-making; computation of the right choice among close alternatives."[27]

A recent experiment (already a "classic"), by ethologist Bernd Heinrich, has demonstrated unequivocally that even small-brained birds are also capable of innovative problem-solving behaviors. The experiment was inspired by a chance discovery among some Scandinavian ice fishermen many years ago. It seems that ravens were stealing the fish caught on their unattended fishing lines, which were dangled through holes in the ice. At first, the fishermen assumed human poachers were the culprits, but eventually a raven was caught red-handed (so to speak). The ravens had learned how to pull up the lines with their beaks and then were using their claws to pin down the loose ends until they could grab the fish.

Heinrich's controlled experiments at the University of Vermont, years later, established that even naive ravens in laboratory cages could quickly learn to perform the same feat with food rewards attached to lines that were dangled from their perches. In fact, they could even discriminate when there were multiple, crossed lines dangling from their perches, only one of which had food; they unerringly picked the right line. So the birds clearly understand the cause-and-effect relationships. Heinrich's delightful 1999 book, *The Mind of the Raven*, provides a treasure trove of both rigorous scientific research and credible anecdotal evidence regarding the mental abilities of these remarkable birds.[28]

Social Learning

Especially important theoretically, because it entails synergies of various kinds, are the many forms of social learning through "stimulus enhancement," "contagion effects," "emulation," and even some "teaching." Social learning has been documented in many species of animals, from rats to bats, to lions and elephants, as well as some birds and fishes, and, of course, domestic dogs. For instance, red-wing blackbirds, which readily colonize new habitats, are especially prone to acquire new food habits – or food aversions – from watching other birds. Pigeons can learn specific food-getting skills from other pigeons. Domestic cats, when denied the ability to observe conspecifics, will learn certain tasks much more slowly or not at all. And, in a controlled laboratory study, naive ground squirrels (*Tamiasciurus hudsonicus*) that were allowed to observe an experienced squirrel feed on hickory nuts were able to learn the same trick in half the time it took for unenlightened animals.[29]

The study of European blackbirds in the 1970s mentioned in Chapter 2 demonstrated just how potent social learning can be. A series of laboratory experiments (reported in the journal *Science*) by ethologist Eberhard Curio and his colleagues showed that the blackbirds' defensive mobbing behavior, for which there is an apparent predisposition, is not transmitted exclusively via the genes but also involves social learning. Naive blackbird "students" were successfully fooled by an ingenious apparatus constructed with mirrors and blinds into mobbing a multicolored plastic bottle, imitating what appeared to be the behavior of their adult "mentors."[30]

True "imitation" (including the learning of motor skills) has also been observed in (among others) gorillas (peeling wild celery to get at the pith), rats (pressing a joystick for food rewards), African grey parrots (vocalizations and gestures), chimpanzees (nut-cracking with an anvil and a stone or wooden hammer), and bottlenose dolphins (many behaviors, including grooming, sleeping postures, even mimicking the divers that scrape the observation windows of their pools, down to the sounds made by the divers' breathing apparatus).

Not surprisingly, the most significant cognitive skills have been found in social mammals, especially the great apes. They display intentional behavior, planning, social coordination, understanding of cause and effect, anticipation, generalization, even deception. Primatologists Richard Byrne and Andrew Whiten, in their two important edited volumes on the subject, call it "Machiavellian intelligence."[31] Social learning provides a powerful means – which humankind has greatly enhanced – for accumulating, diffusing, and perpetuating novel adaptations without waiting for slower-acting genetic changes to occur.

One compelling example is the socially transmitted feeding behavior in black rats, also mentioned earlier. Within the past decade or so, an Israeli population of these rats – which are notoriously opportunistic feeders – have expanded their niche into the young Jerusalem pine forest areas of Israel. However, the only food source available to them is pine seeds, which are enclosed in tough, inedible cones. To get at the seeds, the rats have developed a systematic technique for stripping the rigid scales away from the central shaft, where the seeds are sequestered. Israeli researchers Ran Aisner and Joseph Terkel, in a carefully controlled study, were able to show that this is a culturally transmitted behavior, requiring skills that go beyond simple trial-and-error. The behavior is learned by naive rats only when they are able to observe how an experienced animal does it.[32]

Tool-Using Synergy

As noted in Chapter 2, tool-use is another widespread form of behavior found in nature – from insects to insectivores – and it is utilized for a wide variety of purposes. As Edward O. Wilson pointed out in his comprehensive 1975 synthesis, *Sociobiology*, tools provide a means for quantum jumps in behavioral invention, and in the ability of living organisms to manipulate their environments. Tool-use also represents a new level of synergy in evolution – a class of animal–tool "cooperative" relationships that may result in otherwise unattainable novel effects. Wilson himself provided a broad sampler, compiled from a number of other sources, which I will briefly summarize here. Wilson cited:

* Solitary wasps that use pebbles held in their mandibles to serve as hammers.
* Ant lions that hurl sand at insect prey.
* Archer fish that spit water at insect and spider prey, knocking them into the water where they can be captured.

* Several species of "Darwin's finches" on the Galápagos Islands that use twigs, cactus spines, and leaf petioles as tools for digging insects out from under the bark of trees.
* The brown-headed nuthatches that use fragments of bark, held in their bills, to pry loose other pieces of bark.
* The black-breasted buzzards that drop rocks from the air onto the eggs of other birds to break them open and get at the contents.
* The Egyptian vultures that hurl rocks at ostrich eggs for the same purpose.
* The black cockatoos that use leaves to help them crack open nuts.
* Captive northern blue jays that were observed using strips of newspaper to drag food pellets into their cage.
* The California sea otters that use stones and shells as anvils for cracking open hard-shelled mollusks.[33]

Tool-use is even more diversified and sophisticated in the primates, especially the great apes (chimpanzees, bonobos, gorillas, and orangutans). Chimpanzees are particularly impressive. They frequently use saplings as whips and clubs; they throw sticks, stones, and clumps of vegetation with a clearly hostile intent (but rather poor aim); they insert small sticks, twigs, and grasses into ant and termite holes to "fish" for booty; they use sticks as pry bars, hammers, olfactory aids (to sniff out the contents of enclosed spaces), and even as toothpicks; they also use stones as anvils and hammers (for breaking open the proverbial tough nuts); and they use leaves for various purposes – as sponges (to obtain and hold drinking water), as umbrellas (large banana leaves are very effective), and for wiping themselves in various ways (yes, including chimpanzee equivalents of toilet paper and "sanitary napkins").[34]

Not only are chimpanzees proficient tool-users but they can also make tools. They break off small tree branches and strip them to fabricate ant "wands"; they use their bodies for leverage when they break down larger sticks to make hammers; they work leaves into sponges; and they carefully select stones of the right size and shape for the job at hand and will then carry them to their work-sites.

Chimpanzees are also distinctive (though perhaps not unique) in displaying "true" imitation and the "teaching" of a culturally transmitted tool-using tradition. Zoologists Christophe and Hedwig Boesch have documented that chimpanzee groups in Liberia and the Ivory Coast, but not elsewhere in Africa, use stones and wooden hammers, together with tree-root anvils, to break open the tough *Coula* and *Panda oleosa* nuts that provide the bulk of their nutritional needs in certain seasons. Chimpanzee mothers will also provision their infants for the first few years. But, as the juveniles mature, their mothers eventually start to "wean" them. The youngsters are then encouraged to learn the difficult nut-cracking technique for themselves, under their mother's attentive instruction.[35]

Culture Wars

"Culture" has long been viewed by anthropologists as a unique achievement that, supposedly, separates humans from all other species, not to mention our remote hominid ancestors. But that bastion of anthropocentrism is crumbling. The case against human uniqueness was summarized in a landmark article in *Nature* in 1999 by some of the world's leading animal behaviorists. Under the title, "Cultures in Chimpanzees," the article reported at least 39 different behavior patterns that vary between chimpanzee communities in ways that cannot be accounted for in terms of ecological differences. These range from tool usage to grooming behaviors and even the patterns of courtship. The authors also stressed that the total they cite is probably very conservative: "We must expect that more extended study will elaborate on this picture. Every long-term study of wild chimpanzees has identified new behavioural variants."[36]

Their findings were endorsed in an accompanying article by Frans de Waal on "Cultural Primatology Comes of Age." His overall conclusion was that: "All in all, the evidence is overwhelming that chimpanzees have a remarkable ability to invent new customs and technologies, and that they pass these on socially rather than genetically."[37]

Needless to say, this and other recent writings on culture in animals have provoked a raging controversy. On one side are the many theorists who have accepted biologist John Tyler Bonner's broad definition of culture as behavioral patterns and information that are socially transmitted, rather than being encoded in the genes.[38] Bonner's functional definition leaves wide open the question of which types of behaviors and which particular mechanism(s) of transmission should be included, so the term could be applied equally to birdsong dialects, alarm calls, hunting techniques, and the manipulation of tools. Cultural patterns in these terms have been identified in species ranging from yellow-rumped caciques to rhesus macaques and killer whales.

The naysayers reject all this, of course. Psychologist Bennett Galef called it "misleading to treat animal traditions and human culture as homologous ... and to refer to animal traditions as cultural."[39] Animals cannot learn via true, intentional imitation, as only humans do, Galef claimed. Anthropologist Russell Tuttle was even more emphatic: "I doubt that there is much disagreement among anthropologists and sociologists today that key to the concept of culture is symbols and symbolically mediated ideas, values and beliefs.... No one has shown that naturalistically [reared] chimpanzees have symbolically mediated ideas, beliefs, and values, the sine qua non of culture."[40] More surprising, even ironic, was the objection reported by anthropologist Barbara King. Some cultural anthropologists are abandoning the concept of culture altogether because, they assert, it has been contaminated with politically incorrect innuendos and gratuitous stereotypes. King describes a mock newspaper headline that was making the rounds at a recent symposium: "Apes have culture; humans don't."[41]

In the final analysis, much of the argument about culture (or not) in animals is semantic. Whether it be called "culture," or "traditions," or "socially learned behaviors," the point is that the social transmission of adaptive behaviors is not an exclusively human activity, and that many rudimentary precursors and analogues for our own highly sophisticated cultural skills can also be found in the natural world. As Frans de Waal points out in his insightful new book, *The Ape and the Sushi Master* (2001), culture should be defined in terms of its common denominators and its functional commonalities, not in terms of what is uniquely human. It's rather like confining your definition of an airplane only to a modern commercial jet airliner, while excluding a single-engine Cessna even though it shares many principles and serves the same basic purpose. But whatever you choose to call it, the fact remains that learned, socially transmitted behaviors are not an exclusively human trait. And, more important, they have played a significant causal role in the evolutionary process.

Lamarck's Ghost

Lamarck's ghost haunts the proceedings. And yet, his brooding presence remains in the shadows.[42] Or, to mix metaphors just a bit, he's a skeleton in the closet of evolutionary theory. Despite the crescendo of Darwinized "Neo-Lamarckian" writings in the 1960s, the climactic moment passed without a change in the "habits" of evolutionary biologists. The evidence that behavior, and behavioral innovations, play an important role in evolutionary change is overwhelming, yet the very invocation of Lamarck's name is still treated in some quarters like a form of sacrilege. With some exceptions, conventional Darwinian evolutionary theory generally remains focused on mutations (along with recombination, transpositions, linkage effects, genetic drift, and other molecular phenomena) as the primary source of innovation in evolution. In what John H. Campbell calls "plain vanilla" Darwinism, behavioral influences are not featured and are certainly not associated with Lamarck.[43] Lamarck is still unwelcome at the "high table" of evolutionary theory (as our British cousins like to say). There are several possible reasons for this state of affairs.

Part of the explanation, no doubt, has to do with the taint associated with Lamarck's name. Evolutionary biologists who have been schooled with the standard biology textbooks reflexively associate Lamarck with his theory of inheritance. Perhaps, too, evolutionary biologists are prone to be monotheistic; giving credit to Lamarck might also diminish Darwin's preeminence in the pantheon of evolutionary theory. (Even Darwin's co-inventor, Alfred Russel Wallace, gets short shrift from all but the most punctilious of evolutionists.)

The persistent denial of Lamarck's contribution may also be due, in part, to the Neo-Darwinian love affair with genes. Indeed, the standard definition of evolution tends to be narrow, gene-centered, and circular. Evolution in this paradigm is defined as "a change in gene frequencies" in a given "deme," or breeding population, and natural selection is defined as a "mechanism" which produces changes in gene frequencies. As Campbell put it: "Changes in the frequencies of alleles by natural selection *are* evolution."[44] It follows that mutations and related molecular-level changes are by definition the only important sources of novelty in evolution.

Another way of stating it is that the standard definition equates evolution with genetic changes, rather than viewing evolution more expansively as a multileveled process in which genes, gene complexes, genomes, organisms, and the natural environment interact with one another and evolve together in a dynamic relationship of mutual and reciprocal causation, including (in the current jargon)"upward" causation, "downward" causation, and even "horizontal" causation (i.e., between organisms). The emergence of "multilevel selection theory" in biology during the past few years has been an important step in the right direction.

A further obstacle to the rehabilitation of Lamarck has to do with the fact that his name has long been associated with the "wrong side" (from a biological determinist's point of view) in the hoary old nature-versus-nurture debate. "Habits," after all, imply nurture. Back in the 1950s and 1960s, when the Behaviorist learning paradigm was the reigning model in psychology and the young science of ethology (the study of animal behavior) was confined mainly to a small group of European biologists, the battle lines were sharply drawn. There was an undeclared war between these disciplines and their leading personalities (most notably the prominent Behaviorist B. F. Skinner, the pioneer ethologist Konrad Lorenz, and a few others). One side viewed learning (the "contingencies of reinforcement") as the exclusive determinant of behavior, while the other side regarded behavior as being essentially under genetic control – biologically determined.

Although ethologists (and other behavioral biologists) paid lip service to the idea that both the genes and the environment (both nature and nurture) play a role in producing the phenotype, especially in "higher" animals, much of their work was focused on demonstrating "fixed action patterns," "innate releasers," "imprinting," "canalization," "constraints on learning," behavioral "predispositions," and the like. Likewise, the then-young science of behavior genetics employed hordes of *Drosophila* flies, inbred (and therefore nearly identical) strains of laboratory mice (*Mus musculus*) and human twin-pairs in a research program designed to demonstrate the relationship between genes and behavior. Some of my own research as a young postdoctoral fellow at the University of Colorado's Institute for Behavioral Genetics contributed to this effort. We were able to demonstrate that the genes associated with the expression of four different kinds of aggressive behavior – maternal, predatory, inter-male, and "fear" or threat-induced – were genetically distinct from one another. The genetic substrate for each behavior, at least in *Mus musculus*, sorted independently among different inbred strains.[45]

The terminology may have changed in the past few decades, but the research agenda remains essentially the same for the current generation of evolutionary psychologists, behavior geneticists, and brain scientists; in the age of molecular biology and brain-scanning technologies, the search is being pressed for the biological bases of behavior. In the process, the tide of sentiment – certainly in the life sciences and among many social scientists as well – has turned in favor of assuming genetic influences are involved in shaping animal behavior.

There is even some skepticism these days about many of the earlier claims for learning abilities and the role of learned behaviors in nature. (In truth, some of these claims were inflated.) The milk-drinking blue tits, for example, routinely peel back the bark of trees to search for grubs. So the peeling of milk-bottle caps is seen merely as an extension of a preexisting behavioral proclivity, perhaps through trial-and-error. Likewise, chimpanzees in captivity spontaneously use sticks for poking into holes, even when there are no ants or termites inside; it is evidently a part of chimpanzee nature. And honeybee foragers occasionally use their alternative flower entry technique with other kinds of flowers. In coping with the anthers of alfalfa flowers, it seems the bees were simply applying an existing behavior pattern to the solution of a new problem.[46]

Other Voices

However, there also seems to be growing respect for the role of the "phenotype" as an active agency in evolutionary change. Especially notable are the contributions by Patrick Bateson, Eva Jablonka and her colleagues, Terrence Deacon, Judy Stamps, Beren Robinson and Reuven Dukas, and Carel ten Cate. There is also a much greater appreciation today for the role of developmental and life-history influences in shaping the phenotypes of various species – from the hormonal state of the mother to ambient weather conditions in the animal's environment. Terms like "phenotypic plasticity," "norms of reaction," and "reaction ranges" are commonplace, and there are frequent references to the complex interplay of various factors via "epigenetic cascades," "ontogenetic networks," and the like.[47] One important category of developmental influences is referred to as "maternal effects" – parental actions (or inactions) by either sex that can alter the survival chances of the offspring (such as the choice of nest-sites, or the amount and variety of food that is provided to the nestlings). Differential access to information is another major source of epigenetic variation.

Indeed, behavioral innovations, however they may occur, are beginning to look like the explanation for some of the outstanding puzzles in the natural world. One notable example is the extraordinary diversity of cichlid fish in African lakes, like Lake Victoria, which is nevertheless of recent origin. It is now believed that a highly malleable morphological trait in these creatures, namely their jaw structure, interacted with a great profusion of new food "habits" – specializations on different resources – to accelerate the evolutionary process.[48] Likewise, the approximately 110 species of anole lizards in the Greater Antilles islands correlate well with the diversity of the "niches" that they occupy on different islands: grass, tree branches, tree trunks, etc. Here again, a great range of phenotypic plasticity may have contributed.[49]

Evolution "on Purpose"

An underlying issue, really the linchpin in the theoretical debate concerning the role of behavior in evolution, is the reluctance of biologists – even now that biology is a mature science and is arguably the new "queen" of the sciences – to recognize and incorporate into the core of evolutionary theory the fundamental "purposiveness" and partial-autonomy of living systems. Physics, long the "model" science, has traditionally pursued the search for impersonal, mechanistic "laws" that are applicable anywhere and any-when. Teleology has no place in this enterprise, needless to say, and biology slavishly followed suit in the early years of this century. But in the 1960s and 1970s a number of voices were raised against this ultimately deficient view of how living systems work. The case was argued most eloquently, perhaps, by the influential geneticist/evolutionist Theodosius Dobzhansky:

> Purposefulness, or teleology, does not exist in nonliving nature. It is universal in the living world. It would make no sense to talk of the purposiveness or adaptation of stars, mountains, or the laws of physics. Adaptedness of living beings is too obvious to be overlooked ... Living beings have an *internal*, or natural, teleology. Organisms, from the smallest bacterium to man, arise from similar organisms by ordered growth and development. Their internal teleology has accumulated in the history of their lineage. On the assumption that all existing life is derived from one primordial ancestor, the internal teleology of an organism is the outcome of approximately three and a half billion years of organic evolution ... Internal teleology is not a static property of life. Its advances and recessions can be observed, sometimes induced experimentally, and analyzed scientifically like other biological phenomena.[50]

These days, the term that is most often used by biologists to characterize the internal teleology of living organisms is "teleonomy." Originally coined by biologist Colin Pittendrigh in connection with the 1958 conference on behavior in evolution, the term connotes that the purposefulness found in nature is a product of evolution and not of a grand design (see also the classic discussion of this issue by Ernst Mayr).[51] Teleonomy in living systems is today accepted without question. Yet few theorists take the next step and draw out the implications for evolutionary theory. Teleonomy puts living organisms into a unique category. An organism cannot be reduced to the laws of physics, or be derived from those laws, because its properties – inclusive of its structure, its behavior and its historical trajectory – cannot be predicted from those laws. For instance, the laws of physics are silent about the phenomenon of "feedback" – a fundamental (informational) aspect of all living (cybernetic) systems (see Chapters 5 and 9).[52]

Many theorists skirt this issue by treating organisms as mere vessels (rudderless rafts, not motor boats) that are controlled by "exogenous" factors – genes and the environment. For instance, Edward O. Wilson speaks of behavior as something that is "induced" by environmental forces and "epigenetic rules." The genetic "leash" may be very long, in Wilson's metaphor, but it is still a leash.[53] Others describe evolution as a process in which organisms "track environmental changes." And geneticist Richard Lewontin, in a much-cited essay on the subject that he might word differently today, asserted that "the external world sets certain 'problems' that organisms need to 'solve,' and that evolution by means of natural selection is the mechanism for creating those solutions."[54] In other words, the purposiveness of living systems is irrelevant to the causal dynamics of evolution. Purposiveness may be a product of evolution, but it is not even treated as a bit player or a spear carrier in the process, much less having a starring role.

Chance, Necessity, and Teleonomy

In the tradition of Lamarck and a long line of distinguished dissenters, I disagree. The purposiveness of living organisms is a major causal agency in evolution. Whether or not there is a purposiveness or directionality in the process as a whole is beside the point. That is a question best left to the theologians, and the complexity theorists (see Chapter 9). From an evolutionary perspective, the natural world displays "the piling up of little purposes," as Margulis and Sagan put it. The Nobel Prize–winning geneticist Jacques Monod famously characterized evolution as a process that is governed by two great influences, "chance and necessity."[55] However, Monod's slogan is inadequate; it leaves the organism out of the picture and includes natural selection only by implication. His slogan should be reformulated. The evolutionary process is in fact shaped by four broad classes of influences – *chance, necessity, teleonomy*, and *selection*. Actually five classes. Add synergy.

"Chance" – meaning unpredictable "historical" contingencies – is a factor at every level of life, from mutations to monsoons. "Necessity," too, is found at all levels, from the underlying laws of physics and chemistry to the "frozen accidents" of the genetic code, as well as the emergent properties of the physical and biotic environment.[56] (In truth, chance and necessity sometimes cohabit, but that's another story.)

Teleonomy, it appears, is also found at many levels, but its influence is more diverse. Teleonomy is embedded, so to speak, in the genome of every living being and plays itself out in the process of morphogenesis. We noted earlier that the "selfish genome" is more than a metaphor; novel traits are, by and large, selected in relation to how they serve the needs of the organism as a whole; and the genome is by no means a passive participant in the process.

Teleonomy is also evident in the construction and behavior – the phenotype – of each organism, which imposes many functional imperatives and constraints on how an animal goes about earning its living. To use an obvious illustration, bacteria and blue whales, by their very nature, have rather different nutritional requirements and very different sets of options available for how to satisfy those needs. The common denominator is their purposiveness, their active search for food, and, ultimately, the consequences of that activity for survival and reproduction. By the same token, animals frequently modify their environments in purposeful ways – from beaver dams to chimpanzee sleeping nests, mole-rat burrows, and the ornate courtship bowers that distinguish the bowerbirds.

In the jargon of decision science, living beings are value-driven decision systems. Much of the research relating to this model of behavior these days goes under the heading of "cognitive ethology," and there is currently a lively debate about the nature of animal intelligence and whether or not animals have "minds." These issues aside, the common thread is the assumption that animals are "intentional systems" (in philosopher Daniel Dennett's term). It is taken as a given that there is at least a degree of autonomy – some degrees of freedom – in the shaping of animal choices.[57]

"Downward Causation"

Much of this purposiveness can be said to be a product of upward causation – the expression of the genes and the genome. But much also entails "downward causation," from the whole to the parts. Here

Lamarck's insight – properly modernized and Darwinized – comes into its own.

Although the term "downward causation" is often associated with the psychologist and evolutionary epistemologist Donald T. Campbell, who used it in a 1974 article (and may have developed it independently), the term was actually coined years earlier by the Nobel Prize–winning psychobiologist Roger Sperry with reference to higher-level control functions in the human brain. Sperry was fond of using the metaphor of a wheel rolling downhill; its rim, its hub, all of its spokes, indeed, all of its atoms are compelled to go along for the ride.[58] (For the record, a similar concept, termed "supervenience," was put forward in the early years of the twentieth century by the proponents of "emergent evolution," most notably Conwy Lloyd Morgan.[59])

From an evolutionary perspective, downward causation/supervenience refers especially to purposeful activities at higher levels of organization in living systems (the phenotype) that differentially affect the survival and reproduction of lower-level "parts" (including the genes). Recall the terms "synergistic selection" and "Holistic Darwinism" that were mentioned earlier. Organisms do not adapt to their environments in a random way as a rule (although specific behaviors, like evasive maneuvers, may have a random aspect). An organism's time and energy resources are limited and must be used efficiently – economically – or else. Even trial-and-error processes are purposeful. They are shaped by evolved, preexisting search and selection criteria, namely, the adaptive needs of the organism. These behavioral "choices" may then affect the course of natural selection.[60]

Remember also our discussion in Chapter 5. Natural selection is not really a mechanism but a way of characterizing the functional consequences for survival and reproduction of significant changes in the relationship between an organism and its environment. It is the functional effects of these changes that matter. Thus, a change in an animal's "habits" (or its habitat) may have no significant effect, or it could drastically change the odds of its survival. In the process, this new habit/habitat may alter the context for the selection of various structural modifications.

One famous example – mentioned earlier – involves the remarkable tool-using behavior of the so-called woodpecker finch. *Cactospiza pallidus* is one of the numerous species of highly unusual finches, first discovered by Darwin, that have evolved in the Galápagos Islands, probably from a single immigrant species of mainland ancestors. Although *C. pallidus* was not actually observed by Darwin, subsequent researchers have found that the woodpecker finch occupies a niche that is normally claimed on the mainland by conventional woodpeckers. However, as any beginning biology student knows, C. *pallidus* has achieved its unique adaptation in a highly unusual way. Instead of excavating trees with its beak and tongue alone, as the mainland woodpecker does, C. *pallidus* skillfully uses cactus spines or small twigs held lengthwise in its beak to probe beneath the bark. When it succeeds in dislodging an insect larva, it will quickly drop its digging tool, or else deftly tuck it between its claws long enough to devour the prey. Members of this species have also been observed carefully selecting "tools" of the right size, shape, and strength and carrying them from tree to tree.[61]

For our purpose, what is most significant about this distinctive behavior is its "downward" effect on natural selection and the genome of C. *pallidus*. The mainland woodpecker's feeding strategy is in part dependent on the fact that its ancestors evolved an extremely long, probing tongue. But C. *pallidus* has no such "structural" modification. In other words, the "invention" of a digging tool enabled the woodpecker finch to circumvent the "selection pressure" for an otherwise necessary structural change. This behavioral "workaround" provided a selective "shield," or "mask."

Behavioral innovations can also produce degenerative effects, as Terrence Deacon has pointed out. He provides an elegant example. Most birds, fruit bats, guinea pigs, and primates have lost the ability to synthesize ascorbic acid (vitamin C) internally. The common cause of this "regressive" change, evidently, is that all of these species adopted the "habit" of being prolific fruit-eaters. One consequence, of course, is that exogenous sources of vitamin C are now an essential part of our diet – another example of the paradox of dependency (Chapter 4). On the other hand, this behavioral shift also created a new selection pressure for the primates. In order to become proficient as fruit-eaters, they had to re-evolve a previously lost trait – color vision.[62]

There are many other examples of "downward causation" in living systems, from the DNA damage repair and error-correcting mechanisms in the genomes and cells of complex organisms to the synchronization of reproduction in symbionts (including the mitochondria in eukaryotic cells), the differential reproduction of different "castes" (matrilines) in honeybee colonies in response to the needs of the hive, and the policing of social relationships in mole-rat colonies.

The concept of downward causation can also shed further light on the controversies regarding the celebrated behaviors of blue tits, honeybees, and chimpanzees. The fact that there is evidence of biologically based pre-dispositions in each case does not contradict the fact that learning is *also* involved; it is not an either–or situation. Suppose that blue tits were still preying upon milk bottles. (It has been said that milk bottles eventually adapted to their predators – so to speak – with better caps and protected en-closures, and that the practice of door-step deliveries has become "extinct" in many areas.) Over time, the birds might have evolved longer, stronger beaks, in order to breach the tougher caps and reach deeper into the bottles, and perhaps also a stronger sucking action to draw up the cream. In any case, this novel behavior represented another example of an existing skill (I still like the term "pre-adaptation") that was applied to a new problem, just as Lamarck, Darwin, Mayr, Waddington, and others have proposed.[63]

Likewise, the honeybees' strategy for exploiting the alfalfa flowers that they normally shun involved the application of a preexisting skill to a new situation. It is also possible that some subtle genetic changes may have been selected in the process. For instance, honeybees have relatively poor eye-sight, yet they are readily able to detect which alfalfa flowers have already had their anthers tripped, and which have not. We do not know how the bees make this discrimination.

In the same way, chimpanzees may well have a predisposition for pok-ing sticks into holes, but the skilled fabrication and use of ant and termite "wands" is a much more complex activity which, in fact, many juveniles learn in part from observing experienced animals. Moreover, this behavior did not evolve spontaneously. It was "reinforced." The "rewards" or pay-offs were the proximate cause of its eventual establishment as a widespread behavior pattern in chimpanzees.

Selfish Memes?

We need to make a brief digression at this point to deal with an alternative paradigm known as "universal Darwinism" – especially the currently "hot" concept of "memes." According to the promoters of universal Darwinism, any form of evolutionary change may be viewed as Darwinian in character if it exhibits three key properties: (1) a system of "replicators" (genes are the model, of course), (2) variations among the replicators, and (3) differen-tial "selection" among the variants in each generation via "competition." Some adherents to this paradigm also espouse a fourth, sometimes implicit assumption, namely that the replicators have a degree of autonomy that allows them actively to pursue their selfish interests while the selection process itself is viewed as a purely impersonal, amorphous (mindless) process. Accordingly, in universal Darwinism the replicators are often touted as the primary actors. The father of this paradigm is, of course, Richard Dawkins in *The Selfish Gene*.[64]

In the latter part of his famous book, Dawkins also hypothesized that there may well be an analogue of the genes in cultural evolution, which he called "memes." Memes, he suggested, include whatever we imitate – information, ideas, behaviors, habits, traditions, even artifacts. Moreover, Dawkins speculated, memes may be "active agents" that have a mind of their own. "A cultural trait may have evolved in the way that it has simply because it is *advantageous to itself* [his emphasis]."[65] Dawkins's dangerous idea (to borrow a line from Daniel Dennett) was slow to propagate itself (an argument against the claim that the potency of a meme is independent of the minds it "infects"). However, Dennett gave it new life when he enthusiastically endorsed the concept in his influential 1991 book, *Consciousness Explained*. From a meme's-eye view, Dennett explained: "A scholar is just a library's way of making another library."[66] Call him Dawkins' bulldog.

Psychologist Susan Blackmore, in her recent book *The Meme Machine* (1999), argues that memes are real physical entities, like genes (DNA). Moreover, memes have a mind of their own; they compete among themselves *"for their own sake"* [Blackmore's emphasis]. Just as Dawkins characterized organisms as "machines" for making more genes, so Blackmore tells us that every human is "a machine for making more memes.... We are meme machines." Citing the assertion by Stephen Pinker that humans have "surplus" mental abilities (especially imitative abilities) that cannot be accounted for as adaptations for survival and reproduction, Blackmore contends that the selfish interests of memes can explain the evolution of these otherwise inexplicable surplus abilities. Memes have taken control of our cultural evolution, she says.[67] (In fact, Pinker's thesis seems very dubious; such costly anatomical characters would have been subject to stringent adverse selection if they had not in fact been adaptive for evolving humans – see Chapter 7.)

The trouble is, memes don't really exist as a discrete causal agency in evolution, and saying they do won't make it so. As a metaphor for various forms of learned cultural "information," the term might be quite useful. It has the advantage of being more generic than such familiar terms as "ideas," "inventions," "learned behaviors," "artifacts," etc., and it is certainly preferable to such clumsy neologisms as Edward Wilson's "culturgens." But as a shaper of cultural evolution independently of the motivations, goals, purposes, compulsions, and judgments – in short the "minds" – of human actors, memes may rank right up there with the fiery phlogiston and the heavenly aether. There is no way I can conceive of to demonstrate (or falsify) the assertion that memes are real entities that exercise an autonomous influence in human societies. Others, like the anthropologist Robert Aunger, claim that it may be possible to do so, and that we should keep an open mind. However, this core assumption needs to be rigorously tested before the science of "memetics" can be taken seriously. Genes, and the coils of DNA that comprise the germ plasm, have an independent physical existence and known causal influences. Memes are labels that have been given to whatever we learn from one another. We are told that anything we imitate is a meme – hair styles, clothes, applause, dances, cigarette smoking, superstitions, songs, jokes, religion, and democracy, not to mention science, technology, and the very idea of a meme.

The assertion that minds are "robots vehicles" – passive receptacles for various external inputs – vastly oversimplifies both the neurobiology and the psychology of human learning processes, not to mention the dynamics of cultural life. Memetics, as its practitioners like to call their hopeful monster (to revive a venerable term), is a curious throwback to the Behaviorist *tabula rasa* hypothesis – the claim that human behavior is wholly determined by external inputs ("reinforcers"). To the contrary, memes are always embedded in minds (anything external is only a "latent" meme), and it is minds that do the creation, selection, and use of memes. Humans do not slavishly imitate whatever they see, or hear. They are highly selective, and manipulative, both in terms of their personal choices and in what they may attempt to foist on others. Denial of the primacy of human actors in the selection and transmission of social behavior and cultural information is bad psychology – and bad anthropology.[68] I'm reminded of a whimsical old poem about ghosts that I will take the liberty of bowdlerizing: "Yesterday upon the stair, I met a meme who wasn't there. He wasn't there again today. I wish that he would go away."

But can't it also be said that ideas, ideologies, religions, books, music, technologies, etc., "compete" with one another? Yes, of course, but only metaphorically. To be precise, memes are differentially selected by prospective users. Some bad memes do seem to hang around when they are not wanted, rather like parasites. But memes themselves are "powerless," despite the uncharacteristic "hype" of *Scientific American*, which recently featured a promotional article by Blackmore on "The Power of Memes." False analogies can do a lot of mischief, so it is important to keep the meme in its proper place as a term of convenience for a broad category of social phenomena and not as a distinct causal agency. In so doing, we can also lend support to the null hypothesis: we call the shots on whether or not to imitate the purveyors of the meme as, itself, a prime example of a meme.

"Feedback" [*sic*]

One other current buzzword should also be mentioned briefly. Many theorists these days use the term "feedback" to describe the relationships between behavior and evolution. Conrad Waddington, we noted earlier, used the term to characterize a "circularity" in the relationship between an organism and its environment. Christopher Wills, in *The Runaway Brain* (1993), speaks of a brain–body–environment "feedback loop." James Lovelock characterizes the Earth and its biota as being participants in a self-regulating, homeostatic "superorganism" that is maintained by "automatic feedback" (see Chapter 3). Geneticist David Thaler, one of the leading advocates for the notion of an "intelligent" genome, argues that there is, as he puts it, "feedback between the generators of genetic diversity [mutations, transpositions, etc.] and the environment that selects among variants."[69] Meanwhile, at the other end of the evolutionary scale, economist Brian Arthur has developed what he calls a "positive feedback" model of economic evolution.[70]

Perhaps it was unintended, but these and other theorists have used the term "feedback" in inappropriate ways that may obscure and even misrepresent the underlying causal dynamics. Feedback is a technical term in cybernetics and the information sciences. A cybernetic system is, by definition, a purposeful, goal-oriented system that is controlled by information flows, and feedback *sensu stricto* refers to information that is used by the system to monitor its behavior and make adjustments as necessary to achieve its pre-specified goals or maintain a steady state (homeostasis). Positive feedback will tend to encourage more of the same behavior, while negative feedback will discourage, modify, or terminate the action. The classic example is a household thermostat, which senses room temperatures and turns the furnace on or off accordingly. To use the systems scientist William T. Powers's classic formulation, feedback "signals" are compared to the internal "reference signals," and it is the relationship between the two signals that determines what the behavior of the system will be.[71]

Any usage of the term "feedback" that departs from this goal-oriented, information-driven model is at best metaphorical and at worst misleading. Though perhaps inadvertent, it amounts to sneaking a hidden teleology into the natural world. Sometimes the term misrepresents a pattern of interactive, reciprocal causation. Lovelock's "Gaia" is a case in point. Where is the "reference signal" in Lovelock's superorganism? At other times, the feedback processes are inappropriately enlarged to include the inanimate world. When an organism produces changes in the environment, those changes may produce feedback (in the strict sense) to the organism that will influence its behavior over time. But there is no "circularity" involved; it is a one-way street. The environment does not react to feedback, only to the physical actions that produce environmental changes. A reciprocal feedback "cycle" can only occur when two different cybernetic systems respond to each other. Finally, genomes are not modified between generations by feedback; this amounts to old-fashioned Lamarckism in disguise. Genomes change as a result of differential survival and reproduction (natural selection).

To cite one example, when one of Lamarck's giraffes "selected" acacia leaves and initiated a new feeding pattern in that species, the direct cause was the immediate nutritional rewards – the "reinforcements." There was also positive feedback in the strict sense between the giraffe's digestive system and its brain, which increased the probability that the behavior would be repeated. But the feedback was a secondary, indirect cause of the behavior. If the new behavior contributed to the differential survival of longer-necked giraffes, feedback played only a supporting role. It was the nutritional benefits that mattered. The change was due to natural selection, pure and simple.

"Neo-Lamarckian Selection"

The traditional ways of characterizing evolution – "chance and necessity," or "variation and selective retention," or Mayr's "two-step, tandem process" – are inadequate and require modification. Evolution, like a wagon pulled by a team of horses, has harnessed four distinct kinds of influences: "*chance, necessity, teleonomy*, and *selection*." Another way of putting it is that evolution is very often a four-fold process involving (1) genetic variations,

(2) phenotypic variations (inclusive of developmental influences and be-havioral shifts), (3) ecological (environmental) variations, and (4) differ-ential survival (natural selection). Furthermore, the causal arrows between each of these domains go in both directions. In an earlier book, I char-acterized such Neo-Lamarckian behavioral "choices" as "Teleonomic Se-lection," in order to highlight its purposive nature. In retrospect, it might have been more appropriate to call it "Neo-Lamarckian Selection," in con-trast with Darwinian Selection (a.k.a. natural selection). The proposition here is that Neo-Lamarckian Selection and natural selection often work hand-in-hand to affect evolutionary change. Neo-Lamarckian Selection is purposeful (teleonomic) and may therefore have informational, feedback-related aspects, but the focus should be on the economics – the rewards or "reinforcements" for novel behaviors.

Neo-Lamarckian Selection, I hasten to say, has many forebears. The idea is hardly new. At the turn of the century there was "Organic Selec-tion" (Baldwin); in the 1920s, there was "Holistic Selection" (Jan Smuts); in the 1960s, there was "Internal Selection" (Lancelot Law Whyte and Arthur Koestler); and more recently there has been "Psychological Selec-tion" (Mundinger), "Rational Preselection" (Boehm), "Baldwinian Selec-tion" (Deacon), "Behavioral Selection" (various writers), "neo-Lamarckian evolution" (Jablonka and Lamb), and similar terms. There are some differ-ences of emphasis among these concepts, but by any name Neo-Lamarckian Selection is a process that always occurs in the "minds" of living organisms; it is living beings that do the selecting, and it is a "purposeful" process that is intimately related to meeting basic survival and reproductive needs in a given context.[72]

As we noted earlier, many years ago Ernst Mayr drew a useful distinction between "proximate" causes and "ultimate" causes in evolution, concepts that are somewhat analogous to Aristotle's distinction between "efficient" causes and "final" causes. Proximate causes refer to the machinery of devel-opment, and to the physiology, biochemistry, and behavior of the mature phenotype. Ultimate causes refer to the influence of natural selection, or differential survival and reproduction.[73]

In practice, these two forms of causation interpenetrate with one another. Proximate causes associated with what we are calling Neo-Lamarckian Selection may also be responsible for shaping ultimate causes; behavioral choices may be the efficient cause of natural selection. As we noted earlier, Neo-Lamarckian Selection is implicated in habitat choices, adaptive ra-diations, dietary choices, predator–prey interactions, even what Margulis (after Mereschkovsky) calls "symbiogenesis." In fact, the proximate causes of novel forms of symbiosis, from lichens to such evolutionary turning-points as the origin of eukaryotic cells, are the result of various behavioral "initiatives." A long-term laboratory study by geneticist Kwang Jeon illustrates the point. Jeon began by infecting a strain of *Amoeba proteus* with a bacterial parasite that was resistant to the host cell's digestive enzymes. At first, the parasites were burdensome to the host. But after 200 generations, or 18 months, the parasite had evolved into a mutualist that contributed valuable chemicals to the host's operations. Eventually the symbionts became completely interdependent.[74]

The many kinds of "artificial selection" practiced by farmers and plant breeders, not to mention the more sophisticated technologies associated with genetic engineering, can also be redefined as examples of Neo-Lamarckian Selection – purposeful behavioral selections by third parties that shape the course of natural selection in other species. "Sexual selection" is also Neo-Lamarckian. Mate-choices among reproductive animal pairs involve behavioral choices that may directly shape the course of evolution, as Darwin himself first pointed out.

Coevolution – a concept first enunciated by biologists Paul Ehrlich and Peter Raven – may also involve behavioral influences. The term coevolution refers to cases in which an evolutionary trajectory is co-determined by step-wise modifications in two interacting species over multiple generations. The coevolution of butterflies and flowers is the classic example, but the ant–aphid symbiosis described earlier is also a case in point, and there are many others.[75]

The term coevolution is also used by anthropologists, but they have given it a very different meaning. To these theorists, coevolution refers to the "internal" relationship, so to speak, between genetic and cultural evolution in humankind. Robert Boyd and Peter J. Richerson, for instance, have formulated what they call a "dual-inheritance" paradigm, which they elucidate with some elegant mathematical modeling work. As they put it: "Individuals are the products of gene pools *and* cultures; they are the loci of natural selection and decision making ... Put another way, phenotypic characters acquired via social learning can be thought of as a pool of cultural traits that coevolves with the gene pool."[76]

Likewise, anthropologist William H. Durham, in his 1991 volume *Coevolution: Genes, Culture and Human Diversity*, asserts that: "Human beings are possessed of *two* major information systems, one genetic, and one cultural." Durham proposes that there are at least five different "modes" of gene–culture interactions.[77] A frequently cited example is the incest taboo, a biologically adaptive cultural universal that is also (evidently) influenced by a biologically based antipathy toward mating with close kin. Inbreeding in humans, as in many other species, can produce all manner of deleterious effects in the offspring. Another example cited by Durham is the evolution of adult lactose tolerance in populations that adopted milk-drinking as a regular part of their diets.

A Two-Way Street

In case it needs repeating, Neo-Lamarckian Selection is very far from being the exclusive cause of evolutionary changes. Even Lamarck viewed environmental changes as a prior cause. Lamarck's claim for the primacy of habit changes over structural changes was also overstated.[78] If the dynamics of evolution often involves a four-domain process, it is also a two-way street; the arrows point in all directions between those domains. To reiterate, it is a change in the functional relationship between an organism and its environment that produces natural selection. But the causes of these functional changes can vary widely. They might be the result of a change in behavior. They might be initiated by a mutation, a transposition, or some other change in the genome. Or they might be triggered by a significant change in the environment, including possibly a change in one species that affects other species. Often it involves a complex set of interactions among these agencies.

Consider the "raspberry gene" in the fruit fly *Drosophila melanogaster*, a recessive mutation that is responsible for producing both raspberry-colored eyes and a sharp decline in reproductive success when it is present at both of its two gene loci (the so-called homozygous condition). In a classic set of experiments that pitted the homozygotes against heterozygotes (where only one of the two loci had the mutant gene), as well as individuals without the mutation, the selection coefficient against the homozygote was a powerful 0.5. After only ten generations, the frequency of the homozygotes in the experimental population was reduced from 50% to about 10%. The specific mechanism responsible for this decline was suggested in subsequent experiments with another mutant form. It was found that, when the raspberry gene mutation was expressed (in the homozygote), it was responsible for subtle changes in the *Drosophila*'s stereotyped courtship behavior, which in turn drastically reduced its mating success. What this experiment showed was that a genetic change can be a direct, deterministic cause of differential reproduction and, *a fortiori*, of natural selection.[79]

The raspberry gene effect conforms to the traditional "two-step" evolutionary model. The more complex four-domain model can be illustrated with an experiment involving the so-called crossbills, or crossbeaks – birds whose mandibles (the two parts of the beak) seem misaligned because they cross over at the tips. There are about 25 species and subspecies of these birds worldwide, including three in the Galápagos Islands, and it has been known since ornithologist David Lack published his famous study, *Darwin's Finches*, in 1947 that the cross-over trait is actually adaptive. In the Galápagos it allows the crossbills to pry open the tough seed cones of larch, spruce, and pine trees. In fact, the three Galápagos crossbill species are markedly different from one another, with bills that are specialized for each of the different types of seed cones.

The laboratory experiment in crossbills, conducted by Craig Benkman and Anna Lindholm, was described in Jonathan Weiner's wonderful book, *The Beak of the Finch* (1994), which recounts in fascinating detail the long-range study by Peter and Rosemary Grant (and their colleagues and their daughter) in the Galápagos Islands. Benkman and Lindholm, in their experiment, trimmed the beaks of a group of experimental crossbills (a procedure, like cutting your nails, that does not harm the birds). This "treatment" effectively incapacitated the birds; it confirmed that the cross-over alignment of the mandibles is absolutely essential to prying open the seed cones. In other words, the crossbills' distinctive feeding strategy could not have arisen in the first place without a prior structural variation. However, the experiment also showed that, as the birds' bills grew back, even a small degree of crossing-over was enough to allow them to open some of the cones, however inefficiently, and that their performance progressively improved as the beak tips regenerated.[80]

What this experiment indicates is that, just as Darwin suggested, a "slight modification" (step one) may have opened the door to a new "habit" via Neo-Lamarckian Selection (step two), which natural selection subsequently "rewarded" with differential reproductive success (step three). The new habit in turn established a new organism–environment relationship in which the causal arrows then flowed the other way (step four) as subsequent beak variations were "tested" in relation to the new feeding strategy. The new behavior favored a more pronounced (more efficient) crossing-over morphology.

Neo-Lamarckian Selection and Human Evolution

Ever since Darwin it has been appreciated that animals may actively manipulate their environments in various ways and, in the process, change the selection pressures to which they are subject, but its importance as a source of evolutionary change is only now coming into focus. The Neo-Lamarckian dynamic was recently formalized in a set of models developed by biologist Kevin Laland and two colleagues under the heading of "niche construction." As the authors put it: "Organisms frequently choose, regulate, construct and destroy important parts of their environments ..." These models showed, among other things, that cultural transmission of niche-constructing behaviors could swamp the influence of gene-based behaviors. Their conclusion was that "cultural niche construction probably has more profound consequences [in evolution] than gene-based niche construction, and is likely to have played an important role in human evolution."[81]

Anthropologist Tim Ingold, in a special "Epilogue" to a major edited volume on *Tools, Language and Cognition in Human Evolution* (1993), comes to the following conclusion about the evolutionary process, and human evolution in particular:

> What is required, then, is a much broader conception of evolution than the narrowly Darwinian one embraced by the majority of biologists. Central to this broader conception is the organism–person as an intentional and creative agent ... In this account, behaviour is generated not by innate, genetically coded programmes, nor by programmes that are culturally acquired, but by the agency of the whole organism in its environment ... The change of perspective that this implies requires nothing less than a new theory of evolution.[82]

Well, not a "new theory" but certainly a new perspective. I proffer Holistic Darwinism as a candidate label for this paradigm shift. Holistic Darwinism stresses the role of cooperation and "Synergistic Selection" in evolution. But more important, Holistic Darwinism represents a multi-level, four-domain paradigm; it encompasses chance, necessity, teleonomy, and selection at various levels. And it explicitly recognizes the active participation of living organisms as causal agencies in the evolutionary process via Neo-Lamarckian Selection. One consequence of this paradigm, as we shall see in the next chapter, is a radically different theory of humankind. I will propose that our species in effect invented itself. Not only have we become the sorcerer's star apprentices but, in the process, we have deployed nature's magic to conjure our own evolutionary trajectory.

But be forewarned. The subject of human evolution is an intellectual minefield, with almost as many competing theories as there are theorists, or so it would seem. One reason why opinions on this subject are so strongly held is that it involves a matter that is close to all our hearts: Where did we come from? How did we get here? What are we, by nature? The various "scenarios" that have been proposed over the years also shape, and are shaped by, our ideological predilections. We have been called the "killer ape," the "naked ape" (and accused of being preoccupied with sex), the "talking ape," as well as "man the hunter," "woman the gatherer," and even the "selfish ape" (looking out primarily for ourselves and our close kin). I will suggest that we might instead view ourselves as the "inventive ape," and that many of our most important inventions involved new ways of exploiting the synergy principle; we are also, preeminently, the "synergistic ape."

7 Conjuring Human Evolution: The Synergistic Ape

There are many fearful and wonderful things,
but none is more fearful and wonderful than man.
He makes his path over the storm-swept sea
and he harries old Earth with his plough.
He takes the wild beasts captive and turns them into his servants.
He has taught himself speech and wind-swift thought,
and the habits that pertain to government.
Against everything that confronts him he invents some resource –
against death alone he has no resource.

Sophocles
Antigone

Let's begin with a parable. It is early morning on the East African savanna, and a troop of sleeping baboons begins to stir. The troop has been sequestered overnight at a nesting site (the jagged face of an ancient rift) that is well protected from various carnivore enemies. But now the troop – comprising about 30 males, females, juveniles, and infants – is getting ready to split into smaller "clans" (or sometimes larger "bands") and set off together on their daily food quest. (It's dangerous for baboons to travel alone in open country.) Each clan/band will cover several miles, before returning by nightfall to the safety of its sleeping site.

The leader of one of these groups is a large, confident alpha male that has earned his job the old-fashioned way. He bears a number of battle scars from the formidable canines of his rivals, not to mention some violent conflicts with rival troops. As his group forms a wide, irregular semicircle in a clearing below their nest-site, the leader squats in front of them on his hind limbs – erect, calm, and attentive. One by one the other baboons, using body language and tentative movements, cast their votes (in effect) for various alternative foraging routes. Today there seems to be a consensus in favor of a westerly route that had recently yielded an abundance of berries, fruits, nuts, roots, tender leaves and grasses, some easily captured small game, and several watering holes. The alpha male seems hesitant, perhaps remembering a recent run-in with a pride of lions. But finally the leader rises and begins to move, and the rest of the group – including several females with clinging infants and rambunctious juveniles – sets off toward the west in search of food.[1]

Later on that day, as the group is negotiating a long, shallow ravine between two open areas with low shrubbery and lush grassland, one of the females – her infant clutching tightly to her back – becomes distracted by some succulent rhizomes (root-like plant stems that are rich in nutrients). The female stops to feed, unaware of the fact that a cheetah has been stalking the troop from the ridge above. Now the cheetah moves ahead of her, getting into position to cut her off from her companions. At the last minute the infant, already able to recognize a dangerous predator, stiffens and utters a cry. The mother swirls around and shrieks an alarm call. The cheetah hesitates and, in the next instant, several baboon males with canines bared come rushing to the mother's aid. The cheetah, finding the tables suddenly turned, hastily retreats while the baboon mother and infant quickly rejoin the group.[2]

This episode was just a part of a fairly typical day in the life of these large quadrupedal monkeys, with their distinctive dog-like snouts. Life on the ground in East Africa, especially away from the safe haven of tall trees and sleeping cliffs, can be rich in food resources (for an omnivore) but is fraught with life-and-death challenges. Yet these "smart monkeys," as baboon expert Shirley Strum calls them, are resourceful, adaptable, and formidable; they can hold their own. Next to humans, baboons are the most successfully adapted of all the terrestrial primates. They are ubiquitous throughout Africa and the Near East. Chimpanzees, by contrast, have mostly stayed close to the safety of the trees (at least in recent times) and have relied on a narrower range of resources. They are relicts on the verge of extinction.

Although humans and chimpanzees are more closely related to each other biologically than either one is to the other great apes, in some ways humans also resemble the more tightly organized and aggressive baboons. Our baboon parable could also be a tale about our remote hominid ancestors of several million years ago (though there are also significant differences). *Homo sapiens* evolved in a rich but unforgiving environment, which we have ultimately come to dominate. Perhaps the best all-around image that we will find to describe this process is Darwin's "struggle for existence." Sometimes the living was easy, but many times it was not.

A Wealth of Scenarios

The growing research literature on human evolution represents the scientific equivalent of our "origin story" – as various writers have noted. Many disciplines are involved – paleoanthropology, archaeology, primatology, human genetics, paleoecology, behavioral ecology, climatology, and more. Over the years a number of different scenarios have been developed from the gradually accumulating body of evidence – fossils, DNA studies, indirect data about ancient climates, comparisons (and contrasts) with other species, plus a large dose of inference and even speculation. However, there is as yet no definitive version of our Genesis story. Much still depends upon how this assemblage of evidence is interpreted. The result is a wealth of scenarios and a wealth of conflicting opinions. Science writer Roger Lewin, one of the most diligent (and prolific) students of human evolution, has written an illuminating book about the debates among the fossil-hunters called *Bones of Contention* (1997).[3]

The starting-point and touchstone for theorizing about human evolution is Darwin's *The Descent of Man* (1871). People who find fault with Darwin's speculations tend to forget that he was compelled to reason about our origins without the benefit of a single early hominid fossil. He had to rely on the application of his own evolutionary principles, plus comparisons with other species, some ethnographic studies of contemporary hunter–gatherers, plausibility arguments, and common sense. (And, of course, the writings of contemporaries like Spencer, Galton, and Bagehot.) Whatever might be the deficiencies of his scenario from a modern perspective, it was nonetheless remarkably insightful.

Darwin did not attempt to reconstruct a precise historical narrative. He concerned himself with the functional dynamics and the overall logic of the process. In Darwin's view, the crucial first step in human evolution was a shift from the trees to a terrestrial lifestyle by our remote primate ancestors, though he could only speculate about why these creatures had abandoned the relative safety of an arboreal existence. Perhaps, he theorized, it was "owing to a change in [their] manner of procuring subsistence, or to some change in the surrounding conditions."[4]

Whatever the case, Darwin reasoned that this momentous change of venues by our ancestors encouraged the development of bipedal locomotion, which in turn liberated the forelimbs and hands for other uses. In time, this important innovation allowed for the invention of tools and weapons, a sequence rather like the development of wings and flight skills in ancient, bipedal dinosaurs. Tools/weapons in turn were able to offset our ancestors' obvious anatomical disadvantages (notably their reduced canine teeth) and facilitated the adoption of hunting as a major subsistence strategy. Darwin also stressed the role of social cooperation, reciprocity, and "mutual aid" in human evolution, especially in food-getting but also in conflicts with other groups and other species.

In modern terms, Darwin proposed that natural selection operated at three "levels" – between individuals, between "families" of close kin, and between social groups. Indeed, Darwin believed that competition, including warfare, between various "tribes" played a major role in shaping the course of human evolution. "Natural selection, arising from the competition of tribe with tribe ... would, under favourable conditions, have sufficed to raise man to his high position." The tribes that were the most highly endowed with intelligence, courage, discipline, sympathy, and "fidelity" would have had a competitive advantage, he argued. In an oblique allusion to the role of synergy, Darwin observed that:

> Selfish and contentious people will not cohere, and without coherence nothing can be effected. A tribe rich in the above qualities would spread and be victorious over other tribes; but in the course of time it would, judging from all past history, be in its turn overcome by some other tribe still more highly endowed. Thus the social and moral qualities would slowly tend to advance and be diffused throughout the world.[5]

Finally, Darwin viewed increased intelligence and the development of language in humankind as later outgrowths of tool use and social organization. As we shall see, Darwin's logic has much to recommend it; the factors that he identified are still in play among contemporary theorists.

The Killer Ape

Darwin well understood the relationship between cooperation and competition in nature, and he rejected the notion that humans are endowed with a reflexive killer instinct – a blood lust. The basic challenge that we confront, as do all other species, is survival and reproduction. Warfare, or group aggression, is an instrumentality – a means for coping with this problem in various contexts. However, many other theorists since Darwin's day have treated warfare as a prime mover, or directive agency in human evolution. This bias was introduced back in the 1920s by South African anthropologist Raymond Dart, discoverer of what was then the oldest known hominid fossil (*Australopithecus africanus*). Detecting what he thought was evidence of a violent end in his famous Taung child (and in other fossils that he found later on), Dart characterized our ancestors as "confirmed killers." Our bloodthirsty ancestors, he wrote, "seized living quarries by violence, battered them to death, tore apart their broken bodies, dismembered them limb from limb, slaking their ravenous thirst with the hot blood of victims and greedily devoured livid writhing flesh."[6]

Dart's vision of humankind as the descendants of killer apes may have been overdrawn, but the basic idea resonated with a generation that had just survived the first of two savagely destructive world wars. An innate aggressiveness seemed plausible, and many other writers since then have adopted Dart's basic scenario – if not his purple prose style – including Sir Arthur Keith, Robert Ardrey, Konrad Lorenz, Robert Bigelow, Keith Otterbein, Richard Alexander, Richard Wrangham, and others.

However, a theoretical turning point occurred in the 1960s, when Dart's unflattering image of humankind was challenged by an alternative scenario that came to be known as "Man the Hunter." This was the title of a landmark conference on this theme and of a subsequent published conference volume. (Science popularizer Robert Ardrey called it "the hunting hypothesis.") In their introductory chapter for *Man the Hunter* (1968), two of the leading anthropologists of that day, Sherwood ("Sherry") Washburn and Chet Lancaster, characterized big-game hunting with weapons as the "master behavior pattern" of our species. Washburn and Lancaster argued that it was this distinctive adaptation that differentiated humans from other primates and great apes and gave impetus to the development of our unique anatomical features. Therefore, warfare was not the driver. On the contrary, war is a recent derivative from our hunting ancestry. So, if we resemble chimpanzees in many ways, we also bear some likeness to savanna-living social species such as baboons and wild dogs. Like Darwin before them, Washburn and Lancaster emphasized the importance of intermale cooperation and the use of tools. "The biology, psychology, and customs that separate us from the apes – all these we owe to the hunters of past times," they concluded.[7]

Man the Hunter was a popular idea at first, but it was soon opposed by a feminist alternative dubbed "woman the gatherer." Originally proposed by anthropologist Sally Linton, the idea caught fire some years later when it was treated in more depth by anthropologists Adrienne Zihlman and Nancy Tanner in a much-cited 1978 paper.[8] Zihlman and Tanner asserted that meat plays a relatively small role in the chimpanzee diet, compared to fruits, nuts, leaves, and other gathered foods. More significant, hunting and meat-eating is not the dominant mode of subsistence even in contemporary human hunter–gatherer societies; gathering accounts for more than 50% of the total calories consumed by these peoples. Zihlman and Tanner pointed out that the use of tools for food procurement by chimpanzees is principally a female trait; most of the food provisioning for the infants/juveniles is done by the females, they claimed.

Equally important, they noted, was the fact that male chimpanzees do not as a rule play the guardian/protector role that is seen in savanna-living baboons. In the more common forest-dwelling chimpanzees, there is a loose "fission–fusion" pattern of dispersal (for daytime foraging) and aggregation (for overnight nesting) that is, among other things, notable for the frequent separation of the two sexes. (The exceptions, significantly, are those chimpanzees that inhabit more open areas. There the males are both more protective and more possessive of the females.)

Zihlman and Tanner also asserted that the hunting of large animals with crude weapons at close range is a dangerous activity that most likely did not emerge until much later on, when hominid groups expanded into temperate climate zones with more seasonal food sources. In the female gathering scenario, the males were also viewed as being rather superfluous; they had no significant economic role. Zihlman and Tanner speculated that the males might also have been subject to female sexual selection for sociability and docility.

Food-Sharing

A third scenario, viewed by many students of human evolution as a kind of middle-ground between the two supposedly "sexist" models, was advanced in the 1970s by the anthropologist Glynn Isaac. It has been labeled the "food-sharing hypothesis."[9] Isaac pointed out that food acquisition is typically a "corporate responsibility" in modern hunter–gatherer societies, such as the !Kung San of the Kalahari Desert in Southern Africa, and that the sharing of both plant foods and meat at designated "home bases" is a distinctive human trait; both males and females usually play an important economic role, in contrast with chimpanzees and the other great apes.

As for tools, chimpanzee males do, after all, use sticks and stones for threat displays and occasionally for more violent aggressive purposes (though with limited skill). So do capuchin (*Cebus*) monkeys in the New World. More important, Isaac found what he believed was evidence at hominid fossil sites in East Africa, dating back more than 2 million years, of the systematic butchering of large animals, using stone tools that had been imported for the purpose from some distance away. Isaac reasoned that meat in large quantities was being carried from butchering sites to home bases for later consumption. He also envisioned a division of labor in which the hunting and butchering of meat was complemented by female gathering activities.

While Isaac's scenario was attractive, it was battered by a succession of critical papers and, more important, careful micro-analyses of the bone assemblages. First, Lewis Binford made the argument that the bones found by Isaac were more likely to be natural accumulations of debris (caused, for example, by episodic local flooding) and that the bones might have come from a mix of hominid and carnivore activities.[10] Binford also suggested that ancient hominids were more likely to have been scavengers than big-game hunters.

Binford's reasoning gained support from the studies of Richard Potts and Pat Shipman, who deployed scanning electron microscopes to undertake a minute examination of the bones.[11] These researchers found that the bone collections Isaac had associated with human activities showed a complex mixture of carnivore tooth marks and human cut marks. In some cases, the cut marks were overlaid on the tooth marks, a clear indication of scavenging by hominids, while in other cases the reverse pattern was found. Also, Potts and Shipman found no evidence of "disarticulation" at the joints of these animals, a likely occurrence if meat was being butchered for removal to other sites.

Though it's clear that hominids were using crude stone tools to acquire meat more than 2 million years ago, the current consensus seems to be that this was likely to have been an opportunistic behavior pattern. It does not necessarily support the assumption of home bases and food-sharing. Or rule it out for that matter. Shipman and others have pointed out that the argument about butchering is really about timing; there is much stronger evidence of systematic butchering and meat transport by 1.5 million years ago. Whether or not the meat was acquired by scavenging or hunting remains uncertain, and debatable. Chimpanzees scavenge only occasionally, and several theorists have pointed out that scavenging in competition with other large carnivores can also be a dangerous activity (but see below).

The Nuclear Family

A more elaborate argument for food-sharing and male provisioning, one that did not rely on hunting/scavenging, was developed by anthropologist Owen Lovejoy in the early 1980s. His scenario is often referred to as the "nuclear family model."[12] Lovejoy's reasoning was based on what he perceived to be a major "demographic dilemma" (as he called it) for our remote Pliocene ancestors. Like the other great apes, these early hominids were a "*K*-selected" species, meaning that they produced relatively few offspring, each one of which required a relatively high level of "parental investment" both for prenatal development and for postnatal infant care. Many theorists have noted that the great ape pattern of a long childhood dependency is the biological foundation for the social learning that undergirds our unique cultural attainments. Primatologist Jane Goodall has documented that, even in chimpanzees, the quality of parenting makes a significant difference in how the offspring perform as adults. However, our vaunted cultural abilities come at a price; the capital costs (so to speak) for reproduction are notably high. Even in chimpanzees, it is 5–6 years.

Lovejoy pointed out that the forces favoring selection for increased parental investment by our hominid ancestors must have come into conflict with the climate changes that occurred in East Africa during the late Miocene and Pliocene – most importantly an increase in the seasonality and "patchiness" of food resources and the need for much larger foraging ranges. Among the primates, it is typically the females that provision the young. But this, Lovejoy noted, would have become increasingly difficult under the twin pressures of a greater "demand" for food provisioning and a more time- and labor-intensive food quest.

The obvious solution to this dilemma was to recruit the males to provide food for the mothers and their infants, as many birds and most canids do. Accordingly, Lovejoy proposed that male provisioning, coupled with the emergence of monogamous pair-bonding (so the males would be assured of their paternity and would not unwittingly provision some other male's offspring) was the key to human evolution. The loss of an estrous cycle (an otherwise universal primate trait) and the permanent sexual receptivity of the females were supportive developments, Lovejoy reasoned.

In many birds and canids, the males provision their young by regurgitating food that has been transported in their mouths back to the nest or den. But for early hominids, the only feasible alternative was carrying. Therefore, Lovejoy argued, it was the need to carry food back to home bases to provision females and infants that created the selection pressure for efficient bipedalism – and everything else that has followed. Whereas Darwin had supposed that the emancipation of the hands was a byproduct of bipedalism, Lovejoy turned the tables. He proposed that the need to use the hands for carrying food from distant foraging sites – not tools or weapons – was the driver for the development of our distinctive hominid traits.

Why should there be sequestered home bases for the females and infants? Baboons and chimpanzees do not follow this pattern; the females quickly return to foraging even while they are nursing. Lovejoy pointed out that the combination of expanded foraging ranges and bipedalism would have presented a major mobility problem for our evolving ancestors. Whereas baboons and knuckle-walking apes can transport infants on their backs, bipedal apes obviously cannot. So bipedalism became both a solution and a problem, and male provisioning was the adaptation that cut the Gordian knot.

Lovejoy's scenario is obviously appealing, but it too has been sharply criticized. First, it assumes a radical behavioral shift away from the typical primate pattern. It is also difficult to document. Though inferences can sometimes be made about ancient behaviors from bones and other material evidence, monogamy and male provisioning do not leave fossil residues. A further difficulty is that monogamous pair-bonding could charitably be called an elastic human adaptation; it does not have the degree of consistency (strong genetic heritability) of, say, bipedalism or even language, which are reliably mastered by the overwhelming majority of modern humans without elaborate instruction or external sanctions. Others point out that, in Lovejoy's model, the males would also be at risk of cuckoldry when they were out foraging. Who would guard their mates? It seems more likely that the nuclear family is of more recent vintage. Not only is there considerable variability even in contemporary societies (polygyny is fairly common, as Richard Wrangham and others point out) but the nuclear family pattern is supported in part by various cultural "reinforcements."

Other theorists have objected that the presumed demographic bottleneck which drives Lovejoy's model may have been overstated. It has also been noted that less radical solutions to the problem are both conceivable and perhaps more plausible – for instance the carrying of infants between food patches by foraging females and/or cooperative baby-sitting by older siblings and other females (perhaps grandmothers in one currently popular scenario). Finally, some feminist writers perceived Lovejoy's model to be almost as patronizing to females as Man the Hunter. Rather than portraying the females as competent and self-sufficient providers on their own behalf, Lovejoy was accused of treating females as passive dependents that survived on handouts from foraging males. A close reading of Lovejoy's original paper, though, shows that he intended no such slur. At most, he could be accused of omitting some implicit details regarding the female role.

On the other hand, at least half a dozen subsequent studies by other anthropologists have supported, enriched, or modified Lovejoy's thesis, and, as the battle over sexism in anthropology has abated, there has been a shift in recent years toward a more complex middle-ground viewpoint. Alan Mann, for instance, highlighted the potential for more systematic exploitation of high-protein food resources (eggs, insects, invertebrates, reptiles, birds, small animals) as a way to augment and enrich the hominid diet.[13] In other words, a dichotomous choice between big-game hunting and plant foods is too simplistic; it is more likely that these creatures were opportunistic omnivores. Duane Quiatt and John Kelso have stressed the potential economic benefits of nuclear family units.[14] Jane Lancaster emphasized the mutual advantages that could be achieved by an economic division of labor.[15] Paul Turke pondered the significance of hidden ovulation and the synchronizing of menstrual cycles in human females.[16] And Milford Wolpoff, citing fossil evidence of a high mortality rate and early deaths among our remote hominid ancestors, suggested that a high frequency of orphaned juveniles would have favored the selection of more elaborate kinship bonds and social support networks, possibly involving male as well as female relatives.[17]

The Synergism Hypothesis and Human Evolution

Prime mover theories of human evolution have their uses. Single-factor explanations may draw attention to important aspects of the process, even if they are ultimately found to be insufficient. Lately, the "climate-change hypothesis" seems to have become fashionable in some quarters, inspired partly by new evidence that climate patterns have been more unstable in the past than we had thought (see below). However, many theorists nowadays shy away from this approach; as our understanding of human evolution has increased over the years, prime mover theories have been less favored.

Yet the ultimate theoretical question still beckons: How do we explain the evolution of humankind? What accounts for our remarkable evolutionary transformation? Is there some underlying principle? Or was it blind luck – a meandering "drunkard's walk" without any particular logic to it? I believe that there was an inner logic to the process of human evolution and that the time is ripe for a fresh approach to the broader "Why" question. Perhaps a frame-shift may be possible at this point that will allow us to view the process from a different perspective. I believe the Synergism Hypothesis has something to contribute.

I proposed in my earlier book (1983) that there was no prime mover in human evolution. Rather, our course was determined by an accumulation of various forms of functional synergy with economic "payoffs" for the immediate problems of survival and reproduction. The process was propelled by a series of behavioral innovations and proximate "Neo-Lamarckian" (teleonomic) selection. In the truest sense, the evolution of humankind involved an entrepreneurial process – a pattern of innovation, trial-and-error learning, selective retention, and cultural transmission, and the subsequent natural selection of supportive anatomical changes. Moreover, much of our inventiveness has involved new forms of synergy. Synergy played a key role in this transformation; it has generated potential bioeconomic benefits of various kinds – synergies of scale, threshold effects, combinations of labor, complementarities, cost- and risk-sharing, and the like. This self-making theory has been reinforced by the extensive additional evidence that has accumulated since 1983, I believe.

Under this hypothesis, it is behavioral changes that were the "pacemakers" (to use Ernst Mayr's term) of various "progressive" trends toward greater functional competence and complexity in hominids. In other words, our hominid ancestors shaped the trajectory of their own biological evolution; to a very considerable degree our species invented itself. As the zoologist Jonathan Kingdon put it in the title of his provocative 1993 book, we are the *Self-Made Man*. (In my own earlier book, I alluded to archaeologist V. Gordon Childe's classic 1936 study of the agricultural revolution called *Man Makes Himself*.) However, this paradigm can also be classified as a "gene–culture coevolution theory" (see Chapter 6); behavioral and biological changes are viewed as having been deeply intertwined.

Two other preliminary points are in order here. One is that the assumptions that are made about the context of human evolution are critically important. The external physical and biotic environment – what Darwin referred to as the "surrounding conditions" – were obviously a key factor. But equally important was the "internal environment" – the array of specific biological survival needs that imposed day-by-day imperatives for our ancestors, as well as the biological endowment (the precursors or preadaptations) that conditioned their efforts to meet these needs. In many scenarios, the full context of human evolution is unstated, or downplayed, or used only selectively to develop a theory. The problem is that biased assumptions can lead to logically impeccable conclusions that are totally wrong.

The other preliminary point is that, in speculating about the many unknowns in human evolution, Occam's razor (or the principle of parsimony) provides a useful analytical tool – a way of choosing among various explanations when "hard" evidence is limited or lacking. To be sure, Occam's razor has a double-edged blade, so it must be used with care. But, all other things being equal, the simplest and most incremental (least radical) alternative seems more likely to be the right one. By the same token, the scenario that implies the greatest rewards, or payoffs for the least possible cost or risk (cost-effectiveness) is more likely to have been the one that our ancestors selected, by and large. Although it is uncertain that we will ever find conclusive evidence for many aspects of human evolution, we may be able to rank-order the different scenarios in terms of their relative plausibility.

What Was the Context?

The first and perhaps most fundamental point about the context of human evolution is that our ancestors had to cope with an environment that was highly variable over time.[18] It is now apparent that there have been a total of 27 ice ages during the past 3.5 million years alone, and that many of these episodes involved major changes in land forms and local ecosystems in various parts of the world. For instance, during the deep freeze that began about 33,000 years ago and peaked about 13,000 years later, sea levels declined more than 400 feet from present levels as a mile-deep mountain of ice became trapped in the expanding northern glaciers. This opened up vast tracts of fertile land for habitation on the continental shelves and altered climate and rainfall patterns in many temperate and equatorial areas. Ultimately, of course, this process went into reverse, with equally profound consequences. In fact, we now recognize that significant shifts in local microclimates, and in the relative abundance of plant and animal species, can occur over periods as short as a few years. Although tropical Africa has long been a fecund and bountiful area, it has not been a free lunch. There have been many booms and busts.

The most important ecological influence on the course of human evolution, it now seems evident, was a major long-term change in the subtropical forest belt that spans the mid-section of Africa, including what is now (principally) Tanzania, Kenya, and Ethiopia, and extending at least as far west as Chad. The key factor seems to have been a shift in global climate patterns. Over time this greatly altered the terrain and the ecology of East Africa. Immense tracts of dense tropical forest areas – safe havens with plentiful resources for arboreal primates – were gradually converted into a more broken, "mosaic" pattern with many "patches" of woodland, an abundance of lakes and streams, large areas of marshland, and more open savanna areas.

These environmental changes created both a challenge and an opportunity – a gradual reduction in the traditional primate homeland, a growing array of more seasonal plant materials and, equally important, the appearance of many animals that were more suited to woodland and open grassland areas. Many of these areas ultimately came to be populated with large herds of herbivores (some 20 different species), along with a formidable number of carnivores – saber-toothed cats, lions, leopards, hyenas, cheetahs, and more.

Some theorists believe that these environmental changes were sufficient to account for the origin of the hominid line; they are said to have "induced" or "driven" or "forced" East African apes to shift from an arboreal lifestyle to terrestrial living and to adopt new survival strategies. As one well-known biologist put it, "the causes were ecological."

Biologist Patrick Bateson calls this the "billiard ball" theory of evolution.[19] For one thing, hard evidence of a close correlation between climate changes and the major benchmarks in human evolution is sparse and contradictory.[20] But more important, the deterministic model obscures the active role of the participants themselves. A gradually shifting resource base may have changed the "payoff matrix" – the relative abundance and the costs and benefits of exploiting various food sources, as well as the intensity of competition for these resources. Likewise, a mosaic of forested and open areas may have created many new ecological opportunities. But exploitation of this new environment required "initiative" – along with mobility, the ability to safely traverse open areas, and (most likely) the capacity to defend resource "patches" against competitors. Some of the eastern primates of that epoch chose to change their "manner of procuring subsistence," in Darwin's phrase. Others did not. Once again, Darwin's reasoning was prescient.

The shift to a terrestrial lifestyle most likely did not happen all at once. For one thing, it involved substantial costs and risks. As foraging ranges expanded, so did the time and energy required to exploit them, and australopithecines (and their *Ardipithecus* predecessors) were imperfect bipeds – competent but not as efficient as later *Homo erectus/Homo ergaster*. More important, the exploitation of a mosaic terrestrial environment introduced new risks from predators and competitor species, not to mention rival proto-hominid groups. Some theories of human evolution have downplayed these threats, but it was in fact a major challenge, with life-and-death consequences (as our baboon parable suggested).[21] It is possible for a large primate to survive for many days without food, so long as fresh water is available, but a single encounter with a predator will very likely reduce its survival chances to zero.

There is ample contemporary documentation and some fossil evidence going back to *Ardipithecus ramidus* some 4.4 million years ago (or more) suggesting that our remote ancestors were indeed subject to "predation pressure," as the ecologists would say. Evolutionary biologist Robert Foley points out that there were no fewer than ten large carnivore species roaming East Africa in those days, compared with just two today. Pack-hunting species like *Palhyaena* would have been particularly dangerous, but even giant eagles were a threat. To borrow a line from anthropologist Charles K. Brain, this scenario could be called "Man the Hunted" (and woman the hunted, of course).[22]

Primate Pre-Adaptations

The proto-hominids that chose to venture into this changing (and hazardous) environment brought with them many important precursors/ pre-adaptations (or exaptations if you prefer) – stereoscopic color vision, a vertical climbing anatomy, dexterous forelimbs with manipulative hands, sociality, a relatively high degree of intelligence, an omnivorous dietary pattern (a major advantage) and – highly significant – a form of social organization that is based on a nucleus of closely related males who are joined by unrelated females. Only 6% of the 167 primate species studied to date have male-based groups, and this may have been one of the keys to the emergence of the hominid adaptive pattern (see below).[23]

Two other precursors, or preconditions, may also have been important. One has to do with the fact that our earliest proto-hominid ancestors were diminutive in stature – less than 3 feet tall. Even the much later *Australopithecus afarensis* of 3.5 million years ago is estimated to have weighed some 15% less on the average than modern chimpanzees. Paleoanthopologist Milford Wolpoff calls these creatures "Miocene midgets."[24] In other words, our remote ancestors were even more vulnerable than Darwin imagined.

A second, more controversial point is that – contrary to Darwin's reasoning and that of many other theorists since – bipedalism may actually have been a pre-adaptation that does not need to be explained. The traditional view is that our ancestors evolved from "brachiators" that used their arms to navigate through the trees, much like children do today on jungle gyms. However, a number of theorists – Morton, Washburn, Tuttle, Kurtén, McHenry, Harmon, Wolpoff, Wills, and others – have espoused the alternative idea that hominids evolved from an above-branch biped, rather like an ancient tightrope-walker, that may have resembled a 10-million-year-old Sardinian primate called *Oreopithecus bambolii*. Bipedalism, in other words, may have been an ancient form of arboreal locomotion that was applied to a new situation (Neo-Lamarckian Selection) and was improved upon over time via natural selection. As we noted earlier, this is a common theme in evolutionary history.[25]

However, the "*Oreopithecus* model" conflicts with the widespread assumption that contemporary chimpanzees represent the most likely baseline for our own evolutionary divergence. Chimpanzees show evidence of very little evolutionary change during the past several million years, it is said. So our common ancestor is presumed by some to have been a knuckle-walking frugivore (fruit-eater) that was very similar to modern chimpanzees. Indeed, primatologist Richard Wrangham developed a detailed list of chimpanzee traits that he supposed would also be found in our common ancestor.[26] And Jared Diamond wrote an entire book based on this assumption called *The Third Chimpanzee* (1992).

The trouble is that we really do not know which model is more likely to be correct, because we have not found the common ancestor. In fact, we have not found any ancient chimpanzee fossils at all. There is also a dearth of proto-hominid fossils dating from the most critical period, from 6 to 9 million years ago. Thus it is conceivable that orangutans, chimpanzees, gorillas, and hominids all diverged from a common bipedal ancestor and subsequently followed very different evolutionary paths. (Primatologist Frans de Waal points out that bonobos – the so-called pygmy chimpanzees – use a bipedal gait more often than do chimpanzees, especially when carrying objects.) To borrow the punch-line from a famous joke, it may have been bipedalism all the way down. But nobody knows.[27]

Australopithecine Synergy

In any case, the crucial development among our proto-hominid ancestors – the "pacemaker" for the path we have taken – was the adoption of new, group-based behavioral strategies. Social organization was the key. We took the path less traveled by, to paraphrase the poet Robert Frost, and that made all the difference.

For our purpose, the assortment of early hominid fossils that have been given different generic names (*Ardipithecus*, *Australopithecus*, and *Paranthropus*) can be lumped together in the general heading of the australopithecine "grade." This will allow us to sidestep the controversies over labels, categories, and sequences and concentrate on the most general and important features. One issue has to do with the tempo of australopithecine evolution. The recent discoveries of several 5.2–5.8-million-year-old *Ardipithecus* fossils in the Middle Awash area of Ethiopia and the 6–7-million-year-old fossil find in Chad indicate that the shift toward *Australopithecus* began far earlier than had previously been supposed. Indeed, some features, from their bipedal gait to their teeth, clearly presage *Australopithecus afarensis* (Lucy and her cousins) some 2 to 3 million years later.[28]

In other words, the australopithecines were not simply a way-station or a transitional ape that was evolving toward something "better." They were – by the evidence both of their longevity and their wide geographic range – an immensely successful genus, that most likely pursued various lifestyles, from leaf-stripping to nutcracking and, perhaps, some scavenging and opportunistic hunting. (To put things in perspective, the average lifespan of a species is estimated to be about 1 million years. Three million years is equivalent to roughly 150,000 human generations.) The key to the australopithecines' remarkable success, I would suggest, was a set of novel behavioral adaptations that were highly synergistic – in fact a nexus of mutually reinforcing synergies.

How did a diminutive ape with constrained mobility on the ground and no natural defensive weapons, but with a relatively large brain (slightly bigger on average then modern chimpanzees), manipulative hands, and an omnivorous digestive system, solve the problem of shifting to a terrestrial habitat, broadening its resource base, and, over time, greatly expanding its range? Social organization – group living – was one key factor. In a patchy but relatively abundant woodland environment that was also replete with predators, competitors, and sometimes hostile groups of conspecifics, group foraging and collective defense/offense was the most cost-effective strategy. There were immediate payoffs (synergies) for collective action that did not have to await the plodding pace of natural selection. (A number of theorists over the years have endorsed the group-defense model, including George Schaller, Alexander Kortlandt, John Pfeiffer, Richard Alexander, and others.) However, it is also likely that the earliest of these proto-hominid pioneers stayed close to the safety of the trees; early australopithecines apparently had very powerful arm muscles (Wolpoff compares them to modern steel-workers), perhaps to compensate for the reduction of climbing ability with their hind limbs and feet.

Note that the group-defense scenario does not assume male provisioning, monogamy, or other more radical innovations. It assumes only that the synergies derived from acting collectively – foraging and reproducing as a group – were both immediate and mutually beneficial. Moreover, the odds of survival were greatly enhanced. There may well have been group selection, but it was not based on altruism. It involved what the economists call "collective goods," or "public goods." And because these groups were formed around a nucleus of closely related males, individual selection, kin selection, and group selection would have been aligned and mutually reinforcing – just as Darwin had supposed. Indeed, this anomalous form of social organization may have been a crucial "pre-adaptation."

Why would the males defend the females and infants? For one thing, the males might not have known the paternity of the offspring if the females followed a reproductive strategy of multiple matings, as chimpanzees typically do. Another factor was that all of the infants would have been closely related – "nephews," "cousins," or even younger siblings. A third point is that, in an extremely "K-selected" species with a very long reproductive cycle and a relatively short life-span, each offspring was commensurately more valuable; the benefits of protecting the mothers and infants in a dangerous environment, and the costs of not doing so, were obviously much greater. Finally, in a tightly organized, interdependent group it was not significantly more costly to defend the offspring of close kin than it was to defend one's own progeny; it was not a matter of altruism, or reciprocal altruism but of teamwork in a win–win (or lose–lose) situation – a synergy of scale.[29]

A Division of Labor?

Contemporary hunter–gatherers, not to mention most modern societies, typically have a division of labor along sexual lines, and it is possible that a rudimentary version of this pattern existed also among the australopithecines. Applying the principle of parsimony once again, it seems likely that the females would have been primarily responsible for carrying the infants and shepherding the juveniles, while the males served as the primary (though not necessarily exclusive) guardians for the group. Some evidence for this can be seen in the australopithecine fossils of 3.5–3.0 million years ago, where there appears to have been a sharp sexual "dimorphism"; the females remained very small while the males grew much larger, closer in size to modern pygmies.

Many theorists have attributed this size difference to sexual competition among the males. It is well documented that "sexual selection" can produce significant anatomical differences between the two sexes. However, a division of labor between males and females can also create a "selection pressure" for dimorphism. If australopithecine males were close kin, this might have dampened status rivalries and mitigated male sexual competition. On the other hand, if the males came to play an important role in defending the group, larger size would have provided a significant functional advantage. An analogue can be seen in savanna baboons, where the females average about 30 pounds and the males about 90 pounds, a far greater difference than is typically found in primates.

We may never know for certain about this and many other details relating to human evolution, but group living/group foraging and a cooperative division of labor allowing for increased access to a more dangerous but abundant environment is likely to have been primordial in the hominid line. It would have involved the most limited, incremental behavioral changes with the most cost-effective payoffs for the participants; it was highly synergistic. Moreover, as time went on the group-living mode of adaptation led to other forms of social cooperation and more elaborate forms of synergy (see below). "Culture," after all, is a property of a group, not an individual.

Tools and Synergy

However, there is also a possible problem with this scenario. If high-quality food resources are not sufficiently abundant, a large foraging group might become subject to disruptive internal "scrambling competition." Other primates, like chimpanzees, solve this problem by "fissioning" into small foraging units during the day.[30] However, they are not vulnerable to intense predation pressure. Accordingly, a second key to the australopithecine adaptive strategy may have been the invention and use of tools (and weapons) – a synergistic "soft technology" of wood and bone implements, and perhaps thrown objects as well. It could be said that the "stone age" was preceded by the "wood age" and "bone age."

There have been many tool-use advocates over the years, from Darwin to Dart, Szalay, Washburn, Birdsell, Coursey, and Mann. However, some theorists claim that the first "real" tools in hominid evolution were the manufactured stone cutters/choppers/scrapers that first appear in the fossil record some 2.5 million years ago. These theorists often downplay the importance of tools and envision early australopithecines as having been minimal tool-users at best. But this view, as Jonathan Kingdon observes, is obtuse. For one thing, the lack of fossil evidence is not evidence of a lack – as the old cliché goes. But more important, tool-use can have revolutionary consequences (as we noted in Chapter 6). It can be the functional equivalent of opening up a new ecological niche, or a whole new adaptive mode. Otherwise unattainable sources of food can suddenly become a reliable, even abundant part of an animal's diet.[31]

The current thinking is that so-called "underground storage organs" (tubers, bulbs, rhizomes, corms, etc.) may have played a critical role in the evolution of the australopithecines. In a more seasonal woodland habitat, they provided a reliable, year-round source of food.[32] But this shift also presented a major "technological" problem, namely, how could these resources be exploited efficiently. Digging sticks may have provided the answer. It seems unlikely that the australopithecines would have adapted successfully to a terrestrial lifestyle and survived, even prospered, for perhaps 3 million years without the skilled use of various natural objects – digging sticks, stone hammers, carriers, and the like. Indeed, we noted earlier that chimpanzees and even capuchin monkeys are frequent users of tools for procuring food, and sometimes in conflict situations as well. It seems very improbable that the stone tools dating back about 2.6 million years sprang from the mind of a "tool-challenged" predecessor.

A more unpopular proposition, but equally compelling I believe, is that "weapons" also played an indispensable part in the successful transition to a terrestrial lifestyle. One can hardly exaggerate the value to a diminutive, relatively slow-moving biped of being able to use a short stick (similar to the modern billy club) or a large femur, or even a well-aimed rock, as a defensive weapon. For the very same reason that policemen around the world still use billy clubs and rioters still throw rocks, the wielder of a weighted object can engage an opponent at much less personal risk (beyond arms' length) and can strike a much more damaging blow. Even crude weapons, if skillfully used, would have served as "equalizers" (as six-guns were called in the Old West). Indeed, it has been pointed out that the same hand-held wooden implements can double as digging sticks and defensive weapons.

Here, again, the principle of parsimony (and synergy) might be invoked. Consider the alternatives. A group of cooperating Miocene/Pliocene midgets without benefit of defensive implements of any sort would have been at a great disadvantage against bigger, faster opponents armed with large canines and foreclaws. Conversely, a lone australopithecine, even armed with a weighty femur, would have been at a great disadvantage against a group of hungry hyenas. But a group of australopithecines traveling together in dangerous or unfamiliar country with weapons carried at the ready (a synergistic package) would have been far more likely to hold their own in any life-and-death situation. These creatures may not always have been subject to predation, but even one incident in a lifetime would have been one too many. It was better to be prepared.

Although I don't relish the idea, it is also possible that the "arms race" in humankind began in the Miocene/Pliocene – as many conflict-oriented theorists over the years have claimed. If groups wielding simple weapons gained "leverage" in confronting predators, the same would have been true in confronting various competitors. Again, the principle of parsimony can be applied. The manufacture and skilled use of tools/weapons by these hominids entailed learned, socially transmitted behavioral innovations, and their application to new contexts would have been an incremental step with potentially large benefits.

Does this mean that the australopithecines were "killer apes" after all? I would argue that this image is greatly overdrawn. Warfare is not an instinct;

it's a cultural invention. If there are deep psychological predispositions for aggressive behaviors, they are also shaped by experience and modulated by higher-level cortical processes. The challenge for our ancestors – as it is for us – was survival and reproduction. Survival was the overall challenge, and this involved an array of continuing, inescapable "basic needs." (A recent article of mine in the *Journal of Bioeconomics* documents no less than 14 "primary needs" for modern humans, from food, water, and "thermoregulation" to nurturing offspring and even mental health.[33])

Violent confrontations with other groups can be very risky, especially when weapons are involved; the potential costs could outweigh any possible benefits. It is more likely that armed group conflicts were related to such vitally important objectives as the acquisition or defense of food patches, water-holes, and sleeping sites. Anthropologists Lisa Rose and Fiona Marshall call it the "resource defense" model, but there must have been an equal number of resource "offenses." It is also very possible that the male chimpanzee practice of raiding other groups to acquire females was common among the australopithecines as well, as Wrangham, Foley, Stanford, and others have suggested. In any case, organized collective violence, especially with crude weapons, represented another form of social synergy.[34]

Hard Evidence for Soft Technologies

There is some "hard" fossil evidence for the group-living/tool-using hypothesis. One source of evidence comes from brain "endocasts" – impressions of the interior surface of australopithecine skulls made by paleoanthropologists Ralph Holloway, Philip Tobias, and others.[35] What these surface impressions, and other evidence, indicate is that, even 3.5 million years ago, australopithecine brains, while still quite small by comparison with our own, may already have undergone some internal reorganization. These changes were of the kind that are associated with skilled use of the hands, more complex social behaviors, and more sophisticated communications abilities – though probably not language ability in our sense of the term (more on language below).[36]

In other words, during its first 2 million years or so australopithecines had moved well beyond the cognitive abilities of modern chimpanzees in ways that were compatible with the scenario described above; they were accomplished group-living, tool-using, loquacious bipedal apes, and the changes in their brain anatomy were a reflection of major changes in their behavior. Jonathan Kingdon has pointed out that these intelligent creatures would have had both the incentive and the ability to make more systematic use of various materials and resources in their environment, and plenty of models to use for inspiration. Like many other species, our remote ancestors no doubt learned from observing nature – spider webs, nest-building techniques, food procurement strategies, the hunting techniques of large carnivores, and the like.[37] Thus, a chance observation that some small mammal was digging up an edible tuber with its forepaws might have led to the deployment of digging sticks to exploit this potential new food resource as well. Again, the "rewards" for adopting these innovations would have been immediate.

A second form of hard evidence consists of the very changes in dentition – tooth size and shape, enamel thickness, and other characteristics – that paleoanthropologists rely upon to differentiate among the various fossil discoveries. Australopithecine dentition was significantly different from that of the chimpanzees. The reduction of their canine teeth and the development of larger molars with thicker enamel implies that australopithecines had adopted a more diversified dietary pattern, with a larger percentage of tough, chewed plant foods. And this, in turn, suggests that behavioral changes (Neo-Lamarckian Selection) might have preceded the natural selection of the relevant anatomical changes.[38]

A third source of evidence is the significant alteration in the hands of the austalopithecines. These "handy apes" already had relatively dexterous hands, compared to other primates and the great apes; their thumbs were shortened and their fingers flattened, although their thumbs were still not fully opposable. The modern precision grip, or grips (there is more than one) was certainly not perfected in australopithecines, so it is unlikely that their tool-making was very advanced. Indeed, anthropologists Nicholas Toth and Kathy Schick, in their important research on early tool-making techniques, showed that even the crude, "flaked" stone tools of 2.5 million years ago required considerable manual dexterity and practice. Among other things, Toth and Schick found that even the famous Kanzi, an intelligent and carefully trained bonobo (pygmy chimpanzee), was unable to master these techniques.[39] Accordingly, australopithecines most likely could do things with their hands that chimpanzees could not, although they were still much less adroit than their successors. Much has been made of the cognitive developments underlying the emergence of language, but the developments associated with skilled use of the hands were equally significant, as Darwin long ago supposed.[40]

One further implication of the group foraging and group defense/offense model is that the carrying of objects may have played an important part in australopithecine evolution after all – as Lovejoy suggested though for very different reasons. As these hominids began to move out into more exposed environments, a tacit division of labor may have developed. The females carried the infants (and perhaps digging tools) to new foraging sites, while the males carried tools/weapons that could readily be used for defense.

At first, no doubt, these forays were very furtive; the groups probably remained close to the safety of the trees. But, as their skills, their size, and their confidence increased, they were able to migrate over longer distances and occupy more diverse environments. By 3 million years ago, at the latest, South African australopithecines – having spread some 2,000 miles from their East African heartland – were thriving in much different climates with more varied seasonal resources. Intensive scavenging and small-game hunting may well have become a winter-season imperative for some of them.[41]

The Social Triad

One other form of social cooperation (and synergy) may also have played an important role in the evolution of the australopithecines. It is well known that human females have a much more difficult birthing process than, say, chimpanzees and that the progressive enlargement of the hominid infant's brain is partly to blame. Less widely appreciated is the fact that changes in the anatomy of the pelvic region accompanying the shift to bipedalism were also a complicating factor. In chimpanzees, the infant's head emerges at birth facing forward. The mother can see the infant's face, can manually aid in the delivery process and can take care of the afterbirth details (clearing the infant's nose and mouth, cutting the umbilical cord, etc.). However, the anatomical changes evident even in australopithecine females suggest that the human birthing process, with the infant facing the mother's back, had already become a necessity; the infant's head had to rotate in order to fit through a more constricted "pelvic aperture." One consequence was that the childbirth process in australopithecines all but required assistance – midwifery.[42]

So, in addition to male–male and male–female cooperation, cooperation between females may have become one of the elements of the australo-pithecine adaptive pattern. In other words, there was a "social triad" – a three-way nexus of cooperation and synergy. If so, this basic social frame-work created the scaffolding for a more elaborate and extensive pattern of cooperation over time; humankind evolved as a closely cooperating, group-living species. And our social "coherence," as Darwin put it, has facilitated many other collective synergies as well.

This is not to say that the influence of individual competition, status ri-valries, internal social conflicts, etc., somehow magically disappeared. Then as now it is likely that there was a sometimes precarious interplay between competition and cooperation, between various individual self-interests and the interests of the group. No doubt differences in individual "personali-ties" also played a part, just as they do today even in our great ape cousins. Indeed, a dynamic tension between individual and group interests is com-mon in other social primates as well.[43]

The key to australopithecine sociality may have been the relative costs and benefits to each individual for cooperation or non-cooperation. Why should the males, even if they are closely related, cooperate with one an-other? Or why should the females help one another if they are unrelated and perhaps rivals for social status and the attentions of the males? Reciprocity and reciprocal altruism may help to explain it. But the benefits associated with being included in the group – and the high cost of ostracism – may also have been a major factor. Moreover, each individual had a stake in the viability and well-being of the group as a survival unit. The "public interest," as it were, was rooted in the group and its potential for collective synergy. It provided an incentive for collective measures to contain conflict and enhance cooperation. For instance, a larger group was more likely – all other things being equal – to benefit from synergies of scale in con-frontations with other groups of predators or competitors (not to mention potential prey). The same principle of collective synergy undergirds human societies today.[44]

"Homo Intermedius"

There has been a tendency of late to characterize the emergence of modern humans, perhaps 150,000 years ago, as a "revolution." However, there was a much earlier revolution in human evolution that was at least of equal importance. The earlier revolution occurred some 2 million years ago with the emergence of *Homo erectus/Homo ergaster* and (presumably) various close relatives. To be sure, this transition required many hundreds of thousands of years and included intermediates like *Homo habilis*. It was the combined result of a synergistic set of incremental changes rather than a "megamutation" or a "punctuated equilibrium." However, some of these inventions were ultimately of far-reaching importance; they "seeded" many of the other changes that have followed.

Paleoanthropologists make much of the sometimes subtle differences between ancient fossils, and much of great value is learned from expert analyses of these material residues. However, the intense professional debate over dates, timing, categories, and sequences can also obscure the underlying functional dynamics – the inner logic of the process. Indeed, as more fossil evidence has accumulated in recent years, some hotly contested distinctions have begun to blur. For instance, new evidence of variability and overlapping traits in *Australopithecus afarensis* and *A. africanus* specimens has led some theorists to conclude that these hominids were merely subspecies rather than sharply divergent lineages. Likewise, the distinctions between australopithecines and early Homo (*H. rudolphensis*, *H. habilis*) do not seem as sharp as they once did. Not only did these ancestral hominids coexist for at least 1 million years but it seems that the earliest systematic use of stone tools may have occurred in the australopithecines. Were each of these separate species or only variations on a common theme – like African Pygmies and Watutsis today? There is no way of knowing for sure.

Nevertheless, over the course of roughly a half million years, a major transformation occurred that set our ancestors firmly on a path that has led to modern humankind. Although there are obviously important differences among these ancient hominids, they also represent to varying degrees an interrelated set of functional changes (both anatomical and behavioral) that were intermediate between those of the diminutive small-brained australopithecines and modern, large-brained, technologically sophisticated, language-using *Homo sapiens*. Accordingly, at the risk of offending the professional fossil-hunters, I will lump these fossils together into a generic category called "*Homo intermedius*" and focus on the "Why" question.[45] Why did an apparently successful small biped evolve into something quite different – something much more suggestive of modern humans?

A Technological Revolution

A benchmark that is sometimes used to denote the transition between australopithecines and the *Homo* line is the appearance in the fossil record of the first "flaked" stone tools that were struck from larger quartzite or lava stone "cores." (Other commonly used criteria are differences in brain size, dentition, and the like.) However, it has recently been discovered that *Australopithecus garhi* (or an as yet unidentified contemporary) at Gona, in Ethiopia, was already adept at making and transporting these implements over some distance and using them for a variety of purposes – chopping, cutting, smashing bones, and perhaps skinning – some 2.4 million years ago.[46]

Some theorists liken these early, multi-use tools to Swiss army knives, but this is not really a good analogy. The Swiss army knife consists of a set of functionally specialized blades and other tools that have been bundled together. A better analogy is the famed Bowie knife of the American frontier. For more than four decades, from the 1820s until well after the Civil War, the backwoodsmen, mountain men, swamp rats, and pirates of that rough-and-tumble era relied on a multi-purpose, tempered steel knife with a gently curved, single-edged blade and a sharp tip. It is said to have been invented by Jim Bowie, the frontier legend who later died at the battle of the Alamo. Bowie knives were so perfectly balanced that they could be thrown with deadly accuracy, and they were variously used for hunting, fighting, skinning, chopping, wood-carving, hammering, eating, and even shaving.[47]

It is common these days for writers on human evolution to belittle the so-called Oldowan tool-making tradition – so named because the first stone tool discoveries were made at the Olduvai Gorge by the famed fossil-hunter Louis Leakey (father of Richard) and his co-workers many years ago. The naysayers claim that the earliest stone tools were not very sophisticated, and there is little evidence of further technological "progress" for another million years. But this attitude is condescending. We have already mentioned Nicholas Toth's demonstrations that the manufacture and use of stone tools is a more highly skilled activity than it appears. More to the point, paleoanthropologist Bernard Campbell has argued that, contrary to the belittlers, the Oldowan tools were so successful, and so well suited to the needs of a highly mobile biped, that there was probably no need or incentive for making major "improvements." Leakey himself demonstrated how efficiently a large animal can be skinned and defleshed with a simple stone flake. Campbell also notes that more elaborate tool-making is a time-consuming activity. The advanced "biface" tools of the later Acheulean tradition (dating back 1.4 million years or more) required some 25 steps, while the more highly diversified and stylized Aurignacian tools of the Upper Paleolithic (within the past 50,000 years) required six separate stages and about 245 steps.[48]

The importance of these "crude" Oldowan tools can hardy be overstated. It really amounted to a technological revolution, because it enabled our ancestors to become systematic hunters (and scavengers) and to exploit the teeming herds of large animals that populated the open grassland areas in East Africa, and beyond. Once stone tools were deployed, moreover, the carcasses of these animals provided "raw materials" – horns, bones, skin, and sinew – for many other uses. Just as digging sticks and hand-held weapons may have played a key role in the success of the early australopithecines, the invention of stone tools vaulted our ancestors into a new ecological niche. Equally significant, this adaptive revolution evidently predated the emergence of "*Homo intermedius*" (a.k.a. *Homo erectus/Homo ergaster*) by several hundred thousand years. In other words, synergistic behavioral changes – technological changes – preceded and supported the major anatomical developments that are reflected in the fossil record much later on.

The specimens of "*Homo intermedius*" that begin to appear in the fossil record about 2 million years ago were strikingly "improved." First and foremost, they were twice the size of Lucy and her cousins (and even greater than that by weight), or approximately equal in height to modern *Homo sapiens*. Their brains were also more than double the size of the early australopithecines and showed evidence of further reorganization and a greater degree of lateralization, or asymmetry (associated with handedness and various cognitive specializations). Biologist Robin Dunbar makes a compelling case for a close functional relationship between neocortex size, group size, and language development in human evolution.[49] "*Homo intermedius*" also had longer legs (and shorter arms) and had perfected the striding gait of modern humans. Not only was this more energy-efficient (by some 30%) than the knuckle-walking technique of chimpanzees, but it was especially well suited for longer-distance travel. Whereas chimpanzees in open areas may have ranges of 200 square miles, human hunter–gatherers typically exploit ranges of close to 700 square miles.[50]

Another striking change in "*Homo intermedius*" was a sharp decline in the size difference between the two sexes to approximately modern human proportions (about 15% compared to 25% in chimpanzees). This suggests to some theorists that sexual competition among the males had declined; perhaps the nuclear family and more permanent male–female pair-bonding had emerged (as Lovejoy proposed, only much later and for other reasons).

But it is at least as likely that there were significant functional advantages to having the females be relatively larger. One advantage would have been the ability to travel at a faster pace and not hold back the rest of the group. Another would have been the ability of the females (and the group) to cover more ground and greatly expand their day ranges. Still another advantage might have been the ability to carry heavier infants over longer distances. Larger size also implies a larger birth canal and pelvic aperture, an important consideration in giving birth to larger infants. Certainly it would also have been an advantage for self-defense, when the females were foraging independently or were sequestered at "home bases." In any case, the decline in sexual dimorphism was very likely another example of how behavioral changes were the "pacemakers" for the anatomical modifications that occurred.

A final distinguishing feature in "*Homo intermedius*" – at least in the more delicately proportioned "gracile" forms – was a significant reduction and refinement of the teeth and mandibles, and changes in wear patterns, as well as a reduction in the size of the gut. These changes suggest yet another behavioral pacemaker – a dietary shift to one that required less chewing of coarse plant foods, greater consumption of meat, and, some argue, cooked foods (see below). In turn, these anatomical changes facilitated the later development of the vocal tract and the progressive evolution of language skills (more on language below).

A Neo-Lamarckian Process

What is the meaning of all this? The most plausible explanation, I believe, is that a major adaptive shift – a Neo-Lamarckian evolutionary change – had taken place. In the half million years after stone tools became a standard part of their tool-kit, our hominid ancestors made the transition from a pint-sized forager that only opportunistically hunted and scavenged meat to a systematic hunter and confrontational, or "power" scavenger that relied on meat (along with an array of plant foods) to provide a more stable, abundant, high-quality, cost-effective food supply. "*Homo intermedius*" had joined the ranks of "top carnivores" and could hold its own in confrontations with other carnivore competitors – not to mention potential predators. This conclusion is not original, of course, and there are alternative theories (as we shall see). But I would argue that it is the most parsimonious explanation for the anatomical changes that occurred.[51]

A "trigger" for these changes may have been a significant shift in global climate to a more variable, "oscillating" pattern between 2.8 and 2.5 million years ago.[52] One consequence for various East African inhabitants was a more arid, more seasonal environment with relatively less abundant (more widely scattered) plant foods and relatively more meat on the hoof. A shift to more systematic exploitation of large animals was a cost-effective response to these changes. Moreover, a shift of emphasis to hunting–gathering, versus gathering–hunting, did not require a great adaptive leap; early "*Homo intermedius*" were already terrestrial, group-living, opportunistic hunters. (By contrast, the more "robust" hominids of that era, with a distinctively different anatomy and dentition, may have deployed a feeding strategy that relied on a broad range of lower-quality plant foods.)

Nor was big-game hunting necessarily a dangerous activity – a "contact sport." There are a variety of less hazardous alternatives. As Kingdon notes, coordinated tactics of various kinds can be used to ambush prey, or to panic and drive animals into mudholes, swamps, cul-de-sacs, "deadfalls," or even (in later times) into prepared trip lines, nets, and traps. Later on fire brands were probably also used for driving potential prey, as well as for deterring carnivore competitors. In other words, intelligent problem-solving most likely played a major role in the momentous change to a hunting/scavenging way of life.

Some indirect evidence for this adaptive shift can be found in the very anatomical changes that distinguished "*Homo intermedius*." Their larger size and bigger brains ("expensive tissue") implies that they had commensurately greater nutritional requirements. Paleoanthropologists William Leonard and Maile Robertson have estimated that, compared to the australopithecines, *H. erectus* needed 40–45% more energy to support its increased size alone, and that larger day ranges would have increased total energy needs by as much as 85% (at least judging by contemporary human foragers).[53] The females were especially burdened by the increased nutritional demands of their larger but quite helpless neonates, while the more lengthy period of childhood dependency imposed greater constraints on the females' mobility and foraging ability. Anthropologists Leslie Aiello and Catherine Key calculate that *H. erectus* females may have required some 60% more calories for reproduction than did their australopithecine predecessors, which they believe implies a nutritional "revolution."[54]

There are only so many hours in each day that can be allocated to food acquisition, so the anatomical changes that are evident in "*H. intermedius*" put a premium on obtaining large quantities of high-quality food as efficiently (and dependably) as possible. To put it in economic terms, an increase in the demand required a corresponding increase in the supply. Large-scale hunting/scavenging and provisioning of the group was at once the cause and the ultimate solution to this problem. Meat provides twice as many calories as fruit and ten times as many calories as leaves, not to mention other nutrients. And meat can also be obtained in large packages that are susceptible to collective acquisition, bulk transport, and shared consumption. To borrow a phrase, "Man the Hunter" was an idea whose time had come. However, this does not discount the provisioning role of the females. Indeed, anthropologist Sarah Blaffer Hrdy makes a compelling case for cooperative "allomothering" in evolving hominids as the demands of child-rearing progressively increased.[55]

Anthropologist Craig Stanford, who has done much to revive the case for hunting and meat-eating in human evolution, has developed a novel explanation for why hunting/scavenging became an important part of the hominid food quest. Stanford bases his thesis on (1) data from chimpanzees and modern hunter–gatherers showing that meat provides a relatively small share of their overall food supply, (2) research in chimpanzees indicating that the energetic cost of obtaining meat is often not fully repaid (especially where small game are involved), and (3) studies which suggest that there may be biological limits to how much meat we can consume (about 50% of our caloric intake, according to archaeologist John Speth). It seems that too much protein can be toxic. Therefore, Stanford argues, hunting/scavenging in evolving hominids was motivated more by power politics and sexual competition than nutritional needs. Meat is commonly used by chimpanzees as a form of patronage, or "pork barrel," in an effort to support social and political objectives and win sexual favors. Stanford believes the same would have been true for our hominid ancestors.[56]

However, there is reason to be guarded about this Machiavellian explanation. Social objectives were likely to have been secondary to nutritional requirements for evolving hominids, I would argue. For one thing, modern hunter–gatherers most often inhabit marginal areas. So the data on their nutritional patterns may understate the role of meat-eating in the Pleistocene, when East Africa was teeming with large game animals. Second, if these animals were hunted or scavenged and processed efficiently, they would most likely have provided a highly cost-effective resource; the hunting of small game by chimpanzees in forest areas is not comparable. Third, some human populations in historic times, like the Karimojong and the Mongolian steppe herdsmen, have relied very heavily on meat and other animal products – milk and blood – and have thrived; it is not at all certain what the nutritional constraints might have been for our evolving ancestors. But most important, the nutritional demands of evolving "*Homo intermedius*" put a premium on obtaining large quantities of high-quality food resources for everyone – namely, meat. Finally, the overriding need for close cooperation in a very demanding and dangerous world would likely have restrained pork-barrel politics.

Other scenarios are also possible, of course, but the hunting/scavenging/foraging and food-sharing/provisioning scenario seems most consistent with other evidence – tooth-wear patterns, tool-use patterns, and the anatomical changes that occurred in *H. erectus/H. ergaster*.[57] Over the course of time there were also progressive improvements in tool-making skills (as reflected in the Developed Oldowan and Acheulean traditions), plus more selective use of raw materials, more complete "processing" of animal carcasses and evidence of more specialized tools for different uses, such as wood working, skinning, and plant food processing. It is also likely that there was an increased need for transporting food, stone "cores," and, very likely, water supplies over long distances. Indeed, water became a much more critical resource with the adoption of hunting/scavenging activities in an open savanna environment. As anthropologist Rosemary Newman observes, we evolved into a "thirsty, sweaty animal" that needed as much as two quarts of water per hour while engaged in the hot pursuit of animal prey on a hot day.[58]

A Hotbed of Synergy

"*Homo intermedius*" was the product of an interrelated package of synergies that laid a foundation for the many improvements that have followed. For starters, there was a synergistic relationship between an improved, more efficient bipedalism, increasingly skillful, manipulative hands, and the increased mental powers needed to use them in more productive ways. There was also the social triad – the framework of cooperative relationships between the males, between males and females, and between the females. Over time this framework allowed for the elaboration of the group as a unit of collective adaptation, with greater social organization, more coordination of activities, and especially a division (combination) of labor.

One important example of this trajectory was the adoption of consistently occupied "home bases," or encampments. The evolutionary significance of this development was that it resulted in a significant improvement in economic efficiency for the group as a whole, because it allowed for a more elaborate "combination of labor." Resources as needed – meat, plant foods, stone tool "cores," animal skins, water, firewood, etc. – could be carried to a safe haven (vitally important) and then shared and utilized through a network of reciprocities. Both Kingdon and Wolpoff have pointed out that the ability to carry various materials and artifacts over a long distance is an underrated technological achievement, no doubt because it involved another "soft" technology. Nevertheless, it was a key innovation.

Even more underrated, perhaps because it is "old news" and a veritable cliché about human evolution, is the adoption and controlled use of fire. "Revolutionary" is by no means too strong a word to use for the consequences of this multipurpose invention. Kingdon notes that natural fires caused by lightning or the combustion of gas seepages, etc., had been a part of our ancestors' environment for many millions of years. The earliest hominids no doubt observed the effects produced by fire, especially as an instrument for driving other animals and as a means for killing and cooking potential prey. It's likely that Lucy and her relatives also learned to scavenge at fire sites, just as their carnivore competitors and predators did. Moreover, fire may have begun to play a major part in our evolution much earlier than we thought. The so-called Karari sites analyzed by anthropologist Randy Bellomo suggest that hearths were used, most likely by "*H. intermedius*," at least 1.6 million years ago.[59] And Richard Wrangham and his colleagues go back even further. They believe there are indirect "signals" of controlled fires as early as 1.9 million years ago. More important, they theorize that it was not meat-eating *per se* but the introduction of cooked foods that may have been responsible for the transition to *H. erectus/H. ergaster*.[60]

In any event, over the course of time fire came to have many valuable uses – defense against predators, chasing competitors away from carcasses, tenderizing meat, killing harmful bacteria, breaking down toxic chemicals in the many plant foods that could not be eaten raw, hardening wooden tools, drying skins, deterring insects, providing warmth (especially in colder, temperate climates), and even facilitating long-distance signaling and communications. Later on, our ancestors also learned how to use fire to condition the environment – to clear land and stimulate the growth of favored species and to prepare garden plots for planting. Not only was the controlled use of fire an immensely potent, synergistic technology – one without precedent in evolution – but it became an indispensable adaptation, like our bipedal gait. We are also, in a very real sense, "*Homo pyrotechnicus.*"

These and many other soft technologies – nets, weirs, traps, spears, shelters, containers, rafts, and more – most likely long predated the vaunted achievements of modern *Homo sapiens*. The long-term significance of these synergistic inventions was that they enabled "*Homo intermedius*" to generate surpluses, which allowed for the growth of bigger, more mobile groups of physically larger, longer-lived individuals, and, equally important, their more numerous offspring. Indeed, the population growth curve in humankind may have started to rise as early as 2 million years ago. The evidence for this (admittedly circumstantial) is the fact that, at a relatively early date, "*Homo intermedius*" began to migrate out of Africa and "colonize" other tropical and subtropical areas at great distances from their East African birthplace. Hominid fossils tracing back to between 1.8 and 1.6 million years ago have been found in Indonesia, and some hominid remains of that vintage have also been found in the former Soviet Republic of Georgia (north of modern Turkey and western Iran) and at Ubeidiya in Isreal. These very long-distance travels (on foot, needless to say) were no small achievement; every new environment presented many novel challenges and opportunities, and many unforeseeable contingencies.[61]

Culture, Language, and Synergy

The invention of a more efficient technology was only half the story, however. "Culture" – the accumulated know-how and experience of the group – also became an indispensable part of the package that enabled "*Homo intermedius*" to prosper and colonize new environments. Larger cooperating groups were able to exploit many new opportunities for social synergy, including the sharing of costs and risks, pooling information, a combination of labor, and, not least, many synergies of scale against competitors, predators, and prey. Likewise, mutual aid, or "succorant behaviors," could increase the odds of surviving an injury or illness, while the joint policing of "free riders" and parasites could serve to reduce internal conflicts, as Peter Richerson and Robert Boyd (and others) have pointed out.[62] Anthropologist Christopher Boehm has also shown that cultural processes like group decision-making can play a significant role and become a locus of group selection. (Elliott Sober and David Sloan Wilson make a similar argument.)[63]

As noted in Chapter 6, some theorists claim that culture only evolved in humankind much later on, with the emergence of art, rituals, symbolic language, and the like. But if culture is defined more broadly as a body of socially transmitted knowledge, skills, and artifacts that are passed between generations via learning and teaching, then culture was already a vital part of the hominid lifestyle more than 2 million years ago. After all, the very existence of a tool-making "tradition" implies the ability to transfer the requisite tool-making skills between generations. And this was surely the tip of the iceberg. For instance, modern hunter–gatherers carry with them "mental maps" of hundreds of square miles and can recall the precise locations of literally thousands of water-holes, plants, animals, natural hazards, and other landmarks, most of which they learned from their forefathers. And this says nothing about socially learned subsistence skills.

Beginning with "*Homo intermedius*," culture became cumulative and an increasingly potent adaptive tool; new ideas and inventions were not only preserved and communicated to subsequent generations but were refined and improved upon over time. The group as a whole became a transgenerational repository of adaptive information and an engine for the invention of more synergies. Spears, for example, came to be made of better raw materials; they were more finely shaped and balanced; their tips were fire-hardened; barbed tips were added to increase their penetrating and holding power; wooden spear-throwers were invented as a way to increase their range, striking force, and accuracy; finally, bows and arrows were invented as a lightweight alternative that could increase the hunter's range and precision, and (not least) multiply the hunter's supply of "ammunition." Each of these inventions represented a major (cumulative) economic advance. More food could be acquired more dependably with less time, effort and collective risk. Primatologist Carel van Schaik argues that socially based learning and inventiveness, and the intelligence to exploit it, would have been powerfully favored by natural selection; primatologist Chistophe Boesch and anthropologist Michael Tomasello call it the "ratchet effect."[64]

The backbone of modern human culture is, of course, symbolic language – the ability to communicate complex information among members of the group both rapidly and in great detail. Language gives us an enormous collective advantage that no other species enjoys. Moreover, it is probable that language was a major social tool long before the full flowering of symbolic thought and articulate speech. From a very early date, some form of proto-language was very likely a major facilitator for social organization, and social synergy. There is also good reason to believe that language is another cultural invention that was progressively improved upon over time, with both anatomical changes and further inventions – much like the incremental improvements in bipedalism, our skillful hands, our cognitive abilities, and our tools. (It is increasingly evident that our cognitive and linguistic skills overlap but are separate.[65])

Linguist Philip Lieberman, in his recent book *Eve Spoke* (1998), points out that the vocal tract that makes fully articulated modern speech possible (a late development) involves some major biological costs, most notably physical constraints on our jaws and teeth and a serious risk of choking. "There would have been no increase in biological fitness for the human vocal tract unless a human brain adapted for speech had existed in our immediate African ancestors."[66]

This view of language is hotly debated, of course, though some of the heat is the result of conflicting definitions of what constitutes language, and its key properties. For instance, some theorists apply the term only to communication that employs arbitrary combinations of symbols (morphemes). Others believe that thinking and talking must be identical ("isomorphic"). Still others see language as an extension of our ability to use our hands, and of hand gestures in particular. I use the term "language" in a broad functional sense; language is fundamentally a behavioral means for communicating information; it is not about "thinking," or symbols *per se*, but about intentional "signs" – to use the organizing concept of the science of semiotics.

In this definition, many other species, especially our primate cousins, have at least rudimentary proto-language skills. Vervet monkeys, for instance, have an array of specific "words" to differentiate among various categories of objects, from predators to fellow vervets. Our closest relatives, the chimpanzees, are especially impressive communicators; their social interactions are shaped by a sophisticated range of vocalizations, facial expressions, gestures, and body language. Another example is the elaborate system of vocalizations in mole-rats, described in Chapter 3, which plays an essential role in organizing their social life. In fact, psychologist Charles T. Snowdon has documented that "multi-modal" communications systems are ubiquitous in other social species. He found that 11 of the 18 language criteria developed many years ago by the linguist Charles Hockett are routinely utilized by other animals, and that the remaining seven can also be seen (though not together) in various primates.[67]

"Homo Symbolicus"

Neurobiologist Terrence Deacon, in *The Symbolic Species* (1997), argued compellingly that the biological basis for symbolic language in humankind – which involves an enormously complex set of anatomical changes – could not have burst forth full-blown from some megamutation. (Jared Diamond also likens our language-making equipment to a finely made Swiss watch with many interacting parts.) In fact, the emergence of symbolic language may provide yet another example of hominid self-making and Neo-Lamarckian Selection. Deacon calls it "Baldwinian selection," after psychologist H. Mark Baldwin's "Organic Selection" theory (see Chapter 6), but the thrust of his argument is identical. Language is the product of a coevolutionary process in which behavioral changes were the pacemakers that induced supporting anatomical developments. (For the record, Deacon's argument was anticipated by the "positive feed-back" scenario of anthropologist Ralph Holloway, dating back to the 1960s.[68])

Some theorists – most notably Noam Chomsky and Stephen Jay Gould – argue that human language did not have a functional (adaptive) origin. In their view, it is an "exaptation" – an accidental consequence of the emergence of our large brains. Of course, our large brains are not exactly "accidental." But that aside, our success as a socially organized species has long depended upon social communication. It is inconceivable that australopithecines and (even more certainly) *"Homo intermedius"* could have developed a highly organized pattern of cooperation and group adaptation, including a division of labor, increased mobility, and, ultimately, long-distance migrations, without communications skills that, at the least, greatly exceeded the repertoire of chimpanzee vocalizations (some 30 calls altogether). Like other unique human traits, the functional advantages even of rudimentary verbal communications – perhaps somewhat like modern pidgin language – likely preceded and "induced" progressive anatomical improvements, including development of the vocal tract.[69]

In other words, even very limited language skills provided important functional advantages for hominid groups. And these immediate advantages were the very cause of the natural selection of improvements in the anatomical features that supported improved language skills. Recall the crossbeaks example in Chapter 6. To thoroughly mix some well-known metaphors, Richard Dawkins's "blind watchmaker" (the title of his popular book about natural selection) could not have produced Jared Diamond's "Swiss watch" without the guiding hand of purposeful behavioral changes by our resourceful, inventive, loquacious ancestors.

Some indirect evidence for this Neo-Lamarckian scenario can be found in *Homo erectus/Homo ergaster* fossils. Jeffrey Laitman and his colleagues believe there is evidence for changes in the vocal tracts of these hominids that facilitated a more elaborate form of sound production. By the same token, both Ralph Holloway and Dean Falk, though they disagree about earlier changes in australopithecines, interpret changes in the neuroanatomy of *H. erectus* as suggestive of greater language skills. Holloway, Deacon and others see further improvements in this direction throughout the Pleistocene epoch (most of the last 1.5 million years).[70] These changes do not, of course, tell us much about the causal relationships, but they do support the argument that improved language skills were an integral and necessary part of the "*Homo intermedius*" package.

Language as a Prime Mover?

Some students of human evolution have elevated language acquisition to the status of being a prime mover – or at least *primus inter pares* – in human evolution, especially in the final transition to *Homo sapiens*. For instance, Philip Lieberman (seconded by Ian Tattersall) argues that symbolic language, or rather the lack of it, played a decisive role in the final demise of the Neandertals. It was their limited language capabilities that "did them in." How so? Neandertals still had a more prognathous face (with a forward-projecting jaw) than *H. sapiens*, which Lieberman believes would have hindered their ability to speak clearly. (The problem had to do with the size relationship between the mouth and the pharynx.) To Lieberman this "slight difference" in the ability to talk would have made a decisive difference in the Neandertals' ability to compete with modern humans. Lieberman does not explain exactly why.[71]

In the same vein, Jared Diamond asks rhetorically, what was the "missing ingredient" in human evolution – the key that can account for what he (and others) characterize as a Great Leap Forward during the past 100,000 years? The answer, he says, was "the perfection of spoken language." This is what opened the floodgate of human inventiveness and liberated our big brains from their cranial cages.[72] Tattersall, likewise, calls language "the fount of our creativity."[73]

Finally, Terrence Deacon relates language to the emergence of the monogamous pair-bond and the need for social constraints on male competition in evolving human societies. Deacon gives symbolic language the credit for undergirding our unique "social contract" (philosopher John Locke's term but with a Neo-Darwinian twist). To boil down a lengthy explanation, Deacon argues that a major impetus for the evolution of language was that it provided a means for articulating and making public our personal commitments (and sexual prohibitions) through symbols and public rituals. We are, says Deacon, "*Homo symbolicus.*"[74]

I subscribe to a somewhat different perspective on the role of language in human evolution. Although it is obviously a uniquely human "soft technology," it is functionally related to primate and other animal communications systems. To put it baldly, the purpose of language is to facilitate social communications processes of all kinds. It evolved as an instrumental, multipurpose cultural "tool" – one element in the package of anatomical and cultural adaptations that were progressively improved upon over literally millions of years. It was a mover, but hardly a prime mover. Indeed, the immense anatomical complexity of our language system – with literally dozens of tiny muscles, bones, nerves, soft tissue, and, no doubt, many millions of interacting neurons – make it extremely unlikely that it could be of recent vintage. In a very literal sense, language and the anatomical basis for language coevolved.

In this light, let's first reconsider Lieberman's claim. The fact is that the classic stereotype of the Neandertals as dumb brutes that were simply outwitted by more clever, language-using humans is increasingly untenable. For one thing, the Neandertals survived and reproduced successfully through many ice ages and other ecological challenges in northern Europe for hundreds of thousands of years (until about 30,000 years ago), and they were apparently the exclusive occupants of many parts of that region for almost all of that time. In the Middle East, moreover, Neandertals and *H. sapiens* evidently coexisted in close proximity for some tens of thousands of years. In fact, some theorists believe that the Neandertals may have beaten back the earliest attempts by modern humans to migrate out of Africa, about 100,000 years ago.

It is also increasingly evident that the Neandertals were not adaptively locked in place; they also participated in the progressive trends that marked the evolution of modern humans during the Middle Stone Age (from about 750,000 to 250,000 years ago). Over time the Neandertals became, on average, less robust and more human-like in their facial characteristics, jaws, and teeth. Some researchers believe the evidence shows that the later Neandertals did have symbolic language. Supposed differences in their vocal tracts have been disputed, while various endocast researchers have found no significant differences between Neandertal and modern human brain structures, most especially those associated with language skills.

The Neandertals also deployed increasingly sophisticated cultural attainments – from huts to hearths to bone tools, necklaces, crude carvings, burial of their dead, and the development of advanced tool-making technologies (the Szeletian, Levallois, and Châtelperonian). Many of these achievements were previously attributed only to *Homo sapiens*; it was assumed that the Neandertals "borrowed" them from our ancestors. But nowadays this condescending view is less convincing.[75] The debate rages on, but it seems unlikely that an incremental difference in language skills alone would have been "the difference that made a difference," to paraphrase again the mantra of anthropologist Gregory Bateson. More likely it was only one contributing factor.

Jared Diamond's argument is more subtle. He recognizes symbolic language as an adaptation that was preceded by more rudimentary precursors, and he characterizes the achievement of modern language skills as a trait that merely rounded out a "full package" of adaptive changes. But Diamond also gives language credit for being the basic cause of human inventiveness. "With language we can invent. The essence of human language is inventiveness.... For that reason, it is inconceivable to me that those non-inventive humans of 100,000 years ago could have language as we know it today. I cannot avoid the conclusion that the development of inventiveness was linked to the perfection of human language."[76]

Language as an Invention

I see it somewhat differently. Symbolic language is in fact an outgrowth of our inventiveness (and other cognitive abilities), not the other way around. Inventiveness is hardly unique to humankind, though obviously we excel at it. As we saw in Chapter 6, many other animals also display inventiveness. Ravens are especially impressive. Even in the hominid line, the spark of creativity can be traced back at least to the australopithecines. Cultural innovation shows clear evidence of accelerating in "*H. intermedius*" – despite the fact that they had only intermediate-sized brains. More important, the larger, more complex brains of *H. sapiens* evolved well before the final perfection of our vocal tracts and the flowering of culture.

What happened, most likely, was that an increasingly inventive, sapient hominid developed a more elaborate, more sophisticated form of symbolic language on top of an already existing anatomical and neurological foundation and a more rudimentary mode of communication. The invention of writing provides an apt analogy. The basic principle underlying written language traces to the crude record-keeping systems that were devised as an aid to commerce, most likely long before the Neolithic. Over the course of time, the technology of writing progressed from the use of various tokens as counters to marks on clay tablets to pictographic symbols and, finally, abstract symbols.[77]

It seems more plausible that humans invented language – the "morphemes" (the sound combinations that make up spoken words), plus object-naming, categorizing, syntax, morphology, etc., that are crucial elements of most modern languages. A more effective system of verbal communication was just one of many technological improvements that began to occur with increasing frequency in emerging *H. sapiens*, once the underlying principle of making progressive cultural improvements became an established part of our adaptive strategy as a species. In other words, we invented inventiveness. Just as mounted ("hafted") stone axes were discovered to be vastly more efficient than their hand-held predecessors, so a more efficient form of verbal communication became an essential organizational tool for larger groups with a more elaborate division of labor, greater need to plan and coordinate efforts, and the need to share a richer "data base."[78]

The case for this coevolutionary view of human language is buttressed by some indirect evidence. It is significant that the syntax (the relationships between morphemes, or words) and the lexicon (the specific collection of morphemes and their meanings) differs widely among different languages; these important "surface" elements are culturally determined. Indeed, a child can learn any language, but a severe language deficit will occur if a child is isolated from a linguistic cultural environment. Moreover, each new generation of children must learn the specific repertoire of sounds, and the syntax and lexicon for the particular language they are exposed to. In other words language is, inextricably, both biological and cultural in nature.

It is also significant that even chimpanzees, gorillas, and dolphins have limited (non-verbal) symbol-using capabilities. Thus, our big brain is not essential to symbolic communication, though humans can (obviously) learn language much more easily and are vastly more proficient at it. And, of course, we can vocalize in symbols. To be sure, these animals do not display their symbol-using abilities in the wild; this talent only emerges after they are exposed to careful training. But that is precisely the point. Both in other species and in humans, symbolic communication skills depend upon a complex interplay between evolved biological capabilities (a deep logic structure that is even deeper than language) and a supportive cultural environment. As the biologists would say, language is "epigenetic." What we inherit is a strong prepotency (and the necessary anatomical and neurological hardware) for learning and using language, but the final result is also a product of culture. Linguist Noam Chomsky is famous for the much-debated theory that there is a biologically based "universal grammar" underlying the surface features of various languages. If it does exist, we cannot totally exclude chimpanzees, gorillas, or dolphins from this club.

Finally, let's consider Deacon's thesis about language and social bonding. The basic social dilemma that he identifies – namely, the need to reconcile the vital imperative of social cooperation with the tendency for sexual competition among the males – arose very early in human evolution, most likely even in australopithecines and certainly in the more complex and interdependent "*H. intermedius*" societies. The problem had to be solved long before the appearance of socially recognized symbols, rituals, and the like. Indeed, other primates are quite capable of sorting out these problems and can follow rudimentary social "rules" without the benefit of big brains and complex symbolic language.[79]

The Human "Revolution"?

The self-flattering image of modern *Homo sapiens* as the product of a saltatory leap of some kind seems irresistible. Christopher Wills, for example, characterizes our species as "evolutionary speed-demons." He speaks of humans as having a "runaway brain," which he likens to Cyrano de Bergerac's outsized nose. (Pinocchio's more disreputable nose might be an even more apt metaphor.) Other theorists refer to a "creative explosion" (John Pfeiffer), a cultural "revolution" (Ian Tattersall), a "great leap forward" (Jared Diamond), a "punctuational event" (Richard Klein), and the like.[80] However, I believe it would be more accurate to say that there was a quickening of a pattern of progressive functional improvements, rather than a sharp discontinuity or a new "stage." Even the categorical distinctions between the different tool "industries" are not so clear-cut (pardon the pun) or static as was once thought to be the case.

But more important, during this very long formative period in human evolution – upwards of 1 million years – there was a multiregional pattern of both biological and cultural changes that accelerated over time. Moreover, there was no epicenter for these developments; it was more of a mosaic pattern. To be sure, the pattern was far from smooth, and there were regional differences. But in general the evolving "archaic" humans became more "gracile," their teeth and jaws receded as their facial prognathism declined and, of course, their brains got bigger.

Even more significant is the fact that the decoupling of biological and cultural evolution became more pronounced. For instance, the Acheulean tool industry – including a more highly diversified array of biface tools shaped to consistent patterns – emerged several hundred thousand years before a major surge occurred in brain size, beginning perhaps 500,000 years ago. Likewise, the rise of the Mousterian tool industry some 250,000 years ago, and most especially the development of mounted (or "hafted") stone axes, preceded the emergence of fully modern *Homo sapiens* by perhaps 100,000 years. The hunting of large animals with skillfully made spears had also become routine long before modern humans appeared.[81] In other words, the development of new technologies, especially for food-getting, once again may have preceded and undergirded the biological evolution of more costly (brainy) humans.

There were also a great many other cultural developments during this formative period – larger groups with higher population densities, more elaborate shelters, wider patterns of trade, a greater variety of raw materials acquired from greater distances, and the development of special-purpose tools that, among other things, replaced the common hominid practice of using their front teeth as a vice. In short, the "revolution" image is overstated. Both biologically and culturally, evolving humans made further functional improvements throughout the Middle Pleistocene (from about 750,000 to 250,000 years ago). As Wolpoff notes, "modernization ... was an ongoing process in widespread populations and not a single dramatic event in a single population."[82] Richard Klein concedes: "Middle Paleolithic people were advanced over their predecessors in many ways."[83] (It is also probably safe to assume that many more artifacts have long since decayed or are buried under modern towns and cities.) In other words, the momentum toward modern, culturally complex *H. sapiens* was already well established and was accelerating when our most immediate ancestors emerged, perhaps 150,000 years ago.

Two Paths to Modern Humans

A close second in intensity to the debate over the ultimate fate of the Neandertals is the current debate over the crossing of the final Rubicon to modern humankind. One theory, the so-called multiregional model, is associated with Wolpoff and two colleagues, Alan Thorne and Wu Xinzhi.[84] In their view, the trends that were evident throughout the Pleistocene continued uninterrupted into the late Pleistocene/Upper Paleolithic without any sharp transition. The essential unity of the process, they believe, was assured through a pattern of both gene flow and cultural exchanges between regional populations. Wolpoff illustrates the idea with the image of various stones being thrown into a pond; as the ripples from each stone spread outward, they intersect and interact with each other.

The well-publicized alternative to the multiregional model is the currently popular "African Origins" hypothesis. An increasingly compelling body of genetic evidence – mitochondrial DNA and Y chromosome data in particular – indicate that all modern humans trace their lineages back to a common ancestor (a very small population) in East Africa about 100,000–150,000 years ago.[85] The data also suggest that this founding population grew larger over time and began to migrate out of Africa, starting about 50,000 years ago.[86] In other words, various genetic "markers" indicate that there was after all an "epicenter" for the final lap to humanity and that modern humans effectively replaced the other hominids in various parts of the world in short order, including (of course) the Neandertals in Europe and the Middle East.

Both of these scenarios have problems, however. Briefly, the multiregional scenario requires an implausible flow of genes and cultural information over huge distances and diverse populations, while the Out-of-Africa scenario is based on genetic "indicators" that bear no *direct* relationship to any known anatomical differences. The recent discovery of a language-related gene (*FOXP2*) that may have led to the "perfection" of our linguistic skills perhaps 120,000 years ago is a plausible "candidate," but the consequences of such an anatomical change can at this point only be inferred. Anthropologist Richard Klein believes that such a genetic change was decisive: "The shift to a fully modern behavioral mode and the geographic expansion of modern humans were ... coproducts of a selectively advantageous genetic mutation."[87]

A Synergistic "Package"

However, I would propose a slightly different scenario. It is possible that the migrants from Africa had some "slight advantage" (to use a Darwinian expression) which, nevertheless, made a great difference in terms of the "balance of power" between competing populations. I believe that the difference that made all the difference was not biologically determined. It involved some critically important technological and cultural innovations that were deployed with great effectiveness – possibly with the aid of improved language capabilities and skills. But language was not the whole story. In other words, the modern human "revolution" – the explosive growth and world-wide spread of humankind – was a culturally driven process that utilized new forms of synergy.

A number of theorists – Richard Klein, Luca Cavalli-Sforza, Jared Diamond, Christopher Wills, Paul Ehrlich, Ian Tattersall, and others – hold that the emergence of a more advanced technology was an important factor in the modern human diaspora.[88] It is significant that the timing of the African exodus – if true – coincided with the flowering and spread of the Aurignacian industry, which encompassed a range of technological improvements. These included more diversified and specialized (and efficient) tools made from various materials, more skilled manufacturing techniques, better cooking skills, more elaborate shelters, better food-storage capabilities, greater use of marine resources, larger population densities (approximating modern hunter–gatherers), longer occupation of different encampment sites, and greater mobility.

Needless to say, a more advanced cultural "package" would have provided an important economic advantage – namely, the means to support a rapidly growing population in diverse habitats. However, the Aurignacian technology may also have given our East African ancestors a major "military" advantage. It seems likely that the great human diaspora of 50,000 years ago was not a peaceful trek into virgin territory but a more hostile invasion of already occupied lands; the human wave was often (perhaps not always) accompanied by coercion and warfare.

As we noted earlier, this is not a new theory, but it deserves a new look.[89]

I hasten to add that we are not talking here about wars of conquest or imperialism in the modern sense; the terminal Pleistocene humans were not necessarily more "warlike" in temperament, or seeking dominion for its own sake. More likely, the process was driven by a pressing need for resources to support a growing population in a changing environment. (The last major ice age began about 75,000 years ago, intensified about 33,000 years ago, and peaked about 20,000 years ago.) Call it the "resource acquisition model" of warfare – and human evolution.

Prehistoric Warfare?

It should go without saying that a warfare hypothesis in any form is an unpleasant, even distasteful idea, but there are a number of reasons for suspecting that warfare played a major role in the spread of humankind. One is the evidence (based on a large body of research) that ethnocentrism (loyalty to one's own group) and xenophobia (hostility between groups) is an evolved psychological propensity in humankind with a strong biological foundation and deep evolutionary roots.[90] As James Madison put it in the *Federalist Papers* (one of America's founding documents), "the seeds of faction are sown in the nature of man." We noted earlier that xenophobia is also a common trait in social primates, especially the chimpanzees; analogues to raiding and warfare between groups have also been documented in our closest primate relatives. Hence, it seems reasonable to assume that ethnocentrism and xenophobia also affected the relationships between groups of archaic and early modern *Homo sapiens*, just as they obviously do between groups of contemporary humans.

Another reason for suspecting that warfare was involved in the final emergence of humankind is that the expanding human populations were not migrating into uninhabited territories, for the most part. Like the fifteenth-century Europeans who discovered that their "New World" was already populated with "natives," the prehistoric groups that migrated out of Africa found that other hominids had preceded them by hundreds of thousands of years. Peaceful trade relationships can arise when there is the possibility of mutually beneficial exchanges between groups. But an attempt to seize another group's territory and resources is a zero-sum game. The invaders came as competitors who threatened the livelihood of the established residents. They would not have been warmly welcomed, to put it mildly.

The implication of this dynamic seems obvious. The human diaspora most likely involved many episodes of collective violence – the forcible displacement of the prior occupants by modern humans. However, this coercive pattern most likely would not have been successful without the advantage of a superior military technology, along with (probably) superior numbers and (perhaps) improved language and organizational capabilities. There are two strong suspects for this technological advantage: (1) spear-throwers, which greatly increased the range and accuracy of their users, and (2) a more advanced form of hafted stone axes. The immense advantage of the latter over a simple wooden club or a hand-held axe in hand-to-hand combat should be obvious. It is noteworthy that the first evidence for the use of the more advanced stone axe roughly coincides with the estimate for when the migration out of Africa by modern humans began.[91]

There are many examples of human conquests in more recent history that were based on some combination of larger numbers, better organization, and/or a decisive technological edge. Jared Diamond's Pulitzer Prize–winning book, *Germs, Guns and Steel* (1997) is a paean to the thesis that a synergistic package of historically unique ecological and cultural advantages, not biological superiority, was responsible for the rise and spread of food production (agriculture and animal husbandry) in recent millennia, which in turn laid the foundation for the rise of civilization, wars of conquest, modern imperialism, and all the rest. As Diamond puts it, "farmer power" gave rise both to new technologies (especially steel swords, body armor, and guns) and to potent new disease germs that enabled their possessors to displace (or marginalize) hunter–gatherer societies in every corner of the world.

A chilling example cited by Diamond involved the total destruction of the Moriori hunter–gatherer society on the Chatham Islands (in the Pacific) in 1835 at the hands of 900 well-armed Maori agriculturalists from nearby New Zealand. The Maori first learned of the peaceful Moriori from a transient Australian seal-hunter. Excited by the report that the Moriori had no weapons, the Maori immediately organized a seaborne invasion. When the unsuspecting Moriori did not resist, the Maori raiding party slaughtered them with impunity.[92] (Ironically, the two groups traced their ancestry back to a common Polynesian origin, but they had long since lost contact with one another.)

The Rise of the Zulu Nation

Another well-documented example of warfare as an instrument of cultural evolution can be found in the history of the Zulu nation.[93] Until the early 1800s, the people of Bantu origin who inhabited what came to be known as Zululand (part of the modern South African province of Kwa-Zulu Natal) were a disorderly patchwork of cattle-herding and minimally horticultural clans that frequently made war on one another. The most common *casus belli* were disputes over cattle, grazing lands, and water rights, but the ensuing combat was usually brief. For the most part it entailed prearranged pitched battles at a respectable distance between small groups of warriors armed with assegai (a lightweight, 6-foot throwing spear) and oval cowhide shields. Injuries and fatalities were relatively low.

However, the Bantu were hemmed in geographically, and as the human and cattle populations increased over the course of time, they began to experience increased crowding. Anthropologist Robert Carneiro calls it "environmental circumscription" (see Chapter 8). This led to a corresponding increase in the frequency and intensity of warfare between the Bantu clans until a radical discontinuity occurred in 1816 when a 29-year-old warrior named Shaka took over the leadership of the Zulu clan. Shaka immediately set about transforming the pattern of Bantu warfare by introducing a new military technology involving disciplined phalanxes of shield-bearing troops armed with short hooking and jabbing spears designed for combat at close quarters.

Shaka's innovation was as great a revolution in the African environment as was gunpowder, or the stirrup, or tanks in European warfare. After ruthlessly training his ragtag army of about 350 men, Shaka set forth on a pattern of conquests and forced alliances that quickly became a juggernaut. Within three years, Shaka had forged a nation of more than a quarter of a million people, with a formidable and fanatically disciplined army of about 20,000 warriors who were motivated in part by Shaka's decree that they were not allowed to marry until they were blooded in battle. Shaka's domain had also vastly increased, from 100 square miles to more than 11,500. There was not a tribe in all of black Africa that could oppose the new Zulu kingdom, and soon Shaka began to expand his domain beyond the Bantu peoples' traditional tribal boundaries.

The further evolution and ultimate downfall of the Zulu nation at the hands of technologically superior European armies in the latter part of the nineteenth century is another chapter (but with a similar theme). What is significant in this example is that the rise of the Zulu nation illustrates one of the truisms of human history. The "balance of power" between groups, societies, nations, and even competing species, can be tipped decisively by various combinations of larger numbers, better organization (plus intangibles like leadership, strategy, and tactics), and superior technology.[94]

Warfare is, of course, a brutal and destructive business; the days of glorifying it are long gone. Nevertheless, war-making utilizes many different kinds of synergy – synergies of scale, a division/combination of labor, functional complementarities, facilitation effects, human–tool symbioses, risk- and cost-sharing, even convergent "historical" synergies. Military preparedness and planning are, in fact, implicitly concerned with producing various forms of synergy (or negative synergy – from the enemy's perspective). To modify a famous line from General George Patton, compared to war all other forms of human synergy sink to insignificance. This may be an exaggeration, but warfare has undeniably been an important instrumentality historically for accomplishing the collective purposes of human groups. The thesis here is that our propensity for using collective violence has ancient roots. It did not arise only with the emergence of modern civilization, and it endures because it continues to be perceived as being cost-effective – or at least effective at whatever cost. Nor is our species unique in this regard; rudimentary analogues of raiding and warfare have been observed in social insects and many animals, especially our close chimpanzee relatives (though bonobos may offer an important contrast).[95]

Occam's Razor Revisited

The warfare hypothesis also conforms to the principle of parsimony (Occam's razor), I believe. War-making in the Paleolithic represented an incremental adaptive change, the application to other purposes of behavioral patterns that were developed and perfected by hominid groups in relation to group defense and offense against predators and competitors literally over millions of years. (This is another idea that has many fathers, going back even to Aristotle.[96]) Armed collective aggression for group defense/offense and for hunting/scavenging has been a major part of the hominid behavioral repertoire for at least 2 million years and possibly twice that long.

The role of warfare in the human diaspora of 40,000–50,000 years ago does not therefore presuppose the evolution of a "warfare gene," or a "killer ape." Warfare often produces immediate benefits (for the winners). And it is more likely to be rewarding to groups/tribes/nations that hold a decisive military advantage – a preponderance of power. Some indirect evidence that a prehistoric shift in the balance of power played a key role in human evolution includes the wave of "overkills" (and extinctions) of many large game animals, most likely by human hunters, as well as the rapid growth of human populations as they migrated into new areas, and the occupation by human groups of more marginal, even extreme environments (very likely under duress from superior competitors).[97]

There are, of course, other alternative explanations for the human diaspora. One is interbreeding and gene flow between the mitochondrial Eve's descendants and other populations. In other words, there was no replacement, only the diffusion of certain telltale human genes. However, this scenario seems problematic. One difficulty is that the mitochondrial DNA that has served as a "marker" is transmitted exclusively in the female line. In order for gene flow to occur between populations, the females would have to outbreed with the males of these alien groups, and so would their female offspring. This would have been a very atypical pattern, to say the least. (It is far more likely that the invading males interbred with the resident females.) In addition, the subsequent spread of these genes through various fragmented populations in different parts of the world would have required positive selection; it is highly unlikely that these genetic markers would have diffused and become "fixed" by random drift alone. Yet these markers are not known to be linked to any functionally advantageous anatomical change. Such a linkage is not out of the question, but it remains obscure.

Another alternative theory might be called the "germ warfare" hypothesis. It is now well established that the spread of agricultural populations and the wholesale displacement of hunter–gatherer societies in historical times was abetted in a major way by the spread of lethal disease germs.[98] In fact, far more people have been killed by virulent viruses than by all of the wars of human history. It is certainly possible that migrating human groups of 50,000 years ago carried destructive microbes with them, along with their weapons and other artifacts.

However, the modern-day analogue may exaggerate the role of germ warfare in prehistoric populations. The lethal diseases that enabled invading agriculturalists to decimate native hunter–gatherers in historical times were all derived from domesticated animals (and agricultural activities). Smallpox, tuberculosis, plague, measles, cholera, etc., all came from livestock. These germs were inadvertent by-products of the agricultural revolution. Their lethal effectiveness as *de facto* weapons of war lay in the fact that agricultural populations, with a head start in developing immunities, were able to transmit these deadly germs to populations that had never been exposed to them. So far as we know, such potent biological weapons were not available to our Upper Paleolithic ancestors. Indeed, invading humans were equally likely to become the victims of the resident disease germs when they moved into new environments; they would not have arrived with any particular epidemiological advantage.

In short, interbreeding and germ warfare may have been contributing factors, but it seems more plausible that technological advances – potent new forms of social synergy that produced immediate functional benefits – were primarily responsible for the spread of modern humankind out of Africa and around the world. Coercion is very likely to have played a major part in this dynamic, but it would be wrong to treat warfare as a prime mover. The ability to make war was itself the product of a synergistic package of capabilities.

More important, armed conflict is, after all, an instrumentality for attaining various ends; it is not an end in itself. The *casus belli* between these evolving hominid populations were very likely shaped by many factors, including (perversely) our very success in reproducing and expanding our numbers. Violent conflicts are more likely to occur in constricted environments, or when resource limits are reached. They are more likely when two populations are in direct competition for vital resources and there is little basis for cooperation. Even then, bloodshed is not inevitable. The odds of violence are almost always influenced by a more or less explicit calculus of costs and benefits – and risks. A shorthand slogan for this calculus is, again, the "balance of power" (or, more to the point, an "imbalance of power"). But this venerable concept implies a many-faceted analytical process, not a narrow statistical exercise. Indeed, it can be said that every war amounts to an *in vivo* test of the combatants' assumptions about the balance of power.

The Synergistic Species

There are two "common threads" in human evolution, it seems to me. One is that various forms of synergy played a key role in the process. It encompassed many new forms of social cooperation, many new tools/technologies that produced otherwise unattainable synergies, plus more powerful social communications skills, an accretion of culturally transmitted knowledge and an array of anatomical "improvements." Various synergies of scale, functional complementarities, combinations of labor, cost- and risk-sharing, etc., were the underlying functional "drivers" for the directional changes that occurred over time. Furthermore, each of these forms of synergy supported and strengthened one another; there was a synergy of synergies. For instance, the controlled use of fire was a collective good that provided many survival-related benefits, but it also required a division of labor for gathering fuel, for fire tending, for transporting hot coals from one campsite to another and, eventually, for fire-making. This in turn produced a greater degree of interdependency (as always, paradoxically).

In other words, there was no "prime mover" or megamutation that suddenly allowed slow-witted hominids to think, and talk, and invent. Instead, an accretion of many small inventions, both cultural and biological, ultimately produced a synergistic new package. A threshold was crossed, rather like the achievement of fully competent flight skills in evolving birds (Chapter 5). It was another example of the Bingo Effect. This synergistic "package" included efficient bipedalism, highly manipulative hands, large and sophisticated brains, numerous tools and technologies, elaborate patterns of social cooperation, organized collective activities, and a unique capacity to accumulate and use cultural information. (Perhaps over the millennia there were many hominid equivalents of Imo, the inventive young female macaque.) How do we know it was a synergistic package? We need only to apply the synergy test. Take away bipedalism, or our dexterous forelimbs and hands, or language, or even such important cultural attainments as fire. There is not a single one of these distinctive traits that we could do without.[99]

To put this evolutionary transformation into perspective, consider the fact that chimpanzees can do many of the things we are able to do, in a rudimentary way – walk bipedally, make tools, hunt in groups, communicate orally, make war, even invent new techniques for food-getting and other needs. Captive chimpanzees can also use symbols to "converse" and can produce "art." The difference is that we can do all of these things "more better," as my children used to say (when they were small). And this ultimately large difference of degree is the cumulative result of an incremental, interactive, and, to a large extent, purposeful process – an entrepreneurial process.

Accordingly, the second common thread in human evolution has been a dynamic of Neo-Lamarckian Selection. We progressively invented ourselves – though there was obviously no premeditated plan. (None of our ancestors could possibly have imagined where it would all lead. Nor can we even now, for that matter.) The striking biological differences between humans and chimpanzees, or even humans and the Miocene midgets of 5 million years ago, are the result of a multi-million-year process in which new forms of behavior, and new synergies, were the "pacemakers." And each major new invention redefined the context of our evolution – both the "selection pressures" and the ecological opportunities and threats.

This "self-making hypothesis" is not as radical as it may sound. In fact, it highlights what is understated (or implicit) in many other theories. In *The Descent of Man*, Darwin himself suggested that learned behaviors – including new food procurement strategies, the invention of tools, and new forms of social cooperation – played a significant part in human evolution. (Darwin even allowed for the possibility of true Lamarckism – the direct inheritance of acquired characters.) Likewise, in the "Man the Hunter" scenario of the 1960s, the role of a behavioral shift to big game hunting was stressed. The food-sharing and nuclear family scenarios of the 1970s also assumed that behavioral changes were key factors. More recently, C. Loring Brace, Milford Wolpoff, Ralph Holloway, Jonathan Kingdon, Terrence Deacon, and others have made similar arguments about behavioral innovations in human evolution. Robert Foley is even more emphatic: "Technology may be the key."

Furthermore, the role of Neo-Lamarckian Selection as a pacemaker did not stop with the emergence of modern humans. It has continued even in recent millennia. We are in fact still evolving (and imposing evolutionary changes on the rest of nature as well), though our time-horizon is much too short for us to be able to see this clearly. Some evidence can be found in the marked biological variations among contemporary human populations: climate-correlated differences in facial features, subcutaneous fat layers, body proportions, and skin color; also the unique physiological adaptations found in high-altitude peoples and adult lactose tolerance in milk-drinking populations.[100] In each case, Neo-Lamarckian "habit" changes introduced new selection criteria. More important, some human populations have been growing rapidly in recent centuries while others have not. This too is natural selection. Likewise, our often deadly impact on the rest of the Earth's biota is also, for these species, nothing other than natural selection – properly defined (see Chapter 5). In short, we have become the Sorcerer's Apprentice. The fate of the entire evolutionary experiment is now more than ever in our hands.

8 Conjuring History: Does Cultural Evolution Have an "Arrow"?

> Natural selection has no plan, no foresight, no intention.
>
> Theodosius Dobzhansky

Biological evolution may not have a trajectory, or "arrow," but many of us still cling to the belief that cultural evolution is different. Just look at our progress since the Paleolithic. Indeed, the idea of "progress" – material, moral, spiritual – is one of the oldest and most seductive themes in social theory. It seems to tap into a deep-rooted human bias toward optimism and confidence in the future, though this bright sentiment also seems to be eternally at war with "the dark side" – a more pessimistic and sometimes apocalyptic strain. Nevertheless, many social theorists over the years have viewed cultural progress as the ineluctable result of a law-like, deterministic process.

An Ancient Idea

Aristotle was perhaps the earliest and most influential representative of this tradition. Schooled as a biologist as well as a social theorist, Aristotle believed that all of nature – including humankind – is endowed with a set of capacities or forces of growth and development that are directed by their inherent properties toward predetermined ends. Just as an acorn contains within itself the "plan" for growing into an oak tree, according to Aristotle's famous metaphor, so individuals and families are by nature fitted for becoming part of a "polis" – an organized, interdependent community that may grow and develop over time.[1] Aristotle invented the term "entelechy" (or internal telos) to characterize this inner-directed property of life.

The idea of progress was also rampant in the Age of Reason. Sir Francis
Bacon characterized his era as "a more advanced age of the world, and
stored and stocked with infinite experiments and observations."[2] Blaise
Pascal viewed history as a continuous process of human learning, while the
Marquis de Condorcet, one of the so-called "Philosophes," turned the idea
of human progress into a formal sequence of ten developmental "epochs."
Condorcet also had a grand vision – later adopted by Karl Marx – of a future utopia in which material inequalities, class differences, and immorality
would be eliminated. "The time will come when the sun will shine only
upon a world of free men who recognize no master except their reason,
when tyrants and slaves, priests and their stupid or hypocritical tools, will
no longer exist except in history or on the stage."[3] But it was Auguste
Comte, originator of the "Positive Philosophy" and a founding father of
the science of sociology, who developed the idea that there are underlying
"laws" of social and mental development that every society would sooner
or later follow. Comte's notion that human history could be reduced to
scientific "laws" became what Léon Brunschwicg later called the "darling
vice" of the nineteenth century.[4]

Some natural scientists of that era also shared this progressive vision.
The most prominent of these was Lamarck – or to be precise, Jean
Baptiste Pierre Antoine de Monet (he later inherited the title "Chevalier de
Lamarck" from his father). Lamarck adhered to the Aristotelian idea that
nature is endowed with an inner, self-propelled directionality (though late
in life he changed his views). Lamarck has generally been dismissed (if not
demonized) by modern biologists for being wrong-headed, or worse, about
the mechanism of biological inheritance, but he was actually a remarkable
scientist whose impressive attainments were overshadowed by his mistaken
ideas. Born in Picardy in 1744, Lamarck studied briefly for the priesthood,
served as an army officer during the Seven Years War, then began the study
of medicine but soon switched to a career in botany and later zoology.
Ultimately, he won an appointment as a distinguished professor at the
Museum of Natural History in Paris, where he produced several landmark
studies of plants and animals, along with a method for classifying invertebrates, which later aided Darwin in his own work. Though Lamarck
went completely blind in his last years, he persevered in his research with
the help of others. He died in 1829, long before Darwin's great work was
published.[5]

Lamarck is best known today for his reasonable-sounding but now
obviously mistaken theory about the inheritance of acquired traits (see

Chapter 6). Less well known is the fact that he was also the first of many evolutionary theorists to propose what later came to be called "orthogenesis" – evolution via some intrinsic, or internal influence. According to Lamarck, the primary causal agency in evolutionary change is a "natural tendency" toward continuous developmental progress, energized by what he called "the power of life." Likened by Lamarck to a watch-spring, the power of life was seen by him as a purely materialistic agency. It involved a crude notion of an inherent energy (of which Lamarck distinguished two kinds, "caloric" and "electrical") that operates within organisms through the actions of postulated inner fluids, in a sort of hydraulic manner.[6]

This simple, mechanical formulation now seems quaintly naive. But so does Darwin's misguided theory of pangenesis, in which he postulated that migrating "gemmules" might be a means for transmitting information through the bloodstream of an organism to the germ plasm. Indeed, Darwin even cautiously accepted the possibility of Lamarckian inheritance. In this respect, he was just as wrong as Lamarck. So Lamarck at least deserves credit for the enduring idea that energy has played an important causal role in the evolutionary process. But more important, Lamarck also deserves great credit for changing his views in response to new information. As the late Stephen Jay Gould showed in a major reassessment, Lamarck in his last publication (in 1820) abandoned his orthogenetic model and embraced what amounted to a clear precursor of Darwin's own evolutionary paradigm.[7]

The Age of Spencer

Herbert Spencer, considered by many of his nineteenth-century contemporaries to be the preeminent thinker of his age, elaborated on Lamarck's early, unrevised model (along with the ideas of the pioneer embryologist Karl Ernst von Baer) in his monumental, multivolume *Synthetic Philosophy*. In an outpouring of works that spanned nearly 40 years and influenced many other theorists of his era, Spencer formulated a "Universal Law of Evolution" that encompassed physics, biology, psychology, sociology, and ethics. In effect, Spencer deduced society from energy by positing a sort of cosmic progression from energy (characterized as an external and universal "force") to matter, life, mind, society, and, finally, complex civilizations. Spencer defined evolution as "a change from an indefinite, incoherent

homogeneity to a definite, coherent heterogeneity through continuous dif-
ferentiations [and integrations]."[8]

Though an increase in complexity provides functional advantages,
Spencer asserted, the "proximate cause of progress" in human societies
was the pressure of population growth: "It produced the original diffusion
of the race. It compelled men to abandon predatory habits and take to
agriculture. It led to the clearing of the earth's surface. It forced men into
the social state; made social organization inevitable and has developed the
social sentiments. It has stimulated men to progressive improvements in
production, and to increased skill and intelligence. It is daily pressing us
into closer contact and more mutually-dependent relationships."[9]

Though Spencer is often portrayed as a "conflict theorist" (the inspiration for Social Darwinism) and is said to have reduced societal evolution to
a competitive struggle for the "survival of the fittest" (Spencer's term, not
Darwin's), actually he abhorred war and held a dualistic view. He suggested
that societies can be ranged along a continuum between two ideal types (to
borrow Max Weber's famous term), from predominantly "militant" to primarily "industrial" (economic). Where the former type had prevailed in
the past, it was Spencer's view that the latter would do so in the future,
and that the direction of social evolution was toward material affluence,
peaceful integration, personal freedom, and a withering away of the state.
Ironically, Spencer's utopian vision resembled closely that of his nemesis,
Karl Marx, though the two men obviously differed greatly over the causal
dynamics.[10]

A Passel of Prime Movers

Many other nineteenth- and twentieth-century social theorists also embraced the idea of progress – from the Social Darwinists to the Fabian
Socialists. The pioneer anthropologist Lewis Morgan, for instance, saw
a general trend toward betterment in human societies, which he divided
into three broad stages – savagery, barbarism, and civilization. Likewise,
the distinguished nineteenth-century American economist Henry George
formulated what he called "the law of human progress" based on his vision of continuous economic improvement. And in the 1940s, following
a hiatus of many years in which evolutionary theorizing fell into disfavor
in the social sciences (a complicated, ideologically polluted episode), the
anthropologist Leslie White resurrected the Lamarckian/Spencerian idea

that progress is a direct result of the ability to harness and control energy. White described his theory as the "Basic Law of Evolution" and asserted that "culture advances as the amount of energy harnessed per capita per year increases, or as the efficiency or economy of the means of controlling energy is increased, or both."[11] Calling himself a "cultural determinist," White claimed that culture evolves independently of our will: "We cannot control its course, but we can learn to predict it."[12]

White's energy-based theory influenced a generation of social theorists (Richard Adams, Fred Cottrell, and the brothers Eugene and Howard Odum come to mind). For instance, White was quoted approvingly in a popular set of anthropology textbooks in the 1980s: "Adaptation in man is the process by which he makes effective use for productive ends of the energy potential of his habitat . . . Every culture can be conceptualized as a strategy of adaptation, and each represents a unique social design for extracting energy from the habitat."[13] Most of these energy-oriented theorists shied away from White's monolithic determinism, however.

There have been a number of other candidates for cultural "prime mover" over past few decades. Though Spencer's prior claim is seldom acknowledged, population pressure has been one of the more popular ideas. In the 1960s, the anthropologist Esther Boserup proposed that population growth might have played a key role in the development of agriculture. Don Dumond saw a close relationship between population growth and cultural evolution generally. Michael Harner, inspired by the Marxist anthropologist Morton Fried, developed the idea that population growth was a major cause of internal conflict and social stratification in human societies.[14] But it was anthropologist Mark Nathan Cohen, in a closely reasoned book-length treatment, who adopted the most Spencerian stance. Calling population growth the "cause of human progress," he asserted that population pressure is an "inherent" and "continuous" causal agency in cultural evolution. "Rather than progressing, we have developed our technology as a means of approximating as closely as possible the old status quo in the face of ever-increasing numbers."[15]

Cohen also marshaled an array of circumstantial evidence against a variety of other prime mover explanations of human progress. Fair enough, but there is also a serious problem with Cohen's "default thesis." In essence, his causal dynamic, like Spencer's, is orthogenetic; population pressure is simply a given, a prior cause that needs no explanation and is ever-present – an inherent pressure. This formulation may be superficially plausible, but it takes for granted the prior question: What causes population growth? In fact, all species have the potential for exponential growth, and all species ul-timately have more or less severe constraints. Humans are not exempt from these vicissitudes. Indeed, various external factors, from wars to diseases, droughts, and famines have severely affected the size of human populations historically (as Malthus kindly pointed out).

But more important, human populations do not grow in a vacuum; they grow only in favored locations and at propitious times, when the where-withal exists for growth. And this in turn has depended on specific "adap-tations" (human technologies) in specific environmental contexts. In many parts of the globe, in fact, human populations have manifestly not grown larger over the past few millennia but have remained relatively stable, or de-clined. Thus very low population densities are found, and for good reason, in the open grasslands areas with limited water resources that constitute about 30% of the Earth's surface. There are also many cases where human populations have expanded at one point in their history and then later on declined. A reasonable alternative hypothesis is that population growth has been both a cause and an effect of cultural evolution, along with many other influences; the causal dynamic over time has been multifaceted and interactive. We'll come back to this point very shortly.

The Social Conflict School

Social conflict – internal or external – is also frequently touted as the "en-gine" of cultural evolution, and there is certainly reason to believe that violent confrontations between human groups have ancient roots, as we suggested in Chapter 7. Many theorists have claimed that warfare can also account for the evolution of "civilization," from hunter–gatherers to advanced nation-states. Darwin, Spencer, and a host of Social Darwinists stressed social conflict as a factor, but some theorists have gone even further. They attribute cultural evolution to our inherent "aggressive and acquisi-tive instincts" (shades of Raymond Dart). Sir Arthur Keith was probably the first and least-known theorist of this genre, with his *A New Theory of Human Evolution* (1949), while the writings of Robert Ardrey, Konrad Lorenz, and Robert Bigelow, among others, caused something of a furor in the latter 1960s. (Some, like Bigelow, also stressed the complementary role of cooperation.[16])

The well-known biologist Richard Alexander advanced perhaps the strongest position on this issue. In his so-called "balance of power" scenario, the process of cultural evolution is portrayed as being driven by competition between human groups, which in turn is a result of inclusive fitness maximizing behavior. In other words, war is a form of reproductive competition by other means. "At some point in our history the actual function of human groups – their significance for their individual members – was protection from the predatory effects of other human groups ... I am suggesting that all other adaptations associated with group living, such as cooperation in agriculture, fishing or industry, are secondary – that is, they are responses to group living and neither its primary causes nor sufficient to maintain it."[17] While various economic explanations for large-scale societies are neither necessary nor sufficient, Alexander claimed, warfare *is* both necessary and sufficient.

Another example of the social conflict school is the famous (and controversial) theory of political evolution proposed by anthropologist Robert Carneiro some 30 years ago. Carneiro's theory is more subtle (it relies on a functional argument rather than a presumed instinctual urge), but it too is monolithic. "Force, and not enlightened self-interest, is the mechanism by which political evolution has led, step by step, from autonomous villages to states."[18] Though state-level political systems were invented independently several times, warfare was in every case the prime mover, Carneiro asserted.

However, Carneiro's prime mover has a prime mover of its own. He argued that the "mechanism" of warfare is the product of an underlying dynamic that he called "environmental circumscription" – situations in which a population is ecologically constrained by mountains, deserts, limited resources, or even by other human populations. Once a growing, circumscribed population reaches its Malthusian limit, Carneiro reasoned, warfare and conquest may become the only alternative to starvation. So Carneiro's theory is really a theory about a predictable reaction of human groups to population pressures. "There is nothing wrong with a monocausal theory if it works," Carneiro concluded in a related monograph about the evolution of chiefdomships. "The mechanism that brought about chiefdomships is, in my opinion, the same one that brought about states, namely, war."[19]

It is certainly true that organized warfare has been a major source of synergy in evolving human societies. There are, for instance, the synergies of scale and threshold synergies associated with the number of combatants on each side, always a major factor. There are also the appalling number of technologies that we have invented for killing one another. And there are the many examples of a division (or combination) of labor – say the 5,000-member crew of a modern aircraft carrier. Carneiro's argument is also consistent with many other studies of war. The evidence is overwhelming that warfare has played a significant role in shaping the course of human history. For instance, a major analysis of this issue some years ago, which examined 21 cases of state development ranging in time from 3000 BC, to the nineteenth century AD, found that coercive force was a factor in every case and that outright conquest was involved in about half of them.[20]

But is warfare the necessary and sufficient cause of complex societies? If warfare involves grave and possibly fatal risks to the combatants, we need to probe more deeply into why wars occur. In fact, there is a vast body of research on this subject, spanning several academic disciplines, and this research supports at least one unambiguous conclusion. Warfare is itself a complex phenomenon with many potential causes and many different consequences. Wars cannot simply be treated as the expression of an instinctual urge or an uncontrollable external pressure.[21] There are too many exceptions and too many problems with any monolithic theory. Why is it that some quite warlike societies – like the Yanomamö of Venezuela or the Dani of New Guinea – did not evolve into nation-states? Why did some societies achieve statehood and then subsequently collapse or even disappear? And why did the first pristine states appear during a very small slice of time in the broader epic of evolution, within a few thousand years of one another at most? Finally, there are the cases in which population pressures were relieved by increased trade or an intensification of subsistence technologies.

In fact, there are numerous cases in which chiefdomships and states failed to emerge from a circumscribed context – when the environmental vice was in fact too tight for further expansion. Nor is warfare always correlated with population pressures. There is even some evidence of cases where population pressures and warfare were the result of economic and political integration rather than the reverse.[22] Indeed, the most decisive step beyond autonomous villages – chiefdomships – was sometimes the result of internal synergies (economic and political developments) rather than coercive force.[23] Finally, population pressure can no more be treated as a unitary cause of warfare than it can be used to account for the evolution of technology and economic progress. The prior question, again, is why populations are able to grow in some circumstances and not others? It is not a given; sometimes populations do starve without a fight. In other cases, well-fed but intensely competitive societies plunge blindly into mutually destructive wars – like World War One. In sum, warfare may have had an

important influence, but it is neither necessary nor sufficient to account for
the evolution of complex societies.

Technological Determinism

Technology has been another popular candidate for the role of prime mover
in cultural evolution. Nobody would dispute the fact that technology has
played a major role in human history, with synergies that are readily quantifiable. A !Kung San hunter–gatherer living in the African Kalahari Desert
in the 1960s extracted 9.6 calories of energy from the environment for
every calorie expended, according to the classic study by anthropologist
Richard Lee. By contrast, an American of the 1960s returned 210 calories
for every calorie invested. Since Americans worked twice as many hours on
the average as their Kalahari counterparts, they secured 46 times as many
calories per person.[24]

Many other synergies of this kind are documented in the ethnographic
research literature. A native Amazonian using a steel axe can fell about five
times as many trees in a given amount of time as could his ancestors using
stone axes, and chain saws add many multiples to the lumberjack's bottom
line. Similarly, a shotgun is at least two to three times more efficient than
a bow and arrow at bagging game on the hoof. An early farmer with a
horse and a wooden moldboard plow could turn over about 1 acre a day.
His modern-day counterparts, with specially bred "work horses" and steel
plows, can plow at least 2 acres, while a farmer with a tractor and modern
farm machinery can cover 20 acres per day, and sometimes much more.[25]

Adam Smith also provided a "textbook" example of technological
progress in *The Wealth of Nations* (1776). Smith did a comparison between
the transport of goods overland from London to Edinburgh in "broad-
wheeled wagons" and by sailing ships from London to Leith, the seaport
that serves Edinburgh. In six weeks, two men and eight horses could haul
about 4 tons of goods to Edinburgh and back, Smith found. In the same
amount of time, a ship with a crew of six or eight men could carry 200 tons
to Leith, a load that, if transported overland, would require 50 wagons,
100 men, and 400 horses.[26]

This is the very model of economic "progress" – doing something new,
or else doing the same thing better or more efficiently. Indeed, the ability
of the British to transport their manufactured products relatively cheaply
to far-flung overseas markets gave them a major competitive advantage

during the Industrial Revolution. However, the synergy produced by their merchant ships had many contributing elements: a division of labor (trained seaman), dependable oceangoing vessels (with all of their associated technologies), other essentials like capital, commercial markets, and a navy (to protect the sea lanes), and, not least, an important ecological advantage – namely, the opportunity for waterborne commerce between human settlements that were located, not coincidentally, near navigable waterways with suitable channels, tidal currents, and prevailing winds. Without these additional contributions from nature, the "technology" of merchant sailing vessels would not have been possible.

The accelerating role of technology in shaping the course of our cultural evolution can be seen in microcosm in the history of the great California Gold Rush. Over a 5-year period, from 1848 to 1853, the ontogeny of gold-mining technology in effect recapitulated our entire technological phylogeny up to that time. Contrary to the mythology that has grown up around this renowned historical episode, most of the mining activity was not done by individual prospectors – the legendary "sourdoughs" wading in mountain streams with tin pans. Within the first year, individual panning was largely supplanted by three-man teams using shovels and "rocker boxes," an innovation that increased the quantity of material that could be processed in a day from ten or fifteen buckets to more than 100 buckets, or at least twice as much per man. Shortly thereafter the wooden sluice made its appearance. Though it required six- to eight-man teams, a sluice could handle 400–500 buckets of material per day, or about twice again as much per man as the rocker box. Finally, when hydraulic mining was introduced in 1853, teams of 25 or more men were required to process the materials and manage the water pumps, hoses, etc., that were utilized to blast away the faces of entire hillsides. Yet the amount of material processed daily jumped to 100 tons or more.[27] (This episode also illustrates the fact that almost every advance in technology creates a new imperative for social cooperation and organization, another key point that we will return to shortly.)

A more contemporary example of the impact of technology – documented in some detail by Michael Rothschild in his best-selling book, *Bionomics: Economy as Ecosystem* (1990) – is the declining cost of eggs in the U.S.A. The many progressive improvements that were made over the period from 1910 to 1986 transformed a casual, labor-intensive, backyard egg-collecting activity into a mechanized system of capital-intensive, factory-like operations. As a result, the real price of eggs in constant dollars dropped an astounding 80%.[28] (A similar mechanization may be occurring currently in fish farming, a development with important future food-production implications.)

It's very tempting to view technological innovation as the "driver" of our transformation from small hunter–gatherer bands into large, complex urban societies. Many theorists – beginning with Adam Smith – have embraced this popular view. The legendary economist Joseph Schumpeter gave technology a key role in his magisterial book, *The Theory of Economic Development* (1911). Homer Barnett's *Innovation: The Basis of Cultural Change* (1953), is another landmark statement of this theme. Technological determinism has also been adopted by many historians. The leading contemporary "soft" determinist (so-called because he also recognizes other factors) is the economic historian Robert Heilbroner, whose classic 1967 essay "Do Machines Make History?" was the catalyst for much subsequent debate. In a recent edited volume on this subject, *Does Technology Drive History?* (1994), Heilbroner revisits the issue and concludes that, despite the acknowledged role of many other influences, his basic thesis is sound. "The deterministic view ... [with] its active search for regularities in, and lawlike aspects to, historical change remains the most powerful unifying capability we have."[29]

Management consultant cum economic theorist Michael Rothschild adopts perhaps the boldest position on this issue. Technology, Rothschild claims, holds "the center stage" among the forces that propel history. So powerful are its effects that, if given its head in a free-market environment, the "experience curve" of technological innovation will ultimately "obliterate" the "central myth" of the dismal science, namely the Malthusian limits to population growth – not to mention John Stuart Mill's famous law of diminishing returns.[30] I'm reminded of economist Kenneth Boulding's observation, some years ago, that "anyone who believes that exponential growth can go on forever in a finite world is either a madman or an economist."

What Is Technology?

One problem with elevating technology to the status of a prime mover is that it is not a "force," or a "mechanism." It is not even confined to tools or machines. It is really an umbrella concept – a broad label that we use to identify the immense number of novel techniques that we have devised for earning a living and reproducing ourselves. At bottom, the term refers not to "things" but to "activities" involving various behaviors, tools, objects, or even other organisms, that have been developed, fabricated, or appropriated to serve human purposes.

Some technologies are mainly a matter of deploying knowledge and skills. Thus, many agricultural practices – the use of dung as a fertilizer, crop rotation, interplanting, controlled watering regimes, and much more – are very important "technologies." Likewise, many of our common plant and animal food products are the result of countless generations of selective breeding (genetic engineering) for various desired properties – size, texture, color, nutritional content, disease resistance, and the like. For example, the oldest known ancestor of modern corn, teosinte, produced cobs that were less than 1 inch long. Literally thousands of years of artificial selection were required to get to modern corn cobs, some of which are more than 1 foot long.

Similarly, domesticated animals represent – in essence – some of humankind's oldest and most important "technologies." Dogs, for instance, provide us with a diverse range of services. We have employed them as shepherds, sentinels, hunters, warriors, policemen, beasts of burden, draft animals, drug sniffers, eyes for the blind, and much more. (A new estimate, based on molecular genetic data, indicates that dogs may have been domesticated far earlier than previously supposed, about 135,000 years ago.[31]) Birds have also been utilized in various ways by humankind. There are the famed miners' canaries that once served as living "sensors" for toxic air in mine shafts, the carrier pigeons that provided long-distance communications for generations of warriors and journalists out in the field, the African honey guides (mentioned earlier) that even today cooperate with human honey-gatherers, and, not least, the domesticated chickens that obligingly manufacture our breakfasts.

Many other human technologies involve the more or less skillful manipulation of various natural objects. We have already mentioned the role of fire, one of our earliest and still most vital technologies. The many special techniques required to gather, process, and cook various plant foods also played an important role in our evolution. The use of pits, deadfalls, cul de sacs, and other stratagems for capturing game were very likely among our early food-getting technologies. The diversion of water for irrigation purposes was a critically important step in the development of large-scale agriculture. So were dams, walls, fences, weirs, and many other early cultural innovations. In other words, technology is not really some external agency;

it's a synergistic relationship involving human knowledge, human skills, and various external objects.

A second key point about technology is that it almost always requires organized cooperative activities by humans – what Karl Marx called "relations of production." We saw this with the evolution of gold-mining technologies during the Gold Rush. Likewise, Adam Smith's famous pin factory required ten workers in a closely coordinated combination of labor. And the relations of production in a modern industrial corporation dwarf the paradigmatic pin factory. The Boeing Aircraft Corporation, for instance, until recently had 42 major facilities, 200,000 employees, and some 10,000 "suppliers" – many of them major corporations in their own right – that were scattered throughout North America and, indeed, the world. A Boeing 777 is the product of a vast cooperative effort.

A third point is that every technology is embedded in a specific environment. It is enmeshed, so to speak, in the historical flow; it is not a separate, autonomous agency but is always part of a much larger economic and cultural system. More important, both the natural environment and the historical/cultural context exert a major causal influence; they are co-determinants, as we saw with the technology of merchant sailing ships. Recall also our brief discussion of the automobile in Chapter 2, where the point was made that all of our most advanced technologies rely on many preceding technological innovations and arise only in favored historical circumstances. To emphasize this crucially important point, here are two brief examples.

Sunflower Oil

The story of sunflower oil provides a striking example of the interplay between biology, technology, economics, and politics.[32] Although there is evidence that native Americans cultivated and consumed sunflower seeds for several thousand years, when the seeds were first taken to Europe in the sixteenth century they were used primarily for ornamental purposes. However, the oil-producing potential of one variety was soon recognized, and in 1716 an English patent was granted to Arthur Bunyan for a process designed to extract sunflower oil for human consumption. But at that time there was little interest in producing sunflower oil commercially, so it was not immediately exploited. The planting of sunflowers as a crop did not take hold until well into the nineteenth century.

The first country to begin growing sunflowers for human consumption was Russia, oddly enough, and the primary reason was political. The Russian Orthodox Church had forbidden the eating during Lent of foods that were rich in oil, but sunflower oil, being recently introduced and as yet unexploited commercially, escaped the ban. Russian farmers took advantage of this loophole and for many years were virtually alone in growing sunflowers for oil. In the United States, peanut oil, cottonseed oil, and, more recently, soybean oil dominated the marketplace. Sunflowers were simply not competitive with these other crops. Their yields were not high enough; the oil content of the seeds was too low; they matured too slowly and erratically (which frustrated mechanical harvesting); and they were susceptible to rust and other plant diseases.

However, in the 1890s American agricultural researchers began a breeding program designed to improve the commercial potential of sunflowers. The program continued, with many false starts and setbacks, for 70 years. Finally, in the late 1960s, a hybrid was produced that had all of the requisite properties. It matured rapidly and uniformly; it was resistant to disease; its yields were 20% higher than the non-hybrids; its seeds had twice the oil content of its ancestors; and it could readily be harvested with the combines or the sheller–pickers used for corn. In addition, the plant residues could be used as a high-protein meal, as roughage for livestock, and as a constituent in pressed wood logs and fiberboard.

These biotechnological breakthroughs coincided with a rapid growth in the demand for vegetable oils as a shift occurred in consumption patterns toward polyunsaturated fats. The result of this historical convergence (in the truest sense) has been a dramatic increase in sunflower agriculture. Sunflower production over the past three decades has increased from negligible levels to about 2.17 million tons in 1999. It is now the world's third leading oil crop.

The Automobile Revolution

The automobile, one of the preeminent technologies of the twentieth century, provides an especially well-documented example of how technological evolution occurs.[33] The idea was already "in the air" – both in Europe and the United States – in the latter nineteenth century, as the design and efficiency of small engines improved. This led to innumerable experiments with powered road vehicles – Stanley Steamers, three-wheeled electric vehicles that vaguely resembled today's motorized wheelchairs, and many vehicles, like the popular Duryea, that literally looked like a horseless carriage. A major breakthrough came with the development of a practicable internal combustion engine that could run on gasoline. The basic principles and some early experiments date back to the 1860s, but the first successful gasoline-powered cars were produced by Gottlieb Daimler and Karl Benz in Germany in the 1880s. However, there was no rush into the market-place with this new technology. At the beginning of the twentieth century there were still only 8,000 automobiles of all kinds in the U.S.A. They were mainly playthings for the rich – and venturesome.

One reason for the slow start was that each of these machines was hand-made and retailed for $3,000 or more (about $60,000 in 1999 dollars), compared to a range of $600 to $1,500 (at most) for a horse and carriage. Another problem was that cars could not at first outperform horses for basic transportation needs. They were relatively slow (top speeds seldom reached even 20 miles per hour); their engines and brakes were unreliable; starting a car with a hand crank resulted in many broken arms; steering with a tiller was awkward and sometimes dangerous; and the solid rubber tires and stagecoach suspension systems in the early cars gave passengers a teeth-rattling ride. To top it off, the engines had to be drained and the vehicles stored during the winter months.

At first the United States seemed an unlikely candidate for the leading role it was to play in the automobile revolution. Progress in developing automobile technology had been centered mainly in Europe, and American roads were notoriously bad. In 1904, the United States had only 153,000 miles of "improved" roads (gravel or paved) out of a total of 2.1 million. As one early car-maker put it: "The American who buys an automobile must pick between bad roads and worse." On the other hand, American railroad and trolley systems were (then) among the best in the world.

Despite its slow start in developing automotive transportation, during the first decade of this century the Americans seized the leadership. While the European auto industry poked along in making improvements and expanding production, the Americans moved aggressively to realize the full potential of automotive transportation. Competition was keen, and by 1908 some 515 companies had formally entered the business, more than half of which had already failed. (There were also many backyard tinkerers who never actually produced vehicles for sale.)

In this climate, improvements came rapidly, and within a few years the industry crossed the critical takeoff threshold. By 1906, the number of registered vehicles hit 108,000. Six years later, the total had surpassed one million (thanks in large part to Henry Ford), and by 1925 there were 17.5 million cars on the road. Only the Great Depression of the 1930s and World War II temporarily slowed further growth until the peak year was reached in 1978, when Detroit produced 12.9 million cars in a single year.

The explanation for why the automobile revolution occurred when it did, and where it did, involves a unique combination of factors, a synergistic nexus that included the following:

(1) *The internal combustion engine.* Though obviously important, its superiority was not at first clear-cut and there was much more to it than that. In fact, as late as 1900 the number of steam and electric vehicles in the U.S.A. outnumbered gasoline-powered cars by 3:1.

(2) *Technological refinements.* Ultimate success also depended on a series of inventions which converted the automobile into a fast, reliable, all-weather means of transportation. These inventions included, among other things, more powerful and efficient engines, lightweight (but strong) vanadium steel for making enclosed car bodies, inexpensive, fast-drying paints for protecting car bodies from rust, better brakes, the electric self-starter, headlights, steering wheels, and, not least, antifreeze.

(3) *Balloon tires.* This was a major factor. Originally developed for bicycles, the use of balloon tires in cars eliminated a serious constraint on performance by permitting much higher speeds and a much smoother ride. In fact, some of the earliest car-manufacturing companies were spinoffs from the bicycle-making business.

(4) *Petroleum.* The availability of an abundant, efficient, and easily transported fuel supply was another major factor, along with improvements in gasoline-cracking methods during the critical early years, which more than doubled the yields per barrel and dropped the price dramatically.

(5) *Mass production.* The development of assembly-line production methods was also an important factor, because it drove prices steadily downward to as low as $290 for one of Henry Ford's Model T cars (about 10% of the average 1900 price). Ford's pioneering role in mass production was a reflection of the distinctively American development strategy, which emphasized low-cost transportation for the masses rather than status symbols for the rich. However, this strategy was in turn dependent on the fact that the U.S.A. already led the world

in manufacturing standardization, and in the use of interchangeable parts.

(6) *Raw materials.* Also important was the fortuitous fact that alone among the industrialized nations of that time America was blessed with huge quantities (at cheap prices) of such key raw materials as iron ore, coal (for steel making), wood, and especially petroleum.

(7) *Geography.* Equally fortuitous was the fact that the city of Detroit was close to all of the necessary raw materials (which were then concentrated in the Midwest) and was already a manufacturing center for all of the component technologies: carriage-making, bicycle-making, and the production of gasoline engines for various industrial uses, along with such ancillary trades as leather-working, upholstering, carpentry, and machine-tool making.

(8) *Government assistance.* Road improvements and other measures undertaken by the government also played a major role. For example, the budget for the Federal Office of Public Roads jumped from a meager $87,000 in 1908 to $500 million by 1925. (Contrary to the mythology of capitalism, government and industry were, as usual, partners in developing this major new technology.)

(9) *The marketplace.* As it happened, there was an astonishing increase both in population and in national wealth during the very period when automobile technology was reaching the breakthrough stage. Not only did the population of the United States grow by some 30 million people between 1900 and 1920 (40%), but national income rose from about $14.5 billion per year at the turn of the century to $79.1 billion in 1920. While the European nations were squandering their wealth in the carnage of World War One, America was getting rich – in part by serving as a supplier to the warring nations. (Even allowing for some inflation, per capita income rose by one-third during this 20-year period.) Furthermore, the distribution of income was much more equitable and the general standard of living in America was higher than in Europe. The result was that many more people had surplus income available to invest in a new mode of personal transportation. To put it in economics terminology, a vast potential "demand" for mass-produced automobiles was there waiting to be tapped.

(10) *Promotion and cultural influences.* Finally, automotive transportation was heavily promoted in the United States and attracted many investors and entrepreneurs. Even before the full benefits of mass production were realized, cars were perceived to be cheaper to

operate than horses and carriages. They took up much less space on congested city streets and, believe it or not, were less prone to collisions. They were also viewed as requiring less upkeep and as being cleaner. (Horses produced some 22 pounds of manure per day, which posed monumental refuse and health problems for any large metropolis.) Then there were the psychological factors: the sense of adventure, freedom, and independence that automobiles increasingly provided for their owners.

When all of these factors and more were combined, the result was a mode of transportation that was much cheaper, faster, more efficient, and even more comfortable than the horse and buggy. A doctor with a car could make five house calls when before he could make only one. A local department store could reduce the cost of making deliveries by more than 50%. A farmer could reduce the cost of shipping his produce to market from about 30 cents per ton to 14 cents per ton, and ship it over a much longer distance to boot (when the roads were good).

Clearly, there was no "prime mover." The causes of the automobile revolution were manifold. It was a synergistic convergence. Moreover, it was the combined effects – the payoffs to the users – that, in the final analysis, were responsible for determining the success of this historic development.

Technology and the Synergism Hypothesis

What all of these and many other technological innovations have in common are: (1) they arise from human needs and human purposes in a specific historical context; (2) they utilize but also modify past cultural and technological attainments; (3) they are interdependent parts of a larger synergistic system; (4) they involve highly purposeful, goal-oriented development processes, as well as many progressive improvements over time; and (5) they are subject to a cultural, Neo-Lamarckian Selection process; the outcomes are ultimately epiphenomena – the combined result of many individual user choices among the available options. Indeed, there is at least a tacit benefit–cost calculation associated with each of these individual decisions, though many other cultural influences may also contribute.

The last point above is critically important. It provides the linkage back to the Synergism Hypothesis and our theory of human evolution. In the final analysis, it is the synergies that determine the "emergence" and diffusion of a new technology; it is the payoffs that "induce" the positive selection of an innovation, in accordance with the backwards logic we talked about in Chapter 5. The wellspring of cultural innovation is organized intelligence, but it is the functional effects – the bioeconomic synergies – that shape the selection process. So the Synergism Hypothesis is applicable to the on-going process of cultural evolution as well. There is nothing law-like and predestined about this process, any more than there is a deterministic direc-tionality in the natural world. Nor is it produced by self-serving "memes." Moreover, each succeeding generation in effect reasseses the technologies that it inherits. A given technology is sustained over time by a cultural analogue of what is known in population genetics as "stabilizing selec-tion," just as various functional improvements over time are products of "directional selection" within and across each new generation of users.

By the same token, the decline of a technology – a very real and in-creasingly common phenomenon – is the result of negative, or adverse selection. To illustrate, in 1900 there were fewer than 10,000 automobiles in the U.S.A. compared with about 25 million horses. But by 1960, ap-proximately twelve horse and automobile generations later, the number of automobiles in the U.S.A. had surpassed 75 million and the horse popu-lation had declined to about 5 million. Today there are about 200 million cars and trucks on the road in the U.S.A., while horses are now used mainly for recreation and sport. And, of course, the horse and buggy is only one example of a host of "outmoded" technologies that are now found – if at all – in museums or in recreational settings. For instance, how many of us (except perhaps a few sentimental antiquarians) still use a Smith-Corona, or Royal, or IBM Selectric typewriter – an important industrial technology until 20 years ago? There are also water wheels, spinning wheels, javelins, bows and arrows, wind-up clocks, sailboats, dirigibles – the list is very long.

So technological evolution is not a one-way street. Nor do all cultures necessarily participate in the process. The automobile, for example, was developed and deployed initially only in a handful of industrial societies, not among the Tasmanians or the Inuit. Indeed, many simple "folk" societies still don't use automobiles, for reasons that are both historical and context-specific. Cultural evolution is not, therefore, the product of some intrinsic or extrinsic energy, force, pressure, or logic. It involves a dynamic, multi-faceted Neo-Lamarckian Selection process (more commonly referred to as consumers, commercial markets, and government procurement contracts) in certain favored locations. It is one "mover" of cultural change, just as population growth and warfare, or the threat of war, are movers and shakers. But the search for a "prime mover" is destined to fail.

The problem with prime mover theories, in short, is that they don't work. They may highlight important influences in cultural evolution, but they are manifestly inadequate – perhaps necessary but certainly not sufficient. This is especially apparent when you begin to ask historical questions. Why did a particular "breakthrough" happen when and where it did? And why not at some other time or place? Nor can prime mover theories account for the manifest influence of other important "movers."

But more important, societies do not change in some automatic way or follow a unilinear path. Sometimes the path even leads down hill. We talked about cultural "devolution" in Chapter 4, and traced the decline of the Roman Empire as an example. In fact, a great many civilizations – technology and all – have declined or even disappeared over the millennia, as we noted earlier. Consider the findings of a study by social scientist Rein Taagepera some years ago. Taagepera painstakingly collected and analyzed data for more than 50 historical cases of political states and empires. Not only did every one of them eventually reverse direction and collapse but there was a strong mathematical relationship between the "rise time" of the system (from 20% to 80% of its maximum size) and its duration. In other words, the faster they rise, the faster they fall.[34]

Down with Prime Movers!

"Down with prime movers!" declared Elman Service, one of the leading anthropologists of the twentieth century, in a critique of cultural evolution theory some years ago. "There is no single magical formula that will predict the evolution of every society. The actual evolution of the culture of particular societies is an adaptive process whereby the society solves problems with respect to the natural and to the social–cultural environment. These environments are so diverse, the problems so numerous, and the solutions potentially so various that no single determinant can be equally powerful for all cases."[35] Similarly, the economist Joseph Schumpeter, in a direct rebuttal to the Marxian claim to have discovered the underlying "law" of history (at least according to Marx's co-author Friedrich Engels), drew this heretical conclusion (for an economist): "Because of the fundamental dependence of the economic aspect of things on everything else, it is not possible to explain *economic* change by previous *economic* conditions alone. For the economic state of a people does not emerge simply from the preceding economic conditions, but only from the total preceding situation."[36]

One apt historical example comes from some elegant fieldwork many years ago by the anthropologist Charles B. Drucker among an isolated Philippine population.[37] The Igorot occupy a remote mountainous area of Luzon, where for centuries they have practiced irrigated rice cultivation within an awe-inspiring system of earthwork terraces, dams, and canals that were laboriously carved with simple tools out of the precipitous mountainsides. It was once thought that these massive structures, characterized by early explorers as the "eighth wonder of the world," were thousands of years old and had taken thousands of years to build. But in fact they are much more recent – the product of a heroic response to the Spanish conquest in the sixteenth century and the seizure of the choicest lowland and coastal areas, which produced a wave-like flight of the resident natives into the mountains. The Igorot people had traditionally practiced a low-intensity, seasonally shifting form of cultivation called "slash-and-burn" (or swiddening), but the sudden increase in the population density, and the demand for food, in the more mountainous areas prompted a radical change in the Igorot's food-production technology.

However, the introduction of a rice terrace farming system is only part of the story. The remarkable sustained yields achieved by the Igorot's rice paddies also depend on the constant replenishment of soil nutrients, especially nitrogen. Yet the environmental sources of nitrogen in this area are totally inadequate. The solution, and one key to the Igorot's successful adaptation, is the presence in the ponds of a nitrogen-fixing blue-green algae that lives in a symbiotic relationship with the rice plants. Respiration from the root structures of the plants generates the carbon dioxide that the algae need for photosynthesis and nitrogen-fixing. At the same time, the leaves of the plants shade the rice terrace mud, where the algae live, keeping temperatures cool enough for the algae to thrive. This in turn stimulates the growth of the rice plants. The result is extremely high productivity coupled with great ecological stability. For the past several centuries the Igorot have been able to grow almost enough staple food on a single hectare (2.47 acres) to feed a family of five.

But there is one more critical element in the Igorot system. The ancestral Igorot lived in isolated family groups that were well adapted to a shifting, small-scale plant cultivation strategy, but the adoption of the rice terrace technology required these families to coalesce into a large, integrated organization. Sustained cooperative efforts were necessary, first to design and build this remarkable agricultural system and then to utilize it, maintain it, and expand it over the course of time. Indeed, without constant weeding and repairs the system would rapidly deteriorate. Accordingly, the Igorot had to invent a social and political system to coordinate in a disciplined manner the activities of the many individual family groups. The result is that the Igorot today have a cultural system that would be unrecognizable to their pre-Hispanic ancestors. It represents an interwoven set of ecological, technological, social, and political elements – a synergistic system. How do we know it's synergistic? Just remove a single element – say the blue-green algae – and observe the consequences.

A Trajectory without an Arrow

Most evolutionary theorists have abandoned the search for cultural prime movers. Following in the footsteps of such influential mid-century theorists as Julian Steward, Robert McC. Adams, Elman Service, Kent Flannery, and others, the current generation of anthropologists have adopted more complex, interactive paradigms.[38] As the archaeologist Brian Fagan puts it in his well-known textbook, *People of the Earth: An Introduction to World Prehistory* (currently in its ninth edition): "Most archaeologists agree that urban life and pre-industrial civilization came into existence gradually, during a period of major social and economic change. Everyone also agrees that linear explanations rooted in irrigation, trade, or warfare are inadequate. Recent theories of the rise of states invoke multiple, and often intricate, causes and are frequently based on systems models."[39]

The laws of history are an illusion. They are epiphenomena – artifacts of a cumulative trial-and-success process. There is no intrinsic entelechy. To suppose otherwise is wishful thinking. And yet, many theorists retain a deep sense that history does have a meaningful trajectory. Surely there is a vast and significant difference between the small, Paleolithic hunter–gatherer groups (or the few modern remnants) and the complex industrial societies with populations of many millions that span entire continents. If the human epic does not follow a self-directed "arrow," it nevertheless exhibits a number of sustained evolutionary trends. Moreover, some of these trends have been accelerating. There is good reason to believe that humankind is being propelled toward a climacteric, a future that will be quite unlike the past – just as the twentieth century was radically different from all of its predecessors. Some theorists foresee world peace and world government, maybe just around the corner. Others see ecological disasters and growing chaos ahead. But, in any case, the future is more like a blind man's grope than a preordained script, or a drunkard's walk. (We'll come back to this point in the last chapter.)

Is there an inner logic to this evolutionary dynamic? Can we account for these progressive trends within a Darwinian/historical framework? The answer, I believe, is yes. The basic, continuing, inescapable problem for the human species, both individually and collectively, is survival and reproduction. This biological *problématique*, as the French would say, exists whether we are conscious of it, or care about it, or not, and it entails an array of ongoing basic needs. Accordingly, our survival is dependent upon a multifaceted economic enterprise that we tend to take for granted – until something goes wrong. Each human society represents, in essence, a synergistic "package" of adaptations – an array of cultural elements, practices, and technologies that work together within a given environment to meet our basic needs.

Over the course of the past 100,000 years or so, since the final emergence of humankind, there have been many progressive improvements in this package – ranging from new tools to new forms of communication and new modes of social control and "government." Many of these improvements have in turn enhanced other elements of the package. Thus the dual invention of record-keeping and money greatly facilitated trade, which encouraged the production of economic surpluses, which supported crafts specialists that were not directly engaged in food production, etc. In other words, the basic pattern underlying the progressive evolution of large, urbanized, complex societies has been an accretion of positive synergies that in turn benefited the rest of the package; there was a synergy of synergies.

The Cultural "Gear-Train"

Some years ago, Robert Carneiro developed a useful metaphor to characterize this dynamic. It seems worth repeating:

> A sociocultural system may be likened to a train of gears in which each gear represents a different sphere of culture. In the operation of this system the gears are generally in mesh. The gears differ, however. Some are larger than others, some have finer teeth, some turn faster, etc. Moreover, some are drive gears and engender motion to the others, while other gears

are passive and do not impart motion of their own, but merely transmit the motion they receive.

The gears also vary in the closeness with which they engage one another. If the mesh between any two were perfect and continuous, then the movement of one would automatically produce a corresponding and equivalent movement in the other. But in sociocultural systems, the gears never engage perfectly *or* continuously. Now and then a gear slips out of mesh and may move forward half a turn without causing perceptible motion in the others.

Yet, by and large, the train of gears moves together. A certain position of one gear is not compatible with just any position of some other gear. Thus, leaving our metaphor aside and looking at sociocultural systems directly, we cannot imagine, for example, divine kingship fitting with cave dwellings, trial by jury with percussion flaking, parliamentary procedure with human sacrifice, or cross-cousin marriage with nuclear reactors. When culture advances in one sphere, other spheres do no long remain unaffected. They tend to advance ... as a single coordinated system.[40]

Carneiro's metaphor is supported by an impressive body of ethnographic data and analyses showing that various cultural traits do indeed cluster together. The *Ethnographic Atlas*, a statistical treasure trove, contains many examples. For instance, 90% of the hunter–gatherer societies included in the *Atlas* are nomadic, whereas only 4% of the horticultural societies are nomadic. Likewise, no hunter–gatherer society that we know of has ever achieved the level of a chiefdomship, much less a state. There have also been a number of statistical studies of cultural evolution. Carneiro, in an analysis of 100 selected societies, examined the correlations among 33 politically relevant cultural traits, from the presence of a permanent headman to a professional civil service. The results were impressive. The lowest correlation was 0.65 and the highest was 0.87. In other words, various political practices do indeed tend to "go together."[41] (In a related study, a similar correlation was found between the size of a population and its degree of social and political complexity.[42])

However, it should be emphasized that these relationships imply only a loose determinism; human inventions and actions (or inactions) have also been important determinants. Anthropologist Robert A. Hackenburg's analysis of the divergent survival strategies pursued by the Pima and Papago societies of the American southwest provides a classic illustration.[43] In the years before the introduction of wheat cultivation by the Spaniards, these two contiguous societies had followed very different food-acquisition patterns. The Papago chose to remain nomadic hunters and gatherers and had incorporated relatively little in the way of domesticated foods into their diets. The Pima, on the other hand, had evolved toward a more sedentary lifestyle; they relied heavily on domesticates and had formed into a network of permanently settled villages that were marked by a substantial degree of cooperation and a well-developed political system.

Accordingly, when wheat cultivars were introduced to them, the Pima were prepared to exploit this important new food production opportunity. One problem they confronted was obtaining sufficient water to grow the wheat. So the Pima villages agreed to consolidate their scarce water supplies and to construct an inter-village water system under centralized management. The result was an agricultural boom. In contrast, the Papago were completely unable to exploit this opportunity. Indeed, many of them were eventually reduced to becoming hired labor for the Pima.

Synergy in Cultural Evolution

Some hints about the path that the human career would follow can be found as far back as 250,000–300,000 years ago, at prehistoric sites like Terra Amata on what is today the French Riviera and Torralba in Spain. Here and elsewhere human groups were beginning to aggregate seasonally into larger communities to take advantage of the potential for synergy in various cooperative efforts – in this case the coordinated killing of locally dense concentrations of large game animals (elephants, deer, boar, and others). The evidence includes masses of animal bones, fragmentary remains of large shelters, rich accumulations of highly sophisticated stone and bone tools and other artifacts that indicate intensive human occupation. This in turn suggests that there were growing synergies of scale, an expanding division of labor and more potent human–tool synergies.[44]

However, these synergies were not automatic; they were always confined to specific, favored localities. These evolutionary "hotspots" (a term coined by the distinguished science writer John Pfeiffer many years ago) arose only because they possessed all of the resources needed to meet the full range of basic human needs. There were sources of fresh water, ample supplies of firewood, access to raw materials for making tools (some of it perhaps imported) and an abundance of vegetable foods and local game animals that provided nutritional diversity and, more important, dependable alternatives to migratory animals during lean times.

Although the evidence from these ancient archaeological sites is limited and subject to differing interpretations, the ethnographic research literature contains many well-documented modern analogues. The Great Basin Shoshone are a legendary example, thanks to the pioneering research of Julian Steward in the 1930s, as well as several follow-up studies since then.[45] The Great Basin, in the American southwest, is a dry, harsh environment, and the native Americans who inhabited this area until very recently survived mainly by foraging in very small family groups. The bulk of their diet was plant foods – nuts, seeds, tubers, roots, berries, and the like. Opportunistic hunting of occasional large game animals (deer, mountain sheep, antelope, elk, bison) also played a part in their foraging repertoire, along with some fishing and the capture of small animals. But meat probably provided less than 20% of their total calories.

However, there were two important exceptions to this pattern. Sometimes, especially during the winter, several Shoshone families would gather in larger "camps" near a common resource like a water supply or a pine nut grove. Here they would share information, learn skills from one another, and find mates. But more significant, the Shoshone also gathered in larger groups of 75 or more when special opportunities arose for one of their famous rabbit drives (and sometimes antelope drives). These activities involved highly coordinated efforts with huge nets, rather like tennis nets only hundreds of feet long, that were used to encircle and trap large concentrations of prey. However, the successful deployment of this primitive technology also depended upon the "relations of production" – a division of labor between net holders and "beaters" under the leadership of an experienced and trusted "rabbit boss." As Steward truly noted, enough animals were captured by these team efforts to meet everyone's needs, whereas individual efforts would have been largely ineffective. The results were clearly synergistic.

There are many other cases of episodic hunter–gatherer collaborations to take advantage of a specific, local opportunity for synergy. Netsilik, for instance, come together in winter to cooperate in hunting seals, which must be speared (with tethered lines attached) through holes in the ice. Because the appearance of a seal at a given hole is an unpredictable occurrence, a lone hunter might go hungry on any particular day. But when a group of hunters cooperate in monitoring many holes at once, the likelihood is much greater that at least one of them will be successful. Since one seal can provide enough food for a good-sized group of people, close cooperation is synergistic.[46] (Netsilik groups also collaborate in salmon fishing in early summer and in the hunting of migrating herds of caribou in late August.) In much the same way, during the nineteenth century the native American bison-hunters of the northern Great Plains adopted seasonal patterns of aggregation and dispersion that paralleled the patterns of the animals they hunted.[47]

Or consider this contemporary example. The Nganasan hunters of northern Siberia have developed ingenious cooperative stratagems for trapping herds of migrating reindeer, which otherwise would elude and outrun them. One example of their technology was described by anthropologists Allen Johnson and Timothy Earle in their important study *The Evolution of Human Societies* (1987):

> Reindeer prefer to gather near lakes or rivers into which they can flee for safety when attacked by wolves. The Nganasan take advantage of this by using dogs to drive the reindeer into the water, where hunters in dugout canoes spear them. Another strategy uses "flags" made of poles from which strips of leather are flapping. Since the flapping intimidates the reindeer, simply by placing poles every 15 feet or so the Nganasan construct long funnel-shaped "fences" along which they drive the reindeer into corrals where hunters lie in wait.[48]

It is truly a synergistic system. The Nganasan hunters accomplish through technology and teamwork what none of them could accomplish alone.

Synergy in Settlements

A further step in the process of cultural evolution involved the emergence of more or less permanent settlements – networks of families engaged in shared food-acquisition, trade, and/or defense within a defined territory with a home base. The common catalyst for these settlements was a synergistic combination of resources, technology, and social cooperation – and interdependence. The whale-hunting Tereumiut of the North Slope provide a well-documented contemporary example.[49] These hardy people live along the Arctic coast in scattered "villages" of 200–300 members, comprised of a number of family groups. Although they join forces to hunt walruses, seals, and even inland caribou in lean times, their economy is centered on the cooperative hunting of whales each spring utilizing large whale boats. Needless to say, every aspect of this technology is highly skilled and tightly organized under the guidance of high-status boat-owners and expedition leaders called "umealiq." In a good season, a successful village might take 15 or more whales, amounting to hundreds of tons of meat and blubber. No individual could take even a single whale without help, so the results are highly synergistic.

To stretch their seasonal abundance throughout the rest of the year, the Tereumiut also collaborate in digging large underground ice vaults out of the permafrost, where they store enough food to feed themselves and their sled dogs (another vital "technology") through the winter months. Trade with inland groups of Nunamiut, who are closely related, is also an important part of their economy. In exchange for their whale oil, seal and walrus skins, pottery, wooden containers, and other items, the Tereumiut obtain caribou products and fox, wolf, and wolverine pelts for making tents, clothing, and tools. It's a synergistic system, but it's also highly contingent. Major fluctuations in the whale or caribou population are common and unpredictable. So various collaborative strategies are utilized for reducing and sharing risks. In other words, the Tereumiut have also developed a form of social insurance.

Probably the best-known examples of pre-agricultural affluent societies, verging on the regional chiefdomships that became a commonplace with the emergence of organized food production and animal husbandry, are the intensively studied native American communities of the Northwest Coast, stretching from the Olympic Peninsula in the State of Washington to southern Alaska.[50] Although the Gitksan, Nootka, Tlingit, Chilcat, Kwakiutl, and other "tribes" that inhabit this region were/are mainly hunter–gatherers and fishermen (some still follow the old ways, others do not), they live in an environment of extraordinary, yet somewhat erratic natural abundance – a dense chain of super-hotspots that can support a level of surplus wealth and cultural development that may be unique among pre-agriculturalists.

There are no fewer than eleven species of saltwater fish – cod, herring, halibut, flounder, and others – off the northwest Pacific coast, as well as sea mammals (porpoises, sea lions, sea otters, and even some whales), plus numerous shorebirds and a bounty of on-shore berries, nuts, roots, and other plants. Further inland, there are the famous salmon runs, along with numerous large game animals (deer, moose, bears, caribou, mountain goats, etc.), plus wildfowl like geese and ducks and a plethora of plant foods. Nevertheless, this food bounty was/is also somewhat unpredictable. The salmon runs may produce a jackpot one year and "snake eyes" the next.

The local stock of other fish and game animals can vary greatly from year to year, and the winters can be very lean. So the dense concentration of human settlements that have exploited these resources for many generations – sometimes in large villages numbering 500–800 members – were/are always at risk of food scarcity and even famines.

Like many other pre-agricultural societies, the people of the Northwest Coast spent much of the year procuring their food in small family foraging groups. But in certain seasons they aggregated into larger, organized groups, under a local "Big Man," or leader. The purpose was to deploy an array of specialized technologies to exploit various resource hotspots. In the spring many groups came together up-river for the annual run of the candlefish – so called because their high oil content can be rendered and stored for various human uses, including candles. In late summer and early fall the salmon runs began, and various traps and fish weirs were deployed so effectively that the local Big Man might need to intervene to prevent overexploitation. The fish were then dried on drying racks and in smokehouses and stored in earthen storage pits for the winter months. (Some other tribes, like the Nootka, engaged primarily in offshore fishing.) Wild berries were also likely to be abundant at this time of year, and these were gathered in large quantities and then dried and stored in watertight boxes for later consumption.

As winter approached, the families came together in their permanent villages to share the stored food supply and engage in a variety of crafts – building or repairing their houses, maintaining their storage sheds and boats, making tools, harpoons, fish hooks, clothes, and various decorative items, and fabricating the vitally important cedar storage boxes. There was much trade between villages at this time of year, though in the more remote past there were also frequent wars fought with battle-axes, clubs, and a primitive form of body armor made of leather and wood. Sometimes the objective was booty or slaves, but more commonly the unabashed purpose was to take land and resources away from smaller, weaker groups that were not viewed as "allies."

The success of these remarkable societies – some of which have survived right up to modern times – also depended on a highly effective social and political system with no less than five levels, from individual families to regional alliances. The system was organized around fairly intricate patterns of property ownership and economic cooperation, but the key player was the Big Man and his personal authority. It was the Big Man who was primarily responsible for organizing collective activities, for controlling the use of shared technologies, resolving internal conflicts, acting as chief ambassador/negotiator with other villages, and, in earlier times, serving as the war leader. The Big Men were also the prime movers for the famed "pot-latches," ceremonial distributions (and sometimes ostentatious destruction) of surplus wealth – foods, tools, clothing, boxes, and other goods – that, it is believed, fulfilled important symbolic and political functions.

The relative affluence of these Northwest Coast communities was based on an abundance of resources in a uniquely favored environment. But effective exploitation of these resources also depended upon organized cooperative efforts and appropriate technologies. It was really a synergistic "package." There were synergies of scale, a combination of labor, human–tool synergies, risk-sharing, information and skills sharing, and more (see Chapter 2). If it was an historically unique convergence, it was also an example of the general trend in cultural evolution toward more complex and more abundant packages of synergy, culminating in large-scale urban civilizations. Moreover, at each stage in this process, the potential for new synergies also imposed the need for more elaborate and effective forms of political organization and social control – what Peter Richerson and Robert Boyd call "workarounds" – to counteract the endemic tendency of individuals to "cheat" and pursue their own self-interests at the expense of the group.[51]

The Agricultural "Revolution"

About 10,000 years ago, at the tail end of the Upper Paleolithic, there were many signs of accelerating change in southwest Asia. Larger groups were occupying favored sites for longer periods of time. Often these sites coincided with animal migration routes, but also they typically included an abundance of other needed resources – plant materials, firewood, and, almost always, a large spring, river, or lake. These communities also pos-sessed an increasingly diverse array of refined and specialized tools, as well as a number of art objects and ornaments that were either locally made or acquired through exchanges with other groups. Though specialized hunting and eclectic food-gathering still predominated, as the climate in this region became progressively warmer and more variable, some of these groups began to exploit the increasingly abundant wild cereal grasses.

At first, this seems to have represented only a small sideline, but ulti-mately it would become the dominant theme, along with animal husbandry.

Cereal grains had the potential to provide humans with vast quantities of a basic foodstuff that could be domesticated easily and planted near permanent settlements with better soils and more abundant water supplies. Even wild grains could be harvested relatively efficiently with the right tools. (A test some years ago with flint-bladed sickles indicated that enough grain could be gathered in a couple of weeks to last a family for an entire year.) Especially important was the fact that large quantities of grain could be stored conveniently for long periods, an enormous advantage. But above all, cereal grains provided a dependable food resource.

Many progressive technological developments were required to realize these potential benefits, however. Recall Jared Diamond's thesis (Chapter 2), based on his multidisciplinary synthesis, that the breakthrough to modern, urban civilization in the Fertile Crescent about 10,000 years ago was the result of a "package" (his term) of important factors – a synergistic convergence that occurred at a unique time and place. The list of requisites included an array of domesticated, genetically altered plants and animals that provided a balanced diet (sheep, goats, emmer wheat, chick peas, olives, etc.), as well as draft animals (cattle and horses). It also required an interdependent set of technologies for plowing, cutting, threshing, grinding, food transport, food storage, and cooking. It required specialized tools for processing hides and fibers, for sewing, for manufacturing stone, wood, and bone implements, and much more. In addition, the growth of a large sedentary population required a dependable source of fresh water, a constant supply of fuel for cooking, heating, and light, long-distance trade to obtain specialized raw materials and, needless to say, reliable protection against covetous neighbors.

The ultimate result was that more food could be produced with much less time and effort in a much smaller area. One acre could feed 10–100 times more people than could hunting and gathering. Various writers have noted that a sedentary lifestyle also permitted a reduction in the spacing of births, from about four years apart among nomadic hunter–gatherers to, typically, a two-year separation among settled agriculturalists. This resulted in much more rapid population growth.

Three famous archaeological sites in the Middle East document some of the steps involved in what is usually referred to as the "agricultural revolution," though in reality, it was a gradual process of invention and improvement over a period of perhaps two to three thousand years. One site is at Netiv Hagdud in present-day Israel, where a tiny settlement of perhaps 200 people flourished for several hundred years beginning about

7800 BC.[52] Situated on a small tributary of the Jordan River, the village consisted of a number of sturdy, round mud-brick houses. The inhabitants were fully sedentary, yet their subsistence in many ways resembled the traditional hunter–gatherers. They hunted gazelle, fish, and waterfowl, and they collected some 50 species of wild plants, including wild cereal grasses that were harvested with flint-bladed sickles. Yet something new was also going on. In nearby fields that were easily tilled and close to the alluvial plain and a small freshwater lake, the villagers were raising a domesticated, genetically improved barley that was harvested and then stored in small bins for later use. Nevertheless, the population remained small, and eventually Netiv Hagdud disappeared. It is not known why.

Abu Hureyra in present-day Syria tells a different story.[53] It was first settled around 9500 BC, when a small village of several families settled in a wooded area in pit-dwellings with simple reed roofs. Much of their subsistence depended on migrating herds of gazelle (some 80% of the animal bones found at the site), which were killed *en masse* and then processed and stored for later consumption. However, the inhabitants of Abu Hureyra also collected hundreds of different vegetable foods, including no fewer than three different wild cereal grains, einkorn wheat, rye, and barley. As a result of this nutritional abundance, the village grew over time to some 300–400 people. Yet it too eventually failed. By 9000 BC, the village was abandoned. The climate had become more arid. Perhaps there was a drought. There is evidence of deforestation (partly due, perhaps, to heavy firewood consumption). But this is guesswork.

However, Abu Hureyra was settled for a second time around 7700 BC. At first the newcomers also relied on hunting gazelle and gathering plant foods. But around 7000 BC, a radical shift occurred. Within the space of a generation or two the inhabitants switched to herding sheep and goats and growing a variety of domesticated cereal grains and pulses. The village also grew much larger and its inhabitants constructed fairly elaborate rectangular mud-brick houses with multiple rooms, polished plaster floors, courtyards, and even wall decorations. In short, settled agriculture brought affluence. However, we also know that the women worked very hard grinding the grain; their knee joints bear the evidence of arthritis caused by kneeling for long periods of time. Abu Hureyra was abandoned for the second and last time about 5000 BC. Again, nobody knows why.

Finally, there was Çatalhöyük, now a huge mound (and a long-time archaeological dig) on a fertile alluvial plain in central Turkey.[54] First settled about 9000 BC, during the next two to three thousand years Çatalhöyük grew to somewhere between 5,000 and 10,000 people living in mud-brick "boxes" that were packed densely together into a single mass. Since there were no streets or doors, the residents had to climb over their neighbors' roofs to gain entry, through the roof, to their own living quarters. (The reason for this pueblo-like design is not known.[55]) Archaeologist James Mellaart, who was the first to undertake systematic excavations there, suggested that Çatalhöyük may have been the first truly agricultural community; it seemed to rely on cattle and a range of crops (wheat, barley, lentils, and peas). Çatalhöyük was also notable for its finely crafted obsidian tools and for its proliferation of paintings, sculpture, and statuettes, implying a new level of affluence and leisure time.

A new generation of researchers, led by Ian Hodder, have found evidence to suggest that Çatalhöyük residents did not rely exclusively on agriculture. Using newer, more painstaking micro-analyses, these researchers have identified an abundance of wild plant and wild animal residues. Hodder also believes that domesticated sheep may have played a more important role than cattle at Çatalhöyük. In other words, the residents were also energetic hunters and gatherers; the fertile plains and wooded steppes nearby would have been teeming with big game and wild plants (at least initially). Some of the art also depicts hunters carrying bows. Nevertheless, over time the nutritional balance clearly shifted to agriculture; traditional hunting and gathering activities were no longer sufficient to support such a large sedentary population. Why Çatalhöyük was finally abandoned is also a mystery.

All three of these ancient, long-gone settlements can rightly be viewed as trial-and-error experiments, or as unwitting stepping-stones toward larger, more elaborate, and longer-lived social experiments. There was no end or goal in mind as these novel adaptive strategies were tried out, but they set a new course that eventually led to progressive improvements, much like the emergence of sunflower oil and automobiles many centuries later. In other words, a cumulative invention and learning process was at work.

Synergy in Chiefdomships and States

The next major step in this evolutionary process involved the emergence of chiefdomships.[56] Over the course of time, dense concentrations of agricultural villages began to appear in certain favored areas – regional hotspots with especially rich soils, ample water and fuel supplies, and a favorable climate. In some cases, settlements arose close to a valued resource like obsidian (a hard volcanic rock widely used for making stone tools and weapons), or along a developing trade route. As these settlements grew in size and number, they stimulated more commerce – and more conflict. Eventually a new level of political organization (a workaround) was developed to deal with these concerns (much like our example of the Pima in the American Southwest). Regional concentrations of people and their economic activities began to be united under a centralized political system with a chief and/or ruling council and a division of labor among various officials in a specialization of roles that foreshadowed the larger bureaucracies of full-fledged states and empires. Thus, if technological innovation was one of the "motors" of cultural evolution, the emergence of "government" as a distinct level of social organization and integration was equally important.

Ever since Periclean Athens, political theorists have likened an organized society to an organism, with government being seen as analogous to an organism's brain and nervous system. Modern-day theorists have found the so-called organismic analogy to be one-sided, however. There is much evidence, both in archaeology and ethnography, that the chiefs and the "ruling classes" often benefited disproportionately from the taxes and labor that they extracted from the "working classes" (in Karl Marx's shorthand terminology).[57] They were often exploitative. Nevertheless, centralized chiefdomships were tolerated because they were more than merely parasites. They also typically provided many important services. They kept the peace and resolved internal conflicts; they facilitated (and policed) internal and external trade (which became increasingly complex); they organized public works programs; they provided centralized storage and distribution of surplus food (social insurance); and they bore the primary responsibility for defending the community against both free-lance marauders and competing chiefdoms. In other words, centralized governments were/are instrumentalities for achieving various collective synergies. (Even economist Adam Smith – the prophet of free-market capitalism – recognized the complementary role of government, morality, and ethics in a well-ordered society.)

No doubt every chiefdomship had unique elements, but there was one fundamental commonality among them; each one represented a contingent experiment. The pattern was very similar to the early days of the automotive industry. Many chiefdomships arose, survived for a while, and then dissolved for various reasons. But a small number of them were favored and evolved into even larger economic and political entities – the so-called pristine states and, later, the first empires (beginning around 2300 BC). But again, this was a trial-and-success process, not the playing out of a script, or a program, or a prime mover. And again, it involved a synergistic package of developments that typically included the following:

* Rich alluvial soils, especially in river valleys and flood plains that were close to ample supplies of fresh water.
* A warm climate and a long growing season that, in some cases, allowed for multiple crops in the course of a single year.
* Genetically improved crops that had higher yields (domesticated peas, for instance, have ten times the nutritional value of their wild ancestors).
* Irrigation systems that allowed for putting more land under cultivation for longer periods of time.
* Domesticated animals that provided a wide range of "resources" – milk, meat, fibers, tools, clothing, transport, plowing, even fertilizer and fuel.
* Record-keeping systems and a medium of economic exchange (money) that facilitated internal exchanges and external trade.
* A complex political division of labor along functional lines (i.e., tax collectors, police, soldiers, local administrators, managers, record-keepers, religious officials, and more).
* Success in waging warfare and defending the community, with all that this entailed.

How can we be sure it was a synergistic package? Just take away, say, the water supply. In fact, prolonged droughts are now favored as the explanation for the ultimate failure of many early states and empires.[58] Or take away domesticated animals. As Jared Diamond persuasively argues, the delay of several thousand years that occurred in the development of complex civilizations in areas of the world other than the "cradle of civilization" (the Fertile Crescent) and southwest Asia were due precisely to the absence of one or more parts of the total "package." (Diamond calls it the *Anna Karenina* principle, after the famous opening sentence of Leo Tolstoy's classic novel: "Happy families are all alike; every unhappy family is unhappy in its own way.") It was only after the missing parts were either independently invented or, in some cases, "imported" that other regions were also able to develop food production systems and large-scale civilizations. For example, the flowering of North America's Mississippian culture occurred as late as AD 900, only after a variety of corn adapted to a more temperate climate was introduced into the region. Likewise,

Australian agriculture had to await the importation of cool-climate crops from Europe.[59] These transitions were further examples of the "Bingo Effect."

The Paradox of Dependency

The agricultural revolution and the emergence of complex civilizations also illustrate a paradoxical aspect of human evolution, the phenomenon referred to earlier as the "paradox of dependency." The more important and valuable a resource or technology (the greater the benefits), the more likely we are to become dependent upon it; inventions are very often the mother of necessity, rather than the other way around. Thus, early on in human evolution "*Homo intermedius*" came to depend upon stone "cores" that were suitable for making flaked tools. When they adopted big-game hunting and evolved into "thirsty, sweaty animals," they also came to depend upon readily available water supplies – and salt! Later on our fire-using ancestors required a steady supply of firewood, or some equivalent fuel source. And in the Neolithic, sedentary farming populations also needed pottery clays, draft animals, obsidian, metal ores, and more.

The paradox of dependency may also shed some light on why the shift to settled agriculture occurred. It is not at all obvious that food production was a superior survival strategy, at least initially. One possible explanation for this change might be called the "overshoot hypothesis." The very same improvements in hunting and gathering techniques (and war-making abilities) that propelled *Homo sapiens* to the top of the hominid heap may also have stimulated a more rapid pace of population growth. It has been pointed out that a "founder population" of only 100 people can, with a mere 1.1% annual growth rate, produce a population of 10 million people in only 1,000 years. And 1.1% may have been on the low side for the 40,000 years or more that preceded the emergence of food production and urban civilization.[60]

One likely consequence of a rapid, sustained population increase in emerging humankind was that it led in time to more intensive exploitation of large game animals. As we noted earlier, "megameat" (as some theorists call it) became an essential part of the diet. It enabled evolving hominids to provide high-quality nutrition for themselves and their bigger – and bigger-brained – offspring. The hunters of the late Paleolithic were no longer opportunists; they were heavily dependent upon "megafauna," along with a widening array of plant foods and other sources of protein.

The ultimate result of this rapid growth pattern, most likely, was an overshoot – a wave of "overkills" and a world-wide spasm of extinctions among most of the large game animal species that were hunted by Upper Paleolithic humans. It was, in essence, a negative synergy of scale. These species and their ancestors had survived for many millions of years, through many previous climate disruptions and intensive human predation, only to disappear within a very short period of time. It is more than a coincidence that these species went extinct shortly after modern human hunters with superior technologies first arrived in different parts of the world and began to prey on these sometimes naive animals.

As noted in Chapter 7, humans are the prime suspects in this pattern of overkills, as anthropologist Paul Martin and a number of supporters have long asserted.[61] Soon after modern humans first appeared in Australia, diprodonts, giant pythons, marsupial "leopards," and other large animal species disappeared; likewise, moas, dodos, and flightless geese vanished from various Pacific Islands after Polynesians arrived; fossil skeletons of the now-extinct Eurasian woolly mammoth and woolly rhinoceros have been found with arrowheads still embedded in their rib cages; the once-plentiful European moose and bears also made a sudden exit; Asian rhinos, elephants, and tigers are long gone; and in the Americas, the late arrival of the so-called Clovis hunters, probably around 13,000 years ago, heralded the extinction of a plethora of native megafauna: elephants, horses, camels, giant ground sloths, mammoths, lions, cheetahs, and others.

In sum, if we invented the agricultural revolution that led to modern civ-ilizations, it appears that we also invented the problem that precipitated it – an overshoot that required our inventive ancestors to search for alternatives to a way of life that had been increasingly successful for at least 2 million years. Some recent studies of the agricultural revolution have called into question the long-held assumption that farming was inherently preferable. At first, it may well have been the case that farming was less efficient and less satisfying work. It was driven by necessity, not just opportunities. An example can be found in the classic study of the so-called "acorn economy" in prehistoric California by anthropologist Mark E. Basgall. Basgall showed that the consumption of acorns, a higher-cost, lower-quality food, was a late development among certain California tribes, and that new technolo-gies had to be devised in order to make the acorns edible and to provide safe storage. A key invention, therefore, was an efficient method for leaching the objectionable tannic acid out of the pulverized nuts. Basgall concludes, "acorn economies would have emerged from need rather than desire ..."[62]

Upward and Onward?

The rest of the cultural evolution story, as they say, is history. But that's the point. The evolution of human civilization over the past few thousand years has been a contingent historical process – a play that was written by the actors as they lived their parts. It was not the product of an inexorable law, or logic. Some of the story line in this play was shaped by the natural environment and the legacy of our past history, but each generation faced new challenges and new opportunities. If it appears that the human career has been following a preordained script with an intrinsic goal, or entelechy, that is because it is ultimately guided by a common purpose – survival and reproduction – and enjoys a cumulative cultural inheritance that each new generation builds upon. We are fellow-travelers in a collective survival enterprise, and the historical trends we can perceive in hindsight are the result of a 5-million-year process of cultural learning and invention that has led to many new forms of synergy. So the "arrow" metaphor is totally inapposite. A marginally better metaphor, perhaps, is a smart bomb that is tracking a moving target. Moreover, the target is always receding and forever changing its shape even as we strive to close in on it. It will always elude us. It's an unending challenge, a problem that can never be permanently solved. This is our ultimate fate.

9 　The Science of History

I believe in no fixed law of development....
I believe ... in no law of necessary development.

Charles Darwin

Imagine – if you can – an orderly, harmonious, law-governed universe, a stately celestial procession with the Earth and humankind at the very center of it. "The music of the spheres," an ancient and once venerated phrase, evokes a time long ago when this was the prevailing view of the cosmos. Indeed, the image of an overarching cosmic harmony can be traced all the way back to one of the "founding fathers" of Western science, Pythagoras of Samos (a small island in the Aegean), in the sixth century BC.[1] Pythagoras is legendary as the developer of the theorem for right triangles that bears his name ($c^2 = a^2 + b^2$), but far more important was his seminal discovery – now a commonplace – that mathematics can provide a powerful tool for understanding the physical world.

Pythagoras and "The Music of the Spheres"

The story handed down to us is that Pythagoras had a flash of insight one day when he passed by a blacksmith's shop and noticed that the anvils, when struck with a hammer, produced differently pitched sounds. We are told that Pythagoras rushed home to test his hunch. Experimenting with gut-strings and weights, he confirmed that differences among them corresponded to variations in pitch. (It's said that Pythagoras later used an instrument of his own invention called a "monochord" to measure the physical correlates of harmonic intervals with more precision.) In other words,

Pythagoras found that there are mathematical relationships "embedded" in music. More than that, Pythagoras realized that many other objects in the natural world also have mathematical properties. He reasoned that mathematics could be used to illuminate what later came to be called the "laws" of nature.

This fundamental discovery has empowered the entire enterprise of Western science – and much more besides. Pythagoras was the first of many theorists for whom the mathematical properties found in the natural world evoked a kind of religious awe. Later on he developed his important insight into a grand, mystical philosophy in which everything in nature was viewed as part of a universal order. According to Pythagoras – and a dedicated group of followers known as the Pythagorean Brotherhood – mathematics is the key to understanding the natural world. The relationships Pythagoras had discovered in musical harmony were at once a part of this cosmic order and a potent symbol of its character. As the journalist and music critic Jamie James puts it in his informative book on the subject – appropriately titled *The Music of the Spheres* – to Pythagoras and his followers "music *was* number, and the cosmos *was* music."[2] Call it mathematical mysticism.

Like so many other cults that promoted radical ideas in those days, the Pythagorean Brotherhood was eventually suppressed. A local political uprising served as a pretext. The Pythagoreans were targeted as subversives and its members were rounded up and killed. (Nobody knows if Pythagoras himself was among the victims; he may have died earlier, but his ultimate fate is uncertain.) Nevertheless, the Pythagorean vision of an orderly, rational, law-abiding, mathematical universe survived, and so did his ruling metaphor. In time, the music of the spheres became one of the leitmotifs of Western civilization. It graced the scientific work of the Renaissance; it influenced the writings of many of the early Christian theologians; and it found expression in the growing body of church music.

More than two thousand years later (in 1618), the great German mathematician–astronomer Johannes Kepler invoked the music of the spheres in the title of his monumental second treatise on the laws of planetary motion, *De Harmonice Mundi* (The Harmony of the World). Among other things, Kepler discovered that the angular velocities of the planets corresponded closely to musical intervals. In the summation of his masterwork, Kepler even limned a mystical vision that resembled the philosophy of his ancient mentor. "Henceforth it is no longer harmony made for the benefit of our planet, but the song which the cosmos sings to its lord and centre, the Solar Logos."[3] (Though embarrassed biographers have tended to downplay the fact, there are overtones of Pythagorean mysticism in the writings of Galileo and Sir Isaac Newton as well.)

The publication of Newton's majestic *Principia Mathematica* in 1687 was not only a turning point in the history of science; it led to a change in the dominant metaphor as well. In time, the music of the spheres was replaced by the more "modern" image – used by the philosopher René Descartes in his *Discourse on Method* (1637) – of the world as a well-oiled machine. The cosmos was also likened to a clockworks that was governed by the "laws" of nature – read Newton's laws. Today, needless to say, even the clockworks image has become outmoded, thanks to relativity, quantum physics, the "Big Bang," and, not least, the evidence for biological evolution. The music of the spheres metaphor now seems altogether anachronistic. It's generally treated as a concept better left to astrologers, or quasi-religious humanists, or mystically inclined musicians like Paul Hindemith. (It was Hindemith who composed an eccentric opera about Kepler in the 1930s, now almost forgotten, entitled *Die Harmonie der Welt*.)

The Neo-Pythagoreans

But if the metaphors have changed, the dream of a unified, mathematically "governed" universe remains the Holy Grail of many modern theorists, especially in physics; the underlying Pythagorean dream is alive and well. Though these latter-day visionaries may be lacking in musical accompaniment, they could perhaps be called Neo-Pythagoreans. And the high priest of this brotherhood is the Nobel physicist at the University of Texas, Steven Weinberg. In his Olympian book, *Dreams of a Final Theory* (1992), Weinberg claims that all the "arrows of explanation" in science are connected; they converge and are ultimately joined at a common starting-point down at the quantum level of reality. Weinberg tells us that physicists can already "catch glimpses" of this final theory. It will unify all of the laws of physics; it will be logically tight; it will be satisfying in its completeness; it will have unlimited validity. (Some of Weinberg's loose-tongued colleagues call it "a theory of everything.") This was Einstein's dream, Weinberg tells us. "Einstein's struggle is our struggle today."[4]

Weinberg's imperialistic vision has been echoed, surprisingly, by the distinguished Harvard biologist Edward O. Wilson. Wilson is one of the world's leading authorities on insect societies and the founder back in the 1970s of the controversial new science of sociobiology (see Chapter 5). Wilson has a deep understanding of biological evolution, yet in his recent book, *Consilience: The Unity of Knowledge* (1998), Wilson endorses what he calls the "strong form" of scientific unification. His "transcendental world view," as he puts it, is that "nature is organized by simple universal laws of physics to which all other laws and principles can be reduced."[5] Although he recognizes the awesome complexity of nature, his faith is rock-solid: "The central idea of the consilience world view is that all tangible phenomena, from the birth of stars to the workings of social institutions, are based on material processes that are ultimately reducible, however long and tortuous the sequences, to the laws of physics."[6]

Wilson's reductionism is uncompromising. He claims that such "emergent" phenomena as the human mind can, in theory, be fully understood from a knowledge of its constituent parts and their interactions. Wilson concedes that "this would require massive computational capacity," but he derides the claim by some theorists that the human mind and other such "wholes" cannot be understood by reductionist analyses alone. He calls such holism a "mystical concept."[7]

The Quest for Laws of History

Can the laws of physics, or any other deterministic principle, also explain history – the trajectory of time past, time present, and time future (to paraphrase T.S. Eliot)? Many theorists have tried. One obvious case in point is the Second Law of Thermodynamics, which has often been interpreted as a death sentence for the universe. Initially formalized by the German physicist Rudolph Clausius in the nineteenth century, this hallowed axiom of modern physics refers to the tendency for the energy in any defined system, such as a steam engine, to dissipate (it produces "entropy") and eventually to become unusable. (Entropy is really a counterintuitive concept; it measures a *decline* in the ability of energy to do work.) Clausius was the first of many subsequent theorists who freely extrapolated from the behavior of gas molecules in an enclosed box, or water molecules in a pan of heated water, to the dynamics of the entire universe. "The energy of the universe is constant," Clausius intoned in his magnum opus *Abhandlungen über die Mechanische Wärmetheorie* (1864), while "the entropy of the universe tends towards a maximum." In other words, the ultimate "heat death" (*Wärmetod*) of the universe, in Clausius's dour image, is only a matter of time. If the universe were in fact a clockworks, the clock would be slowly winding down.[8] (I will come back to this famous but debatable prediction in the last chapter.)

Biological evolution has also attracted the Neo-Pythagoreans; over the years a succession of theorists, often oblivious to one another (so it would seem), have proposed a variety of deterministic ordering or energizing principles – or "laws" of history. Some go back to the Enlightenment itself. There is, among others, zoologist Jean Baptiste de Lamarck's "power of life" (1809), the English polymath Herbert Spencer's "universal law of evolution" (1857), the French philosopher Henri Bergson's "*élan vital*" (1907), the German philosopher Hans Driesch's "*Entelechie*" (1909), the philosopher–priest Pierre Teilhard de Chardin's "Omega point" (1959), the biologist Pierre Grassé's "*idiomorphon*" (1973), and the child psychologist Jean Piaget's "*savoir faire*" (1978).[9] Most of these grandiose visions have been dismissed by mainstream scientists as "vitalism" – a class of explanations that, like Creationist theology, fall outside the scientific pale. The distinguished twentieth-century philosopher of science, Karl Popper, called it "historicism" and attacked such pretensions in a famous book, *The Poverty of Historicism* (1957).

However, in the past few years the dream of a law-like evolutionary process has been resuscitated inside the pale; a number of scientists (particularly in physics and biophysics) have advanced a variety of order-creating principles that have gained at least a respectful hearing. Backed by appropriate mathematical treatments, these paradigms are variously referred to as "structuralist theory," "self-organization theory," "dynamical systems theory," "entropy theory," and, lately, "complexity theory."

If there was a prophet who inspired this quest, it was the Nobel physicist Erwin Schrödinger in his now-legendary book with the provocative title, *What Is Life?* (1945). Based on a set of invited lectures at Trinity College in Dublin during World War Two, Schrödinger's book portrayed a living organism as being, quintessentially, an embodiment of thermodynamic order. Living systems manage to circumvent the relentless tendency of energy to dissipate (to become entropic) by creating "negative entropy," as Schrödinger called it. Organisms are able to do so, he explained, by "extracting order" from the environment. He described organisms as "sucking orderliness" from the universe, and he identified as their most distinctive feature their "well-ordered" thermodynamic state. Although it is often said that organisms feed upon energy, Schrödinger declared, this is "absurd ... what an organism feeds upon is negative entropy."[10]

Unfortunately, this formulation is equally absurd. The term negative entropy means, literally, an absence of an absence of order. Negative entropy is merely a convoluted synonym for available energy, or energetic "order." In fact, Schrödinger defined negative entropy mathematically as the reciprocal of physicist Ludwig Boltzmann's famous statistical expression for entropy. So the term has no independent status. A more serious objection, though, is that Schrödinger in effect answered his own portentous question about the nature of life by reducing its rich complexities to a one-dimensional problem in thermodynamics and physical order. Energy metabolism is only one facet of the many-sided problem of earning a living (and reproducing) in the natural world. Information, for instance, plays a critical role. So negative entropy is a mathematical caricature. It is not, after all, a more precise, "scientific" description of a living organism.

But this is hindsight. In the post–World War Two era, Schrödinger's dream of reducing living systems to the laws of physics was an exciting prospect; it served as an inspiration for many other scientists. For example, the eminent biophysicist Harold Morowitz, in his influential and still useful volume, *Energy Flow in Biology* (1968), proposed that the evolutionary process has been "driven" by the "constant pumping" of energy inputs, mainly from the Sun. He called evolution the "necessary" result of our perpetual sunbath. "The flow of energy through the system acts to organize that system ... Biological phenomena are ultimately consequences of the laws of physics."[11] (Later on in his book, however, Morowitz conceded that other principles and processes are also important; while energy flows are necessary, they are not sufficient.)

The Law of Evolution?

Another well-known and widely cited Neo-Pythagorean is the Nobel physicist Ilya Prigogine, whose "universal law of evolution" (as he calls it) echoes Herbert Spencer's vision in more than just its name. As noted in Chapter 8, Spencer's "law" postulated a spontaneous, energy-driven trend, a cosmic progression from energy (an external and universal "force") to matter, life, mind, and finally, complex civilizations. "From the earliest traceable cosmical changes down to the latest results of civilization," Spencer wrote in an early tract called "The Development Hypothesis" (1852), "we shall find that the transformation of the homogeneous into the heterogeneous is that in which progress essentially consists."[12] Spencer believed that simple, homogeneous systems are inherently less stable than those which are more differentiated and complex. Although he didn't explain why, Spencer asserted that more complex forms are functionally more "advantageous." (This claim would now be challenged; any advantages are always contingent and context-dependent, and are always net of the costs.)

Prigogine's theory, like Spencer's, is also energy-centered. He characterizes living systems as "dissipative structures" that feed upon and "dissipate" energy in order to maintain their structural order. What Prigogine adds to Spencer's formulation is a proposed evolutionary mechanism. In "open" thermodynamic systems, which are subject to continuous energy inputs, evolution may occur "spontaneously," Prigogine claims. The reason is that, if the energy flows are sufficiently abundant, they may produce structural "instabilities" which may, in turn, produce perturbations or "fluctuations" in the direction of greater complexity. Prigogine refers to this dynamic as the principle of "order through fluctuations," and he characterizes a living organism as "a giant fluctuation stabilized by exchanges of matter and energy." He calls it an "autocatalytic theory."[13]

Dissipative structures certainly do exist in nature. The example cited *ad nauseam* by the followers of Prigogine is the so-called Bénard cell. Named for the physicist who first observed it at the beginning of the twentieth century (Henri Bénard), this well-documented phenomenon involves the spontaneous formation of hexagonal convection cells in a pan of shallow water when it is heated slowly and uniformly. Similar heat-generated geometric patterns occur in a variety of natural settings, like deserts, the arctic, and in the atmosphere. And convection cells are only one example of self-ordering behavior. The photogenic Belousov–Zhabotinski reaction, named for its co-discoverers, is another well-known phenomenon; the spreading concentric waves and expanding pinwheels that occur in certain kinds of chemical reactions are frequently used to illustrate books and articles about self-organization. So too are the non-linear "strange attractors" and "fractal patterns" that can be rendered with colorful computer graphics.

The physicist Herman Haken, founder of the science of "synergetics," has also identified many kinds of self-ordering behavior. One especially important example (described earlier) is lasers, where stimulated photons form "coherent," synchronized light waves that are concentrated in tightly focused beams. The aptly named "hypercycles" – the self-generating (autocatalytic) chemical cycles identified by the physical chemist and Nobel laureate Manfred Eigen – represent yet another example of a self-ordering process.

The problem is, none of these important phenomena, which require a relatively new kind of non-linear mathematics even to describe them, can provide answers to Erwin Schrödinger's tantalizing question: What is life? Many theorists assume that there is no fundamental distinction; living systems are merely an elaboration of these well-documented patterns of physical order. Schrödinger likened the genome to an "aperiodic" (irregular) crystal. Physicist Per Bak likens living systems to sand piles.[14] And Prigogine compares glycolysis with Bénard cells. However, glycolysis is an intricate and highly orchestrated biochemical process. It involves a series of precise sequential steps with multiple energy transfers and nine different reactions, each catalyzed by a specific enzyme. Prigogine also suggests that there is an analogy between Bénard cells and the complex, feedback-controlled behavior of slime molds (*Dictyostelium discoideum*). These one-celled creatures forage as independent units, but on signal they can congregate into a single mass and transform themselves into a unified reproductive "fruiting body."[15] On close inspection, these analogies blur a crucial distinction between physical order and the "purposeful," information-driven, functional organization of living systems – a "combination of labor." (Elsewhere I argue that, to avoid confusion, the term "organization" should be reserved exclusively for such purposive, cybernetic processes, and that everything else in nature should be characterized merely as "ordered" – or "self-ordered.")

Order for Free?

Perhaps the most highly touted and articulate of the new generation of Neo-Pythagoreans is the biologist Stuart Kauffman. One of the leaders of the Santa Fe Institute, the high-powered think tank that serves as corporate headquarters for complexity theorists, Kauffman is inspired by the vision that mathematics will ultimately reveal the underlying dynamics of nature. In his 1995 book, *At Home in the Universe*, Kauffman writes: "I shall argue in this book that ... order is not accidental, that vast veins of order lie at hand. Laws of complexity spontaneously generate much of the order of the natural world."[16] "Order, vast and generative, arises naturally," he claims.[17] He also calls it "order for free."

Natural selection is not irrelevant to the process of evolution, Kauffman says, but it should be pushed into the background as an agency that merely provides "fine tuning" and "modest improvements" in the order that arises "spontaneously" in nature; he calls it a "deep theory" of self-generated evolution. In a comment that could be passed off as a quote from Pythagoras, Kauffman reasserts the mystical notion that the underlying order in nature "is rooted in mathematics itself."[18] Like his ancient forebear, Kauffman proclaims that he is pursuing "the ideal of reductionism in science."

Kauffman's treatment is lively, informative, and elegantly written. His scholarly credentials are also impeccable. Some aspects of his vision are highly original and intriguing, including his "scenario" for the origins of life as a collectively autocatalytic "phase transition." Recall Kauffman's analogy of 10,000 buttons scattered on a hardwood floor that are gradually tied together in pairs until, at some point, the pairs begin to be connected. Then, at some critical stage in the continuing process, the buttons will all be tied together into a single interconnected web. "The analogue in the origin-of-life theory," Kauffman explains, "will be that when a large enough number of reactions are catalyzed in a chemical reaction system, a vast web of catalyzed reactions will suddenly crystalize. Such a web, it turns out, is almost certainly autocatalytic – almost certainly self-sustaining, alive." Accordingly, Kauffman believes that the emergence of life should be "expected ... almost inevitable."[19]

Kauffman's argument is seductive, but it remains a promise unfulfilled. Like Prigogine and many other proponents of the laws of history, Kauffman relies on analogies – the Great Red Spot on the planet Jupiter, Bénard cells, the Belousov–Zhabotinski reaction, Per Bak's sand piles – as well as some examples of self-organizing patterns in biological processes (like heartbeats). Kauffman also finds some "hints" of ordering principles in development, in the immune system, and in ecosystems. But, in the final analysis, these represent only "clues." They do not provide direct evidence that living systems as complex wholes, much less the larger process of evolution, are primarily the product of a law-like autocatalysis. At best, these hints and clues can only whet our appetites for what Kauffman concedes are the "laws of emergent order, *if* they should someday be found" [italics added].[20]

For the record, it should also be noted that, in his most recent book, *Investigations* (2000), Kauffman moves toward the middle ground. He now acknowledges an interplay between self-organization and selection. More important, he focuses on the role of "autonomous agents" – "organizations" of matter, energy, and information – that are active participants in what he characterizes as an emergent, "creative" process. (Sound familiar?) Yet Kauffman is still searching for overarching "laws." He speculates that there may be a "fourth law of thermodynamics" – an inherent trend in nature toward greater diversity and "the persistent evolution of novelty in the biosphere." He also stresses what he calls "collectively autocatalytic" phenomena (a.k.a. synergy).[21]

Self-Organization

A word is also in order here about what has come to be known as self-organization theory. In their new book on *Self-Organization in Biological Systems* (2001), biologist Scott Camazine and his co-authors usefully define self-organization as a term that encompasses a broad array of pattern-forming physical processes – from sand dunes to fish schools. Self-organizing systems are distinctive in that they are *not* organized by some outside force or agency; the causal dynamics are "internal" to the parts (or participants) and their interactions.[22] This seems straightforward enough, and it presents no inherent conflict with Darwinian theory (differential selection among functional variants), or the bedrock principle that coded "instructions" in the genome play a primary directive role in the ontogeny and life-history of living systems. Nor does this conflict with the idea that the genome might use simple rules or procedures, and co-opt basic physical principles, to generate structural complexity.

Nevertheless, many enthusiasts have assumed that self-organization and natural selection are opposed to one another, and that evidence for self-organization somehow diminishes the role of natural selection. Kauffman is one. Another is biologist Brian Goodwin, who points to the marine algae *Acetabularia* as an example of a living organism that seems to follow a simple mathematical principle as it develops a stalk and various branches from a single giant cell.[23] However, as John Maynard Smith has pointed out, genetic "instructions" and "feedback" are also involved in the growth of *Acetabularia*. Moreover, it is likely that the evolution of *Acetabularia* was influenced by differential selection among functional variants. So *Acetabularia* does not undermine Darwinian theory.[24]

In a similar vein, entomologist Robert E. Page and philosopher Sandra D. Mitchell proposed that a division of labor can "emerge" from living systems without the influence of natural selection.[25] This claim overlooks the fact that the emergent effects produced by a division/combination of labor are subject to differential selection in relation to the synergies they produce. They are not exempted from natural selection. Camazine and his colleagues summed it up nicely: "There is no contradiction or competition between self-organization and natural selection. Instead, it is a *cooperative 'marriage'* [their emphasis] in which self-organization allows tremendous economy in the amount of information that natural selection needs to encode in the genome. In this way, the study of self-organization in biological systems promotes orthodox evolutionary explanation, not heresy."[26]

The Law of Higglety Pigglety

The antithesis of the Neo-Pythagorean quest for deterministic laws of history is what has been called historical particularism. History, in this paradigm, is characterized as being a relentless stream of unique events that are often disconnected, and certainly not law-like in their behavior. Sometimes history may even seem like "just one damn thing after another," in Winston Churchill's mocking caricature. The twists and turns of the historical narrative, it is said, are inherently unpredictable. And the results often seem to be purely a matter of chance, or luck, or serendipity. For instance, nobody could have predicted the asteroid (dubbed Chicxulub) that, most likely, wiped out the dinosaurs and many other species some 65 million years ago. And that catastrophe may have been only one of several mass extinctions over the course of evolution.

Darwin's theory of evolution is often accused, especially by its Creationist opponents, of being based on "random" influences. While mutations are not strictly speaking random (they are highly canalized), it is true that biological evolution is an historical process in which the most important shaping "mechanism" – natural selection – is always highly contingent and context-specific. As Darwin himself put it in *The Descent of Man* (1871), any evolutionary innovation depends upon many "concurrent favorable developments" that are always "tentative."[27] Moreover, the products of evolution suggest anything but a harmonious natural order. (One of Darwin's detractors characterized his theory as "the law of higglety pigglety.") Random, even capricious events have greatly affected the course of life on Earth, and we can no longer view our own species as the anointed culmination of the process. The Nobel Prize–winning biologist Jacques

Monod, in his famous book, *Chance and Necessity* (1971), described evolution as "chance caught on the wing."[28] And his Nobel collaborator François Jacob, in an often-cited article in the journal *Science*, characterized evolution as a process that is more like *ad hoc* "tinkering" than preplanned engineering.[29]

In fact, the very notion of some overarching form of "progress" in evolution (in the normative sense) is now widely criticized by biologists. We just happen to be one twig on one branch of a luxuriant bush of living organisms that traces its ancestry back at least 3.5 billion years – an inconceivably long time. Yet, from a different point of view, we also happen to be the winners, for the moment, of a relentless evolutionary steeplechase. To Richard Dawkins, it all adds up to "nothing but pointless indifference." A doleful Jacques Monod concluded: "Man knows at last that he is alone in the universe's unfeeling immensity, out of which he emerged only by chance."[30]

The high priest of this paradigm – also rather surprisingly – is the late Stephen Jay Gould. As we noted earlier, in *Full House* (1996) Gould suggested that the history of life on Earth resembles a "drunkard's walk" – an essentially random process without any inherent directionality. Evolution has been shaped by "the luck of the draw," as he put it. Gould highlighted this disparaging vision with a metaphor. If we could "rewind the tape" of history and replay it, he said, the outcome would be very different. "Our persistent attempts to build abstractly beautiful, logically impeccable, and comprehensively simplified systems always lead us astray.... Evolution follows the syncopated drumbeats of complex and contingent histories, shaped by the vagaries and uniqueness of time, place, and environment. Simple laws with predictable outcomes cannot fully describe the pageant and pathways of life."[31]

Is the "Science of History" an Oxymoron?

I submit that the truth lies in the middle. Science and history need each other. Evolutionary history is not simply an expression of a clutch of deep "laws." Nor, at the opposite extreme, is it an aimless narrative – a disconnected series of entries in a cosmic Day Timer. The vision of a random, chaotic historical process is just as one-sided as is the image of an all-encompassing law, or laws of everything. There is a middle ground between these two paradigms that can perhaps be characterized as the "science of history."

Although this term may sound like an oxymoron – a juxtaposing of two contradictory ways of viewing the universe and our place in it – it most definitely is not. As Jared Diamond (among others) has pointed out, the scientific method is not confined only to the search for deterministic, or even probabilistic (statistical) laws. The term "science" is actually derived from the word *scientia* which in turn traces to the Latin word *scire* – "to know." Science also plays a vital role in trying to understand historical phenomena.

In a broad sense, science is concerned with the pursuit of reliable objective "knowledge" about the properties of the natural world and its multitudinous cause and effect relationships, whether it be about repetitious phenomena or unique (historical) events. (A discussion of this issue can be found in a recent article by the philosopher Carol Cleland.[32]) Evolutionary biology provides a preeminent example of a science that is grounded in history, but so are ecology, paleontology, geology, climatology, and cosmology, among others. The great twentieth-century evolutionary biologist Theodosius Dobzhansky pointed out many years ago that evolution via natural selection is really an "anti-chance" theory. It involves a cumulative historical process in which novelties of various kinds are converted into stable functional designs that persist over time. As Darwin himself expressed it, "when we regard every production of nature as one which has had a history ... how far more interesting, I speak from experience, will the study of natural history become."[33] In effect, evolution has been a cumulative learning process, and there have been innumerable "progressive" improvements over time in relation to the needs and functional capabilities of living organisms. "Progress" in this limited (engineering) sense of the term is endemic in evolution. To cite just one example, there is a potent synergy of scale in relation to energy consumption in living systems. As organisms increase in size, they become ever more energy-efficient; they use proportionately less energy per unit of body mass.[34]

The key point here is that evolutionary history has also played an important causal role in shaping the character of the biosphere. The course of evolution cannot be reduced to a law or laws of history, much less to the laws of physics. To reiterate, biological evolution has been shaped by the combined influence of four distinct influences, chance, necessity, teleonomy, and selection (the differential survival and reproduction of different functional designs, along with the information required for reproducing them). In short, a science of history must also account for the shaping influence of history itself, inclusive of natural selection.

Emergence and the Rules of Chess

"How do living systems emerge from the laws of physics and chemistry?" asks the well-known computer scientist/psychologist John H. Holland in his recent book, *Emergence: From Chaos to Order* (1998). "Can we explain consciousness as an emergent property of certain kinds of physical systems?"[35] Holland, who is one of the leading figures at the Santa Fe Institute, concedes that emergence can be defined in various ways, and that many aspects remain "enigmatic." He confines his attention only to the sub-set of systems that, he says, are "rule-governed" – like chess – systems in which "a small number of rules or laws can generate surprising complexity." Of course, laws don't "generate" anything or "govern" anything. They identify regularities in relationships, and in the patterns of causation in dynamic processes; we characterize these formulations as scientific "laws" because they allow us to generalize and make predictions.

But that aside, Holland's example serves to illustrate the basic problem with the Neo-Pythagorean approach to explaining evolution, and especially biological complexity. Even in chess (to use Holland's example), you cannot utilize the rules to predict "history" – i.e., the course of any given game. Indeed, you cannot even reliably predict the next move in a chess game. Why? Because the "system" involves more than the rules of the game. It also involves the players and their unfolding, moment-by-moment decisions at each choice point among a very large number of available options. The game of chess is inescapably historical, even though it is also constrained and shaped by a set of rules, not to mention the laws of physics.

So too, the evolutionary process has been cumulative and "interactive" – more like a chess game when the system has been properly specified. The "Epic of Evolution" (Edward O. Wilson's inspired term, oddly enough) represents a confluence of many independent forces, preconditions, and past events that have combined to produce an ongoing stream of effects. Even if every component – every variable – in that stream is fully "determined" by non-random material causes, each discrete moment of time is characterized by a unique configuration of influences. The variables combine to produce a distinct causal "event" that is inescapably novel; it is shaped by all the events and interactions that preceded it. Recall how, in Chapter 2, this was characterized as a synergistic "convergence" because the total "package" of determinants may be causally unconnected until the instant at which they interact – rather like two cars in a head-on collision (to use a gruesome example). Engineers refer to such phenomena as being "loosely coupled" as opposed to their engineering designs – "tight coupling."

Wilson claims that we lack only sufficient computational capacity to elucidate the workings of a complex system like the human mind. The problem with this formulation is that the human mind is not a disembodied physical entity, or a mass-produced machine with interchangeable parts. Each mind is also the product of its particular history – its distinct phylogeny, its unique ontogeny, and its ongoing, moment-by-moment interactions with the environment. Molecular biology and neurobiology – however important – can illuminate only some of the levels of causation in the life of the mind. Indeed, the conceit that we could reduce and subdue the tangled skein of history to the laws of physics if only we had sufficient computing power is, to put it charitably, naive. It presupposes omniscience.[36]

A footnote to this point concerns the unprecedented "hype" associated with the recent publication of an ambitious book by the physicist cum business executive Stephen Wolfram. In his bulky, 1300-page tome, *A New Kind of Science* (2002), Wolfram makes some breathtaking claims. His "revolutionary" discovery, as he puts it, is that "rules based on mathematical equations can be used to describe the natural world." He calls it "one of the more important single discoveries in the whole history of theoretical science." And he claims that it solves "the single greatest mystery of the natural world. How come complexity."[37] Simple rules (algorithms) and computations using those rules are the answer, he says. He tells us that he can use this strategy to describe any system.

Bold words, though hardly without precedent. In fact, Wolfram's name can be added to the long list of Neo-Pythagoreans whose visions have ignored, or denied, the historicity of history – its immense, unruly autonomy. The basic problem with Wolfram's approach can perhaps be illustrated with a metaphor. Imagine a huge pond, ringed with people who are all throwing stones into it, each at will and in their own way. If only a single stone were being thrown into the pond, Wolfram might well be able to compute the ripple effect produced by the stone, if he knew the stone's shape and mass, the force exerted to throw it, its trajectory, the altitude and wind conditions, and other variables. He might even be able to calculate the interactions between several ripple patterns if he knew all the details about when, where, and how each stone was being thrown into the water. But neither Wolfram nor anybody else could know in advance how, when, and where all of the many stones will be thrown into the pond and how their innumerable ripple patterns will interact. Such autonomous "interactions" are at the crux of the historical process, and our mathematics is not omniscient any more than we are.

Wolfram claims that Darwin's theory cannot explain biological complexity, while his ideas can. Actually, he's got it backwards. The Synergism Hypothesis provides an explicitly Darwinian explanation for complexity that encompasses the combined influence of chance, necessity, teleonomy, and the differential selection of functional properties. On the other hand, Wolfram's mathematical tools are blind to at least two of these four causal agencies, teleonomy and natural selection. His notion of biological complexity is a zebra's stripes, a formulation that ranks right up there with sand piles, Bénard cells, and other simplistic metaphors for living systems. This will not do. (For more on Wolfram, see the Afterword.)

Molecular Computers

A similar mind-set (meme?) has infected molecular biology, where one eminent biologist, Sidney Brenner, claims that, if he were given the complete DNA sequence for an organism and a powerful enough computer, he could compute the organism.[38] As the distinguished geneticist Richard Lewontin points out in his new book, *The Triple Helix* (2000), the metaphor of a computer, now commonly used both in molecular biology and in brain research, is a throwback to Descartes's machine image. It's "bad biology," Lewontin says. Deterministic, machine-like models of biological processes are fundamentally flawed. Living systems are profoundly shaped by random events, by context-dependent influences, and by the nature of their interactions with the surrounding conditions. For instance, the particular three-dimensional folded structure assumed by a protein, which determines its functional properties, depends upon the specific external conditions it encounters during the fabrication process. In other words, a protein is not completely specified by its amino acid sequence. As Lewontin puts it, biological systems are shaped by a "nexus of weakly determining forces," including such things as its initial conditions, historical contingencies, and reciprocal cause-and-effect relationships between an organism and its environment(s).[39]

In sum, history matters – a lot. Evolution is not simply an epiphenomenon of a few deep laws. (A healthy dose of jaundice, so to speak, can be found in Nancy Cartwright's illuminating 1993 book, *How the Laws of Physics Lie*.) To illustrate, in physicist J. Willard Gibbs's classic formalization of the Second Law of Thermodynamics, no fewer than 11 real-world constraints, including gravity and various other physical forces, have to be set aside. In other words, if you want to use Gibbs's "law" to predict what will happen in the real world, you have to relax all those constraints and deal with the mess that results. That's what engineers, as opposed to theoretical physicists, have to do every day.

Every historical context is inescapably burdened, constrained, and shaped both by the laws of physics and by history. Biological processes – from the intricacies of the double helix to photosynthesis, the homeobox gene complex, the cytoskeletons inside each eukaryotic cell and the elegant electrochemical cascades produced by neurons – are historical "inventions" that, in turn, have shaped the entire future course of evolution. The Nobel biologist Francis Crick coined the term "frozen accidents" to describe these developments, but it might have been more felicitous to call them "procreative accidents." Not only have these ancient inventions been highly "conserved" over time, as the biologists like to say, but they have in turn supported many later inventions (see Chapter 3).

By the same token, even a relatively simple ecosystem is influenced by many interacting co-determinants – geology, climate, local weather conditions, past activities by living organisms, and much more. A coral reef, as we noted earlier, is an historical artifact – an elaborate structure erected in certain favored ocean locations by many generations of coral polyps and their symbionts. The ancestral polyps are long gone. However, their limestone skeletons provide a stable platform for their descendants (unintentionally, of course), plus a habitat for an immense variety of other living species. Some of these species in turn help to support the coral polyps (also unintentionally) by providing a portion of their food supply.

The pioneer geneticist Sewall Wright developed a metaphor for this paradigm many years ago that I will take the liberty of embellishing.[40] There is always a high degree of determination involved in the flight of an airplane, as a consequence of physical laws and aerodynamic constraints, not to mention various engineering requirements and the technological state of the art. At the same time, purposeful elements are involved in the decision to build the airplane in the first place, as well as the precise design of the plane and the specific choices regarding where, when, and how the plane will be flown. The actual flight history of the plane might also be influenced by such unpredictable (or marginally predictable) factors as weather and wind conditions, malfunctioning components, the flight plans of other aircraft, and possibly even the actions of hijackers and terrorists.

So what "causes" a particular airplane to be at a particular airway inter-section at a particular date, time, and altitude? The answer entails far more than the laws of physics. The Pulitzer Prize–winning historian David M. Kennedy wryly observes: "As historians never tire of saying, each historical moment is unique and must be understood in its own terms."[41]

Simple Models and the Hierarchy of Nature

A second problem with the Neo-Pythagorean paradigm is that it discounts the fact that the material world, and living organisms in particular, are organized hierarchically (some prefer the novelist Arthur Koestler's term "holarchy") and that new principles, and new capabilities, arise at each new "level" of organization. The simple, one-level mathematical models that are often used to simulate real-world systems are thus blind to a bedrock property of nature. This problem was discussed many years ago in two classic articles that appeared four years apart in the pages of *Science*, America's premier scientific journal. Both of these landmark appeals for a multileveled scientific enterprise categorically rejected the "strong" reductionist program.

In the first of these articles, "Life's Irreducible Structure" (1968), the noted chemist Michael Polanyi pointed out that the hierarchy (holarchy) of levels in the natural world can be identified in terms of distinct "boundary conditions" that impose more or less stringent constraints on the laws of nature. Each level in this hierarchy works under principles (and "laws") that are irreducible to lower-level laws. Equally important, the principles that operate at higher levels may serve to restrict, harness, and organize lower levels. [42]

Thus the principles of hydrodynamics, which describe the aggregate be-havior of water, cannot be reduced to the principles associated with the covalent bonds of its individual molecules, much less the quantum-level principles associated with their quarks and gluons. Nor can the complex functional organization of a living cell be explained in terms of its DNA, much less its protons and electrons. Similarly, in human language (to use one of Polanyi's examples), the principles of grammar utilize but also sub-sume the principles of phonetics.

Polanyi's argument was seconded and augmented by the Nobel physicist Phillip Anderson in his 1972 *Science* essay called "More is Different."[43] In Anderson's words: "The reductionist hypothesis does not by any means imply a 'constructionist' one: The ability to reduce everything to simple fundamental laws does not imply the ability to start from the laws and reconstruct the universe ... The constructionist hypothesis breaks down when confronted with the twin difficulties of scale and complexity ... At each level of complexity entirely new properties appear ... Psychology is not applied biology, nor is biology applied chemistry."

Anderson used the now familiar term "broken symmetry" to characterize such level-shifts. His examples included complex organic molecules, superconductivity, and crystal lattices. "We can see how the whole becomes not merely more but very different from the sum of its parts." In other words, new levels of organization in nature create new relationships (new synergies) and, in the process, redefine the boundaries of the "system." Strong reductionism is oblivious to such higher-level principles.

The Causal Role of Synergy in Evolution

Needless to say, the one-level thinking that characterizes the Neo-Pythagorean paradigm also excludes by fiat the causal role of synergy in the natural world. Synergy has provided a cutting edge in evolution – and especially the evolution of complexity. Yet the Neo-Pythagoreans, by the very nature of their single-minded quest for "a small number of rules or laws," in Holland's phrase, discount any sources of causation that are irreducible and context-specific – in other words, historical. While some forms of synergy arise from the actions of law-like, deterministic forces or influences, a great many more do not.

The reductionist riposte to the Synergism Hypothesis, paraphrasing Wilson's argument above, is that synergy can be predicted from a knowledge of the parts and their "interactions." One problem with this formulation is that, in reality, the interactions among the parts define the system; systems do not float on top of the parts and their interactions. This is not "reductionism"; it's really systems science in disguise. But a more serious problem with this claim is that it is often not possible to know in advance what all of the "parts" are, much less how they will interact when brought together. Moreover, the interactions between the parts and their environment(s) may be even more important than the interactions among the parts. This dynamic often involves unique historical "events." Though the evolutionary process is constrained and even partially caused by laws (and principles) that operate at lower levels of organization, the course of evolution has not been determined by those laws. New sources of causation (new "degrees of freedom," to use the mathematical term) are created at each new level of organization, and interaction.

Some forms of synergy, as we saw, are produced by the forces that underlie the laws of physics. But many more are historical in nature, or involve some combination of the two. When we fall down, gravity is clearly the winner. But when we pick ourselves up again, a complex set of engineering, bioenergetic, neuromuscular, and even psychological principles come into play. Just as the individual snowflakes that comprise the winter snowstorms in the mountains may contribute to the snowpack that later causes spring flooding in the river delta, in the same way the synergies produced by quarks and gluons, protons and neutrons, or hydrogen and oxygen compounds have produced many other "downstream effects" that cannot be explained in terms of the physics alone. More important, these emergent, synergistic effects are responsible for the evolution of living systems and their parts, not the laws of physics.

Though many other theorists in recent years have embraced this multilevel paradigm, many more, unfortunately, remain dogmatic reductionists. This is what prompted two distinguished physicists, the Nobel Laureate Robert B. Laughlin and David Pines, to publish a frontal assault on reductionism in the *Proceedings of the National Academy of Sciences* (2000) under the title "The Theory of Everything." In their words:

> The Theory of Everything is not even remotely a theory of everything ... The fact that the essential role played by higher organizing principles in determining emergent behavior continues to be disavowed by so many physical scientists is a poignant comment on the nature of modern science ... The central task of theoretical physics in our time is no longer to write down the ultimate equations but rather to catalogue and understand emergent behavior in its many guises, including potentially life itself. We call this physics of the next century the study of complex adaptive matter. For better or worse we are now witnessing a transition from the science of the past, so intimately linked to reductionism, to the study of complex adaptive matter, firmly based in experiment, with its hope for providing a jumping-off point for new discoveries, new concepts, and new wisdom.[44]

It is time to put an end to the vexatious debate over "holism" versus "reductionism" (in reality, both are necessary but insufficient) and shift our focus to "architectonics" – the study of how the world has been "built up" from the novel cooperative effects produced by many interacting parts; we need to focus more intently on the joint effects produced by the relationships that arise between things, or organisms. If reductionism is necessary for understanding how the "parts" work and how they interact, holism is equally necessary for understanding "why" living systems have evolved, and what effects they produce. As I suggested at the outset, the universe can be portrayed as a vast structure of synergies, a many-leveled "Magic Castle" in which the synergies produced at one level serve as the building blocks for the next level. Moreover, unpredictable new forms of synergy, and even new principles, emerge at each new level of organization. This ultimately necessitates a science of history.

An Unfinished Symphony

In its original connotation, the "music of the spheres" is clearly an anachronism, and so is the Neo-Pythagorean quest for "laws" of everything, or even laws of evolution. But perhaps we might be able to reincarnate this musical image from the graveyard of moribund metaphors to serve a different purpose. This ancient metaphor could also serve to highlight the Synergism Hypothesis. As Marcel Proust famously suggested, much can be learned by seeing familiar things with "new eyes."

A revised, updated rendition of the "music of the spheres" can be illustrated with a story from the world of theater. There is a musical play for children, called *Me–Myself the Wicked Elf*, in which one of the musical notes, "Me" (typically acted by an outgoing pre-teenager) happens to be an egotistical malcontent who rebels and leaves his partners – Do, Re, Fa, So, La, and Ti. But soon the joy of his new-found freedom gives way to sadness as "Me" comes to realize that, despite his unique talent, he cannot make music without his companions. "Me" then proceeds to sing a plaintive song, mostly in a monotone, about how he is the only "free" musical note, but at a very high price; he is now a musical mute. Needless to say, it all ends happily. In the last scene, there is a joyous reunion and a polyphonic finale.

This charming children's play could also serve as a metaphor for the human condition. The music of the spheres could be likened to a grand, unfinished symphony. Or maybe it is more like a colossal gig, with enough jazz musicians to fill a stadium. (No metaphor can really do it justice.) At any rate, the music of the spheres suggests an intricately interwoven set of themes, with many different notes played by many different instruments, and it is this ensemble that collectively shapes the progression of the overall process. Moreover, the music of the spheres does not follow a predetermined score. Evolution is an irreducible historical process, and we are helping to write the music as we go along.

One of the major themes of this unfinished symphony is synergy – the magical effects that occur when the individual notes are scored and played in unique combinations. These combinations of "parts" are the very source of nature's creativity, whether it be the photons of sunlight that "co-operate" to produce the warming light of the solar flux, or the subatomic "building blocks" that combine to produce the wondrously different natural elements in the periodic table, or the small alphabet of four nucleotide "letters" that are able to produce the recipes for an immense variety of living systems, or the 26-letter alphabet (plus a handful of punctuation marks) that undergird Shakespeare's plays, or (not least) the notes that provide the score for Schubert's Unfinished Symphony.

The Epic of Evolution is truly a tale of magic on a cosmic scale. We still don't fully understand it. Perhaps we never will. (Some major aspects of the current "standard model" remain deep puzzles.) But, at the very heart of this epic we can now see more clearly than ever the inner workings of what has been a profoundly creative process. To some, this epic amounts to nothing more than the playing out of a preordained, deterministic "script," whether it be "God's Plan" or the fixed "laws of nature." However, I have advanced a very different vision – the notion that we are the products of a truly amazing historical adventure; we are fellow-participants in a creative work-in-progress, the outcome of which we do not know.

Some may choose to view this process as a war, a battleground of ruthlessly competitive "Darwinism" on a grand scale ("nature, red in tooth and claw" in the carnivorous image of poet Alfred Lord Tennyson). But I hold a more positive view of the evolutionary epic. To borrow (again) the slogan of the distinguished paleontologist George Gaylord Simpson, it is more like a tale of "trial-and-success." We inhabit a universe that is still inventing itself, and we are an integral part of this creative process, not outside observers. Indeed, humankind does represent a culmination of a sort. To quote Teilhard de Chardin (who in turn quoted the biologist Julian Huxley), we are "evolution become conscious of itself." Accordingly, we can help to shape its course, for better or worse. The choice is up to us – and our choices matter.

10 Conjuring the Future: What Can We Predict?

Though history doesn't repeat itself, it does tend to rhyme.

Mark Twain

Plans are useless, but planning is indispensable.

Dwight D. Eisenhower

The ancient dream of being able to predict the future of the human career has retained its seductive allure over the years. For example, the well-known twentieth-century anthropologist Raoul Naroll documented the historical trend toward ever-larger aggregates and concluded that the probabilities for achieving a world state ranged from the year 2125 (a 40% probability) to 2750 (a 95% probability).[1] Robert Carneiro came to a similar conclusion but used a different methodology. He projected forward the historical trend toward a decreasing number of independent political units and declared that, provided nuclear annihilation can be avoided, "a world state cannot be far off." It's a matter of centuries or even decades, not millennia, he declared.[2]

Sociobiologist Edward O. Wilson, in a 1970s book coauthored by Charles Lumsden (*Genes, Mind and Culture*), also joined the futurist parade with what he characterized as an "autocatalytic" theory of cultural evolution.

The ultimate triumph of both human sociobiology and the traditional social sciences would be to correctly explain and predict trends in cultural evolution on the basis of their own axioms ... Will the social sciences, using all of the considerable resources at their disposal and designing ever more comprehensive multifactorial models, be able to explain history more fully and perhaps even predict with moderate accuracy? We believe the answer is yes, at least on a limited scale. . . . The prediction of history is a worthwhile venture.[3]

The Logic of Synergy

The latest entry in the Neo-Pythagorean archery contest, surprisingly enough, is the well-known journalist and writer on social theory, Robert Wright. Wright is a card-carrying Neo-Darwinian and a stout proponent of the individualistic, competitive model of human behavior. He's especially enamored with game theory approaches to human evolution. Yet in his latest book, *Nonzero: The Logic of Human Destiny* (2000), Wright embraces a vision of cultural evolution that owes more (unwittingly, no doubt) to Herbert Spencer than to Charles Darwin.

In effect, *Nonzero* is a book-length argument for the proposition (masked by the terminology of game theory) that the direction of human history, our progressive cultural complexification, is due to the beneficial effects produced by cooperation – in other words synergy. Wright seems uncomfortable with the term "cooperation," though he does occasionally use the term synergy. Perhaps this is a reflection of the tradition in evolutionary theory of viewing competition as the "motor" of evolution and cooperation as a synonym for altruism. Accordingly, Wright adopts "non-zero-sumness" as a euphemism. The term refers to the many cooperative relationships that are win–win (or sometimes lose–lose) rather than having zero-sum, beggar-your-neighbor consequences. It is really a book about cooperation and synergy in deep disguise.[4]

Wright's thesis is that non-zero-sumness (read synergy) has been the driver of human history, as well as the underlying source of our progressive cultural development. "The direction of history is unmistakable," he claims. The "trajectory" of our evolution has been determined by "the relentless logic of non-zero-sumness."[5] He speaks of "the power of zero-sum logic," and he calls it "the secret of life."[6] But Wright's theory begs the question: How can "logic" drive anything, or do anything? Game theory is an analytical exercise, a mental operation, not a causal agency. It abstracts and assigns numerical values to the functional effects that are produced by cooperation. In actuality, Wright's logic is "driven" by the synergies that cooperation produces. It's the synergies – and how these are distributed – that determine the players' preferences and the ultimate selective consequences.[7]

It could be said that the distinction between non-zero-sumness and synergy involves only an alternative way of viewing the same phenomena. That's true, but underneath the semantic differences there is a deeper level of disagreement. Wright conceives of non-zero-sumness as a deterministic historical "force" that has followed a predestined, perhaps even purposive path. For instance, he calls the emergence of agriculture inevitable. "Given enough time it was bound to happen."[8] He also asserts that "the caprices of history may alter the tempo but not the direction."[9] When you get to Wright's fine print, however, you find that he relies on technological innovation as his "prime mover" (*déjà vu*). Wright claims that all of our various technologies were bound to arise sooner or later, and he cites some examples of multiple independent inventions in support of his case. He says that these inventions were predestined by the basic human talent for inventiveness.[10] In other words, Wright's theory boils down to a sort of biological determinism – the influence of an innate propensity.

Unfortunately, this propensity is not everywhere in evidence. Technological innovations have not been evenly distributed around the world, needless to say. To the contrary, there has been a shifting geography of cultural "hotspots" over the past 10,000 years and more. Indeed, societies often lose technologies. Perhaps the most egregious example is the hunter–gatherers who migrated to the island of Tasmania in the 1600s. When this long-isolated small population was first discovered by Europeans centuries later, they had lost the use of barbed spears, bone tools, nets, hafted axes, even fire. In sum, there are plausible alternative explanations for the historical convergences identified by Wright (see Chapter 8).[11]

Even more troubling is Wright's assertion that this orthogenetic trend can be projected into the future. "Globalization," he says, "has been in the cards since the invention of life."[12] Using a line that could have been penned by Herbert Spencer, Wright assures us that the current turmoil in the world is leading us ultimately toward equilibrium and relative stability; it's "the storm before the calm." A magnificent new global structure is being erected before our eyes, Wright says. It cannot exactly be called a "law," but it is predictable – that is, overwhelmingly likely (with 99.99% confidence).[13]

The Logic of Failure

The problem is that the technological optimists see only the doughnut, which is getting larger. But so is the hole. And the hole may be increasing faster than the whole. Neo-Pythagorean theorists commonly make what professional forecasters long ago identified as a fundamental logical error. To put it very simply, a trend is not a law. A past trend cannot be extrapolated into the future if any of the variables responsible for that trend (say climate) are subject to unpredictable perturbations. Nor is it likely to continue if any major new, unforeseen influence should enter the picture. Call it synergy plus or minus one.

The fact of the matter is that our knowledge of complex systems is still very imperfect. More important, the greater the complexity of a system, the less likely we are to know beforehand what all of its dependencies and vulnerabilities are. The logic of non-zero-sumness is challenged, and sometimes mocked, by *The Logic of Failure* – the title of a brilliant 1996 book by psychologist Dietrich Dörner, who showed how easily we can become trapped in failure-prone systems of our own creation.[14]

Dörner illustrated his thesis with a planning game involving the Moros, a mythical tribe of West African semi-nomads who wander from one watering hole to another with their cattle herds and who raise a little millet on the side. The object of the game is to introduce various interventions to aid the Moros, such as digging deep wells to expand their pastureland, providing hybrid strains of millet, improving public health services, and other measures. With the help of a computer program that allows the participants in the planning game to observe the results in a matter of minutes, Dörner regularly humbles interdisciplinary teams of experts who introduce well-meaning changes that ultimately produce disastrous consequences. It turns out that many things can go wrong. The solution to one problem may create another problem; a small problem may be obscuring an unseen larger problem; cause-and-effect relationships may be poorly understood; short-term benefits may introduce hidden long-term costs; and so on. Dörner does not have to search very far for real-world examples: the Aswan Dam, the Chernobyl disaster, the introduction of the pesticide DDT, and many others.

The logic of failure also applies to the broader sweep of human history. The Neandertals prospered for perhaps 200,000 years and then vanished rather suddenly around 30,000 years ago. Rome dominated the ancient Mediterranean and much of Europe for more than 500 years, then it was gone forever. During that same period, the grand city of Teotihuacán was flourishing in Mesoamerica. Teotihuacán survived for about 900 years and grew to a peak of about 125,000 people. Yet in due course it too collapsed and the population declined to perhaps one-quarter of its former size.[15]

More recently, the British Empire, the Thousand-Year Reich, the Greater East-Asia Co-prosperity Sphere, and the Soviet Empire (among others) have all become fading memories.

The "caprices" of history are not simply quirks, anomalies, or blips. They are not temporary road-blocks that can be got around. They are an integral part of the process. The optimists' dismissive response is "upward and onward." Despite innumerable setbacks, they argue, there has been a general trend over the past 10,000 years toward larger societies, more advanced technologies, and greater complexity. Maybe so, but this discounts the other half of the story. One could just as readily argue that progressively larger, more colossal failures and collapses are the predominant long-term trend. Pessimists like the biologist Paul Ehrlich (author of *The Population Bomb* and many other books), or the Club of Rome (sponsors of the famous "Limits to Growth" models), or even more recent, more sophisticated doomsday model-builders like Willard Fey (of the Georgia Institute of Technology) and Ann Lam warn that, unless there is a drastic change of course, we may go the way of Rome and Teotihuacán and many other failed civilizations.[16] Indeed, some social theorists portray history as an unending cycle of growth and decay. So which aspect of past history is more predictive? Is it the advances, or the recessions? Or should both be treated as real possibilities for the future? Are we advancing as a species or painting ourselves into a corner? By the same token, in the broader epic of evolution, which "trend" is more predictive for the ultimate fate of humankind – the 99% of all living species that are now extinct or the 1% that remain? Peter Ward and Alexis Rockman, in a new book entitled *Future Evolution*, claim that our species is "extinction-proof" (a boast that is eerily reminiscent of the claim that the Titanic was unsinkable).[17] It's a gamble with long odds that we would be foolhardy to take.

The Perils of Prediction

The fact is that predicting the future remains a perilous enterprise. The historical record is littered with events and developments that were not predicted, while many of the things that have been confidently forecast over the years proved to be wishful thinking.

The year 1900 was a time of millennial optimism, certainly in the United States. "Progress" was the buzzword of the day, and many futuristic fantasies bloomed, from egalitarian communist societies to idealized garden cities with perfect hygiene and abundance for all. Nobody back then predicted any of the major shaping influences of the twentieth century: two horrific world wars, nuclear weapons, the devastating influenza epidemic of 1918, the current AIDS epidemic, or global warming, to mention just a few. Nor were many of the century's most stunning technological achievements anticipated – transoceanic jumbo jets, television, computers, lasers, vaccines, cell phones, the Internet, scanning electron microscopes, birth-control pills, and a host of other inventions. On the other hand, there have been many famously wrong technological predictions over the years. Many of them circulate in e-mail messages these days:

* After evaluating the new "telephone" in 1876, Western Union concluded that "the device is inherently of no value to us."
* Lord Kelvin, the chemist and President of the Royal Society, in 1895 declared that "heavier-than-air flying machines are impossible."
* French Marshall Ferdinand Foch (just before World War One) concluded that "airplanes are interesting toys but of no military value."
* Radio pioneer David Sarnoff was told by potential investors in the 1920s that "the wireless music box has no imaginable commercial value."
* Or remember the skeptical Jack Warner, President of Warner Brothers movie studio in 1927, asking "Who the hell wants to hear actors talk?"
* Then there was Thomas J. Watson, Chairman of IBM, in 1943: "I think there is a world market for maybe five computers."
* And Bill Gates in 1981: "640K ought to be enough [memory] for anybody."[18]

The World Future Society, founded in the 1960s, provides a public forum for people who are professionally involved in "forecasting, planning or other future-oriented activities" (as their publicity material puts it). The Society now has some 30,000 members in 80 countries and can boast of having an impressive roster of heavyweight directors and council members. Its distinguished journal, *The Futurist*, is packed with leading-edge trend analyses and projections. However, *The Futurist*'s articles almost always feature qualifying words like "could," "may," "might," or "is likely to." Professional futurists know that unequivocal predictions are always risky and very often wrong. Their abundant use of various "weasel words" represents a tacit acknowledgment that we often do not know in advance all of the factors that may affect future outcomes; we are not omniscient. Equally important, we cannot anticipate how human decisions and actions – by individuals, business organizations, or governments – may shape the outcome. Indeed, many of the futurists' forecasts are deliberately designed to be self-negating prophesies – catalysts for remedial actions that will falsify them. As one of *The Futurist*'s own authors noted: "Predicting the future has been one of Man's less successful ventures as the future rarely turns out as forecast."[19] Similarly, a staff writer for *The Economist* recently observed: "It is one of the great mysteries of economics: why do forecasters get things so wrong, so often?"[20] And, in a delightful article about the age-old demand for prophets, writer Barbara Holland wryly commented: "The term 'foreseeable future' is an oxymoron."[21]

Replacing Forecasts with Strategic Plans

The truth is that we cannot predict with any assurance what the world will be like even 50 or 100 years from now, much less in the year 3000, 5000, or 10 000. Are we, then, at the mercy of unforeseen contingencies and capricious influences? Are we adrift in a sea of unpredictable forces? The answer is a resounding no, but we need to climb inside of this answer and inspect it more closely.

Most major business enterprises, at least in the U.S.A., engage in "strategic planning," often with the help of professional management consultants. What undergirds and guides each of these exercises is the basic goals/objectives of the organization and its overall strategic situation. There is a coherent focus to the strategic planning process. Moreover, the most successful businesses generally try to keep Dwight Eisenhower's dictum in mind. They also prepare for the unexpected; they develop alternative future "scenarios"; and they make contingency plans.

The same kind of strategic planning is urgently needed for the larger problem of survival and reproduction for the human species. Recall the discussion in Chapter 7 about the ongoing research effort related to basic human needs, and the inescapable fact that our survival necessarily involves a multifaceted enterprise. The "survival indicators" project in particular has identified an array of some 14 basic needs that must, by and large, be satisfied continuously.[22] Each of us requires a synergistic "package" of resources and compatible environmental conditions. For instance, we cannot for very long do without fresh water, or waste elimination, or adequate "thermoregulation," or sleep. A society that can reliably provide the requisites for the full package of basic needs will thrive. But if even one of these needs is not satisfied – if any part of the package is missing (synergy minus one) – the entire enterprise is likely to be threatened. Accordingly, each society, and the global community as a whole, should have a strategic plan that is focused on how to provide for our basic needs under an array of uncertain future "scenarios." Of course, this is easy to prescribe but very difficult to develop, much less to implement. Nevertheless, it is vitally important that we should try. The issue is far too important to be left to the marketplace, or to the complacent fantasies of wishful thinkers.

There are a number of considerations that should guide such a strategic planning process, it seems to me. One has to do with the double-edged role of technology. As we saw in the last few chapters, technology has been a major element of our adaptive strategy as a species for millions of years, and the pace of technological innovation has greatly accelerated in recent centuries. Many of these technologies have enhanced our ability to provide for our basic needs, and much more besides. But the downside is that we have also become the captives of these technologies; we are now deeply dependent upon them (the paradox of dependency). As a result, many of our technologies represent "instrumental needs" that are equally vital to our survival. For instance, the very idea of a modern industrial society without electrical power, high-speed transportation, and long-range communications is unthinkable. Though we tend to take it for granted, we are also completely dependent upon a vast, complex, technology-intensive food production system. We cannot turn this technological clock back, but we can certainly make it run a lot more smoothly.

Our bedrock biological needs (and our various instrumental needs) define the strategic problem that we confront as a species. Unfortunately, we cannot make unqualified forecasts about the future because we do not, and cannot, know in advance all of the variables that will affect our course, including the role of human knowledge, human creativity, and human folly. However, we can make a great many contingent, "if-then" predictions. If, for example, we were ultimately able to detect potentially destructive asteroids from outer space and could reliably destroy them before they impact the Earth, then we might conceivably eliminate one contingent long-term threat.

Likewise, we can safely predict that there will someday be an upper limit to population growth. As the economist Herbert Stein put it: "If something can't go on forever, it won't." However, we can only guess what that population ceiling will be, much less when it will be reached in each country.

If by some miracle we were able to institute rigorous world-wide population control measures, our ability to predict global population size would
measurably improve. Yet we still might not be able to anticipate the downside effects of such looming megathreats as global warming, pandemic
diseases, monster earthquakes, or climate-altering volcano eruptions. For
instance, when Mt. Tambora in Indonesia erupted in 1815, it did so with
such enormous force that its dust cloud affected weather conditions – and
crops – for many months afterward in places as far away as New England.
(The locals referred to it as "the year without a summer.") Then there is
the threat associated with weapons (and wars) of mass destruction. It is
too late to put this genie back in the bottle, and we have recently gotten a
whiff of what could happen if the gods of high-tech war are unleashed.

Some Contingent Trends

We can also observe many contingent negative trends (various forms of
black magic) that represent potentially serious survival threats. As the
philosopher John Locke truly observed: "Hell is the truth seen too late." If
major steps are not taken to address these challenges, the ultimate consequences are predictable. To cite just a few examples:

* Global warming should be at the top of everyone's list. Alaska is thaw-
  ing; the Greenland and Antarctic ice sheets are shrinking; global temperatures have already increased an average of 1 degree Fahrenheit during
  the past 100 years and another 2 to perhaps 11 degrees are projected
  for this century. This temperature shift compares, ominously, with the
  temperature range for the last ice age (5–9 degrees on average), though
  in the opposite direction. Drastic, possibly rapid climate changes and
  significant shifts in sea levels are increasingly likely.[23]
* Megadroughts are a thing of the future, with or without global warming.
  Over the past 10,000 years alone, there have often been prolonged periods of drought in various regions – a result of global climate changes –
  that have far surpassed anything the world has experienced in recent
  centuries. There is also increasing evidence that climate patterns can
  change very quickly.[24] The collapse of many ancient civilizations may
  well have been caused by severe, decades-long climate perturbations, as
  we noted earlier.[25] Yet we are totally unprepared for a recurrence of this
  likely event.

Consider this: the state of California, with its rich soil, salubrious climate, and the longest growing season in the world, produces 90% of the apricots, 87% of the grapes and avocados, 86% of the peaches, 83% of the lemons and strawberries, 80% of the artichokes and lettuce, 73% of the broccoli, and 53% of the cauliflowers grown in the United States, along with about one-third of the cherries and pears and a significant percentage of the nation's oranges, wheat, rice, and other crops.[26] California also currently has a population of about 35 million people and is projected to grow to 50 million by 2025. Unfortunately, California is one of the areas that has been susceptible to severe megadroughts in the past. If another such megadrought were to occur in this region, many other countries besides the U.S.A., could be hit with soaring food prices and food shortages. The dark side of the global food economy may be the paradox of dependency writ large.

∗ As *The New York Times* noted in a recent editorial, "the most precious fluid on earth is not oil, but water." With or without global warming and megadroughts, the world stock of fresh water is being depleted at a rate that will soon threaten our food supply, because a major share of the world's agriculture depends upon artificial irrigation. Drinking water and water for sanitation and industrial uses are also threatened. Even now, some 1.3 billion people (20% of the global population) do not have safe drinking water, and at least 4 million people die each year from waterborne diseases. Within the next 50 years, population growth will create the demand for a 50–100% increase in fresh water supplies, a staggering challenge. Yet we are currently depleting many of the lakes, rivers, and aquifers that serve existing populations. For instance, the great Ogallala Aquifer, an underground river in the American southwest that was once the size of Lake Huron, will be gone within 20–30 years.[27]

∗ In the twentieth century alone, smallpox killed more humans than all of the wars and other epidemics combined, an estimated half a billion people all told. The suffering caused by smallpox is beyond our comprehension. Of course, the smallpox vaccine has now reduced the incidence of this once dread disease to minuscule proportions, but we also became complacent about it. Smallpox vaccinations are no longer universal even in the developed countries, and vaccine supplies were allowed to dwindle and their quality to deteriorate. After the 9/11 terrorist attack, when it became apparent that smallpox is top-rated among the weapons of choice for bioterrorists, we were finally alerted to the danger in continuing to follow this course. Now steps are under way to confront this risk. However, smallpox is only one of the Malthusian scourges that are still a menace, despite the dream of someday wiping out diseases. Tuberculosis, for instance, is once again on the rise. AIDS afflicts up to one-quarter of the population (and half of the youths) in some parts of Africa, with serious economic consequences for the rest of the population in these nations. As of 2001, 21 million people had died of AIDS and another 36 million were infected. Equally ominous is the fact that various malevolent microbes have been able to evolve new, resistant strains at a rate that exceeds our ability to develop new defenses. We are losing the battle. To make matters worse, an unintended side effect of world-wide air travel and trade is that it provides more efficient ways to spread disease germs. Then there is the once and future role of abrupt climate changes and extreme weather patterns in spreading diseases. The potential for future pandemics is high.[28]

* Today there is a fifth horseman of the apocalypse, unforeseen by the biblical writers or Thomas Malthus. The very complexity and interdependence that has brought so many economic benefits to the modern world has also become a growing threat. Even as technology has enhanced the power and effectiveness of the human survival enterprise, it has also greatly increased the potential for doing mischief – the paradox of dependency. For instance, the many failures associated with what are referred to as "Large Scale Engineering Projects" (or LEPs) have been analyzed and chronicled in various recent publications. "Welcome to the world of high-risk technologies," says Charles Perrow, author of *Normal Accidents*. "There is a form of accident that is inevitable ... even 'normal.' This has to do with the way failures can interact and the way the system is tied together." He calls it "interactive complexity."[29]

A prime example is the Y2K computer "bug," a narrowly averted disaster. Instead of being able to greet the new millennium with confidence in our technological progress, we approached this historic turning point with grave apprehension and a vast world-wide effort to avert what would otherwise have been a widely destructive example of synergy minus one. As we noted earlier, the possibility of a millennial nightmare was very real. Y2K involved a unique technological glitch, to be sure. But computer viruses represent a continuing threat. In 1994, the number of viruses averaged about one per thousand computers per month. In 1999 the number had grown to 7.5 per thousand per month, even as the total number of computers in service continues to mount.[30] Within the past three years alone there have been several highly destructive epidemics. One, dubbed "Melissa," infected 19% of the nation's large computers and caused some $393 million in damage. More recently, the so-called "Love Bug" caused an estimated $10 billion in damages world-wide. Among other things, the Love Bug bit the Pentagon and forced the Ford Motor Company to shut down its internal e-mail system for several hours.

Nor are viruses the only vulnerability in an increasingly computer-dependent economy. For instance, it is estimated that about 13% of the total electrical power usage in the U.S.A. is related to computers and the percentage is still growing. Yet major power outages are becoming increasingly likely as the electrical utility industry in the U.S.A. falls further behind in building new capacity to keep up with demand. It is estimated by the U.S. Department of Energy that demand will increase 17% by 2007, but capacity will increase by only 4%.[31] The potential consequences of this negative synergy are obvious. To be sure, the soaring prices and rolling blackouts that occurred in California, home of Silicon Valley, during 2001 also involved market manipulation by energy providers, but they are only a foretaste of what's to come.

There are many other opportunities for making gloomy "if-then" (or if-not-then) predictions in our current environment. Some 69% of the commercially exploitable fish stocks are in urgent need of action to conserve their depleting numbers, according to the Food and Agriculture Organization of the U.N.[32] Amphibians have also been drastically declining worldwide, even in some protected areas.[33] Neotropical migrant bird populations are shrinking in numbers; coral reefs are deteriorating; there has been a rash of puzzling mass deaths in sea mammals; deforestation continues to occur in many areas of the world at an alarming rate; soil erosion and water quality declines are proceeding apace.[34] Half the people in the world have no sanitary way to dispose of human wastes, and some 90% of the contagious diseases in the developing world are waterborne.[35]

Perhaps most disturbing is the synergistic nexus – dysergy on a global scale – between population growth, economic development, advancing technology, resource depletion, and environmental pollution. This is unsustainable. In an exhaustively researched and heavily documented book on the environmental history of the twentieth century, *Something New Under the Sun* (2000), historian J. R. McNeill notes that the future may be inherently uncertain but that, most likely, "sharp adjustments will be required" in our current global life style.[36] Indeed, political scientist Thomas F. Homer-Dixon and his colleagues in a major project on Environmental Change and Acute Conflict some years ago, which brought together a team of 30 specialists, came to a profoundly disturbing conclusion: "Scarcities of renewable resources are already contributing to violent conflicts in many parts of the developing world. These conflicts may foreshadow a surge of similar violence in coming decades...."[37] (Recall the discussion in Chapter 7 concerning the "Resource Acquisition Model" of warfare in human evolution.)

Cassandra Meets Pollyanna

Yet there are also some positive signs in the current picture, if not yet enough. One obvious example is the Internet. In 2000, the web had grown to a total of some 1.5 billion pages and was increasing at more than 2 million new pages per day. Meanwhile, advances in computer technology continue to offer more for less. The cost of manufacturing hard disks plunged from $11.5 per megabyte in 1988 to $0.04 per megabyte in 1998. By the time this book is published, it is estimated that the cost will decline to a fraction of a penny per megabyte.[38] Miniaturization also continues to amaze. The computing power that once filled a large room now fits in the palm of your hand, and there's much more (for less) to come. The long-term potential for productivity improvements and cultural synergies have only begun to be tapped.

The banner headlines in mid-2000 announcing that two rival teams of molecular biologists had completed a "map" of the human genome was only the beginning, a milestone in a longer-range effort that will ultimately provide many new tools for aiding human health and welfare. Each human genome contains two sets of 23 giant chromosomes, and each set contains more than 3 billion base pairs of DNA. We now know the sequence, the order in which the base pairs are arranged, but we still understand very little about exactly how the map actually produces the organism; the key to it all still eludes us. The molecular biologist and Nobel Prize–winner David Baltimore predicts that there is still another century of work to be done. Nevertheless, we can glimpse the road ahead. There is great promise, as well as risks, in genetic engineering.

Another hopeful trend involves advances in alternative energy technologies. Solar power is becoming increasingly competitive, and so are windmills. In the 1980s, when there was a burst of experimentation with new energy technologies, the most advanced windmills could produce energy for about 38 cents per kilowatt hour. Now the cost is down to 3.5–5.0 cents per kilowatt hour, and wind farms have been cropping up in such unexpected places as the Iowa corn belt and West Texas, not to mention continental Europe. Within the next decade wind-farming will become a major industry.[39]

These and other prospective developments are fairly safe bets. But if we let our imagination range more freely, we can also conjure more dramatic changes as improved technologies are applied to some of our pressing problems. For instance, the day may come in the not too distant future when vast arrays of solar- and wind-powered energy generators are coupled to improved water desalinization and purification systems. What now seems like a major threat, a growing shortage of fresh water, in due course could evaporate (excuse the pun).

Or consider this modest fantasy. As the highways and the skies become ever more crowded with fossil-fueled vehicles – and people – the technology required for the "teleportation" of real-time, interactive motion-picture-quality video images of people, places, and events is becoming ever better and cheaper. And a whole lot faster. It is a good bet that we are only at the beginning of a transportation revolution in reverse as virtual transportation becomes an attractive alternative to the real thing in many cases – from visiting grandma in Peoria to making a business presentation to a client in Paris. And, in the bargain, we may be able to gain some purchase on such fossil-fueled problems as global warming and urban smog.

There are many other technologies that hold promise for the future: fish-farming, hydrogen-fueled automobiles, seawater-tolerant crops, superconductivity, and tilt-rotor aircraft, to name a few. But there is much more to be done, including more effective implementation of the technologies we already have. The good news is that these are choices we have the collective power to make.

Synergy and the Fate of Humankind

The future, in short, is not preordained by any law, or an inherent logic. It will be co-determined by the combined influence of chance, necessity, teleonomy, and differential survival/selection. It will be partly determined by what we make of it – by our creativity and our choices. But it will also be shaped by forces that we can still only partly predict; nature will continue to have many surprises in store for us. The Neo-Pythagorean fantasy is bad science.

A frequent cause of grief in modern societies is our failure to understand the full dimensionality of our "systems." Like the ill-defined chess game, or the ventilator dust in the air traffic control center, or the flawed baggage handling system at the new Denver Airport, we are constantly challenged to expand our understanding of what all of the "parts" are in our systems, and to cope with the patterns of interdependency that these systems have created. There is an ever-present danger that our short-sightedness will produce unpleasant (or fatal) surprises. By the same token, we must learn to develop more sophisticated ways of understanding the larger, systemic consequences of our actions and our technologies – the "externalities," as the economists would say. The idea is captured in biologist Garrett Hardin's well-known law: "You can never do just one thing."

However, systems thinking involves rigorous, disciplined, even tedious work. It runs against the grain of a hip-shooting, quick-and-dirty culture that is addicted to technological innovation, ready or not. The chief scientist at Sun Microsystems, named Bill Joy paradoxically, made headlines a while back when he issued an apocalyptic public warning about runaway technology: "The 21st-century technologies – genetics, nanotechnology and robotics – are so powerful that they can spawn whole new classes of accidents and abuses."[40] Joy's message was that we must slow down and think through the consequences of our headlong rush into these and other cutting-edge technologies. The market-driven pressure for rapid innovation runs diametrically counter to this objective. And the dynamic of Neo-Lamarckian Selection in the marketplace is almost always focused on the short-term benefits, especially if the long-term costs are obscure or fall in somebody else's yard. These consequences may only become apparent later on when the "failure analysis" experts are called in to explain what went wrong.

Four key concepts developed in this book should be central to a more systemic view of the future:

(1) Almost every important innovation or change is part of a very complex, synergistic system, whether we are aware of it or not, and will likely have an impact on that system.

(2) Major innovations are unavoidably subject to the paradox of dependency; the synergies are contingent upon the functioning of all the parts in the particular historical context, and the cost of losing it will be directly proportional to the benefits.

(3) The technological and economic systems that we depend upon for our very survival are vulnerable to the synergy plus or minus one phenomenon; the whole can be undermined or destroyed by an unforeseen intrusion or the loss of a major part.

(4) The synergies of scale that can accumulate over a time-course often have a double edge. They can produce the positive synergies associated with any more-is-better dynamic, but they can also produce negative synergies (black magic) where more is not only less but becomes a serious threat.

There are many useful ways of viewing the world, each with its merits and weaknesses. The synergy paradigm and the Synergism Hypothesis is obviously only one alternative, but it deserves to be utilized more aggressively, I believe. "Complexity" is a buzzword in academic circles these days, but an equally important concept is that of the human survival enterprise as a purposeful, interdependent, synergistic system that is, very possibly, more than 5 million years old. And the key to our survival has always been cooperation – and synergy. Adam Smith's observation in 1776 is even more true today: "In civilized society [man] stands at all times in need of the co-operation of and assistance of great multitudes ... man has almost constant occasion for the help of his brethren."[41] To put a modern spin on it: Nowadays, it's better to be in the "diamond lane" than the "fast lane," but you have to cooperate if you want to use it (legally).

Tying Up the Loose Threads

Stepping back to view once again the broader Epic of Evolution and the paradigm that I call "Holistic Darwinism," several major themes should be reiterated:

* It is time to replace the simplistic, one-leveled Neo-Darwinian selfish gene paradigm, a prime example of reductionism and the search for deterministic "mechanisms" in evolution. The alternative is a multi-leveled model – from genes to ecosystems and, indeed, the entire biosphere – a model in which the survival units at each level are functionally

interdependent "wholes." To paraphrase the poet John Donne, no gene is an island.

* The evolutionary process can best be characterized as a day-by-day dynamic in which there is a continuing intersection between chance, necessity, teleonomy, and differential selection in relation to survival and reproduction. All of these factors jointly influence the process.

* "Synergistic selection," the differential survival of interdependent, synergistic wholes, has led to the emergence of higher-level self-interests (selfish genomes, selfish individuals, selfish groups, selfish nation-states) whose "purposes" transcend and may sometimes conflict with the interests of the parts. "Cooperation" (and synergy) is not, therefore, an anomaly or a minor theme in evolution; it is the *raison d'être* for biological (and social) complexity.

* Higher-level biological systems can themselves become a cause of evolutionary change via the process we called "Neo-Lamarckian Selection." (Recall our key argument against the extreme reductionist position; the parts can be treated as a whole if they act as a whole.) This is also true of "superorganisms" – organized societies.

* Human creativity, human technology, and our collective, "political" actions are neither *sui generis* nor separated from the evolutionary process. Humankind has merely amplified a deeper, longer-term evolutionary trend. For better or worse, we are now a major cause of evolutionary change, both in ourselves and for the rest of the planet. We are the Sorcerer's Apprentice.

* The future of the human species, and of complex modern societies, is not predetermined; it is up for grabs. Some theorists think we are destined to have world peace and global prosperity, that our destiny will be determined by political innovations, high technology, the learning curve, or whatever. Others, like the writer Kevin Kelly, former editor of the *Whole Earth Catalog*, think our species is already *Out of Control* (1994).[42] Alan Lightman describes our situation as being like a speeding train without an engineer. On the other hand, we still have the capacity to make informed collective choices and to pursue a different course. We can let the negative trends in the world continue without our conscious intervention, or we can take arms against our ocean of troubles (to update Shakespeare's metaphor). We can – indeed we must – undertake strategic planning for the global survival enterprise.

In sum, we cannot predict the future, but we have the capacity to influ-ence its course and to increase our degree of control over the challenges we face. But this will occur only if there are new forms of cooperation and collective action – new levels of social synergy. World government (with teeth) may not be a foregone conclusion, but it is not a bad idea.[43] At this critical juncture in our evolution, our long-term survival (and that of our descendants) may very well depend upon it. Walter Lippmann, one of the most distinguished political analysts and newspaper columnists of the twentieth century, summed up our current predicament in a post-retirement interview with the journalist Henry Brandon a few years before he died in 1974. I first quoted from it in the conclusion to my 1983 book. It is still relevant today:

> *Lippman:* This is not the first time that human affairs have been chaotic and seemed ungovernable. But never before, I think, have the stakes been so high. I am not talking about, nor do I expect, a catastrophe like nuclear war. What is really pressing upon us is that the number of people who need to be governed and are involved in governing themselves threatens to exceed man's capacity to govern. This furious multiplication of the masses of mankind coincides with the ever more imminent threat that, because we are so ungoverned, we are polluting and destroying the environment in which the human race must live.
>
> *Brandon:* Where does this lead us?
>
> *Lippmann:* The supreme question before mankind – to which I shall not live to know the answer – is how men will be able to make themselves willing and able to save themselves.[44]

Revisiting "The Sorcerer's Apprentice"

Unfortunately, we still don't know the answer. This remains the great unre-solved challenge for the twenty-first century. We need to take very seriously the moral lesson in Goethe's original tale about "The Sorcerer's Appren-tice." The "bottom line" in this classic story was that the Apprentice did not know how to reverse the Sorcerer's spell. After the Apprentice had transformed a broom into a water-carrier, the drone dutifully began to haul buckets of water from a nearby well but could not be stopped be-fore the Sorcerer's house had become completely flooded with water. (The situation was saved when the Sorcerer returned.) In other words, a little sorcery can be a dangerous thing, with unforeseen and possibly uncontrollable consequences. Garrett Hardin – who is famous for his classic article on "The Tragedy of the Commons"– many years ago penned what remains a profoundly important truth about the human condition:

> We cannot *predict* history but we can *make* it; and we can *make* evolution. More: we cannot avoid making evolution. Every reform deliberately instituted in the structure of society changes both history and the selective forces that affect evolution – though evolutionary change may be the farthest thing from our minds as reformers. We are not free to avoid producing evolution: we are only free to close our eyes to what we are doing.[45]

The evolution of our species is preeminently a story of empowerment, a story in which the Apprentice has been gaining mastery over more and more of the Sorcerer's secrets. Now we are gaining control of the ongoing epic itself. Witting or not, we control the Sorcerer's magic. So the near future, for better or worse, will be shaped in large measure by what we choose (or fail) to make of it.

But what about the long, long term? What about Clausius's famous "heat death" hypothesis (see Chapter 9)? Is life in the universe ultimately doomed to suffer the cosmic equivalent of running out of gas/petrol? Several generations of physicists – and the many philosophers who have taken their cues from these scientists – have forecast a cold and lifeless fate for the universe as the energy available to do work gradually dissipates. Indeed, the recent discovery that the universe is actually expanding at an accelerating rate suggests to some cosmologists that the end is inevitable – though perhaps not for another 100 billion years, give or take.[46] (Though our species might not be around to witness this finale, perhaps others will; one guesstimate puts the total number of possible civilizations in the universe up in the quadrillions.)

However, this death sentence is highly debatable. For one thing, we know that Clausius's grandiose prophesy was based on a false analogy and some flawed assumptions about the properties of the universe.[47] But more to the point, our entire picture of the universe and its dynamics may be changing as a result of the recent, stunning realization that the missing (unaccounted for) mass of the universe, a staggering 90–95% of the total, may consist of a vast ocean of "dark energy" and "dark matter" – so-called because we are unable to detect it. This incomprehensibly large energy pool may be responsible for driving the acceleration of the universe and for the mysterious (and powerful) gravitational influences that seem to be shaping the puzzling behavior of entire galaxies.[48] We may well be on the verge of some very exciting and positive new discoveries in cosmology.

Many years ago, Albert Einstein, in a letter to a child who had written to him with a query about the fate of the universe, gave this response: "As for the question of the end of it, I advise: Wait and see." That's still very good advice.

Afterword

Winston Churchill – a prolific author (among his other accomplishments) – once remarked that you never finish writing a book, you merely abandon it. In an age when scientific developments seem to come with breathtaking speed and new books and journal articles flood in daily over the transom, it seems harder than ever to let go of a manuscript.

In Chapter 5 (above), in the course of re-introducing the "Synergism Hypothesis," I took the liberty of borrowing a timeless metaphor from Shakespeare. If there is indeed a tide in the affairs of men, it may account in part for the vicissitudes of this theory since it was first introduced in 1983; it seems that the Synergism Hypothesis may be an idea whose time has finally come. But, to borrow another nautical metaphor, "the tide waits for no man." This "Afterword" is a concession to that imperative. I will mention just a few of the developments – and hindsights – that would otherwise have missed the outgoing tide.

The Bioeconomics of Evolution

One important development/hindsight that deserves to be emphasized concerns the growing relationship between biology and economics. The Synergism Hypothesis is, quintessentially, an economic theory of cooperation and complexity in evolution. To repeat, the key theoretical claim is that the "payoffs" produced by various kinds of synergy are responsible for the evolution of complexity in nature, and in human evolution. Thus, it is indeed a positive development from my point of view that an economic view of the evolutionary process has been emerging as a major theme, both in biology and economics. I mentioned earlier the rise of such interdisciplines as bioeconomics, evolutionary economics, behavioral ecology, and ecological economics. Early landmarks in this vein included the writings of biologist Michael Ghiselin, the series of edited volumes by John Krebs and Nicholas Davies, two books by paleontologist Niles Eldredge, the textbook by Robert Ricklefs, and others.[1] Economists, who have been a bit slower to leave the starting gate, can nevertheless point to the work of Sidney Winter, Nicolas Georgescu-Roegen, Gordon Tullock, Jack Hirschleifer, Kenneth Boulding, Robert Costanza, Ulrich Witt, Geoffrey Hodgson, John Gowdy, and others.[2] Kurt Dopfer's new edited volume on *Evolutionary Economics* (2001) provides a useful, up-to-date overview of this developing field. But Haim Ofek's important book on *Second Nature: Economic Origins of Human Evolution* (2001) provides something more – another major voice, and additional documentation, for my theory of humankind as a self-made species (Chapter 7). Ofek emphasizes especially the role of exchange and trade, as well as the division of labor in evolving human economies. Ofek's argument is seconded by science writer Matt Ridley in his newest book, *Nature versus Nurture: The Origin of the Individual* (2003).

Cooperation: Another Idea Whose Time Has Come

Another tide also seems to be reaching the flood stage. Since the manuscript for this book was completed, many more publications have appeared that, in effect, have reduced the theoretical obstacles to cooperation in human evolution. To cite a few of the relevant items: There is the work by Herbert Gintis, Sam Bowles, Ernst Fehr, Simon Gächter, and others on "strong reciprocity" as a cooperation-enhancing mechanism; as the term implies, strong reciprocity is cooperation that is egoistic, not altruistic, and is therefore dependent upon aggressive punishment to prevent cheating.[3] Closely related to this is the expanding body of work on "fairness" as a facilitator of cooperation.[4] Also important is the work by Robert Boyd, Peter Richerson and others on the role of group-serving norms in securing cooperation.[5] New insights into cooperation among chimpanzees has been provided by Christophe Boesch, and by Sarah Brosnan and Frans de Waal.[6] And more evidence has been accumulating that cooperation, much of it altruistic, is widespread in foraging societies and in small-scale agricultural societies (for instance, Kim Hill reports 10% of the foraging activity is altruistic on the average, and sometimes the percentage is much higher, among the Ache of eastern Paraguay).[7] What all of this adds up to is a strong affirmation for the centrality of cooperation and (needless to say) the role of synergy in human evolution.

Human Nature *Redivivus*

It was never really dead, of course. But the social sciences and their camp followers in the humanities and political theory denied its existence during an ideologically tainted era that encompassed much of the last half of the twentieth century. This episode was discussed briefly in Chapter 5. It is worth emphasizing here that, as the evidence in favor of the concept continues to mount, the few besieged voices that defended the concept in the 1980s have now become a large chorus of vociferous advocates.[8] Human nature is another idea whose time has come – though in this case, it's a second coming. For one thing, the new interdiscipline of bioeconomics utilizes a paradigm that is subversive to the classical economists' notion that human "wants" and "preferences" are infinitely variable. On the contrary, as evolved biological creatures, we are constrained by an array of at least 14 specific categories of basic needs – imperatives for survival and reproductive success. These needs are detailed in a recent paper of mine.[9] Accordingly, a bioeconomic perspective on human nature requires us to reformulate our conception of an organized society; it is at bottom a "collective survival enterprise." This bedrock premise has both empirical and normative implications. Among other things, it greatly influences the content of our "preferences."

A second facet of human nature that is coming into much better focus is the psychological substrate for our social and ethical life. Among other things, it is increasingly evident that a sense of fairness is part of human nature. There are pathological exceptions, to be sure, and a common propensity for rationalizing unfairness away when it suits us. Nevertheless, a psychological predisposition toward fairness represents one of the foundations of human nature – and of social life. Indeed, there is a growing body of research in psychology and anthropology that contravenes the classic model of "economic man." We are not all coldly rational, calculating egoists. Fairness also has a strong, if imperfect, pull on our preferences and our conduct – even to the point that we may be willing to make personal sacrifices for its sake. Moreover, our innate sense of fairness – enshrined in the Golden Rule – is a cultural universal.[10]

How can this be? What adaptive advantage could a sense of fairness have bestowed on our remote Pleistocene ancestors, such that it was "blessed" by natural selection and incorporated into the undergarments of our evolving human nature? The most likely explanation, in a nutshell, is that the principle of fairness came to play a central role in reconciling conflicting claims of self-interest within the groups/bands/tribes that were indispensable to our ancestors' survival and reproductive success over many thousands of generations, as we discussed in Chapter 7.

The Science of History (Continued)

A critique of Stephen Wolfram's book, *A New Kind of Science*, by the physicist Steven Weinberg (the "high-priest" of the Neo-Pythagoreans) in a recent review essay is especially noteworthy.[11] Weinberg does not criticize Wolfram's reductionist stance, needless to say, but he does dispute Wolfram's fundamental, unproven claim that space and time are particulate, not continuous, as well as Wolfram's failure to deliver on his inflated promises. On the other hand, Weinberg speaks approvingly of the quest for "laws of nature, the rules that govern all phenomena." However, Weinberg's argument is surfeited with non-sequiturs and implicit contradictions. He makes a huge theoretical concession by suggesting that natural selection might favor some rules over others – rules that "generate the kind of complex systems that improve reproductive fitness." In other words, natural selection is a superordinate causal agency in nature. Equally telling is Weinberg's "deterministic view" of free will: "We have no mental experience that tells us our decisions are not inevitable consequences of past conditions and the laws of nature." Of course, "past conditions" means past "history." So the laws of physics are not omnipotent after all! Weinberg concedes that what he calls "free-floating theories" (like chaos theory) may be "interesting and important," but they "are not truly fundamental, because they may or may not apply to a given system. . . ." To the contrary, the unique contingencies, synergies, and artifacts of history are equally "fundamental." Physicists Nigel Goldenfeld and Leo P. Kadanoff, writing in a special issue of *Science* (1999) devoted to "complexity," drew the following conclusion: "Up to now, physicists looked for fundamental laws true for all times and all places. But each complex system is different; apparently there are no general laws for complexity. Instead one must reach for 'lessons' that might, with insight and understanding, be learned in one system and applied to another. Maybe physics studies will become more like human experience."[12]

Evolutionary Ethics?

In an essay of mine that was published a few years ago, the title asked, rhetorically, is evolutionary ethics an idea whose time has come?[13] In fact, evolutionary ethics is a subject that has been debated ever since Darwin's day. The basic issue is whether or not human ethical systems can be explained – and justified – in terms of evolutionary principles. In recent years there has been a significant upsurge of publications devoted to this issue, including many new books (as well as a number of books on Darwinism and religion) and countless journal articles. Indeed, an Internet search under "evolutionary ethics" yielded 65,400 citations.

Darwin himself answered in the affirmative. He believed that socially organized groups – and the moral systems that gave them "coherence" – played a key role in human evolution. Organized human societies are thus not simply cultural artifacts; they are products of our evolution as a species and have played a vital role in the success of our ancestors. At the height of Neo-Darwinism, when group selection was rejected as an evolutionary agency, Darwin's scenario seemed problematical. But now, as we are in the process of recovering Darwin's Darwinism, his explanation has regained credibility. And as the evidence of our duality, as both egoistic and social (even moral) animals continues to mount, it may well be that the time has come for evolutionary ethics. Indeed, the relevant question may be, are we yet ripe for it?

Acknowledgments

I have intellectual debts that go back to the 1970s. However, a number of people have helped in various ways to bring this particular work to fruition. Early on, I was most grateful for the support and encouragement of John Maynard Smith, Eörs Szathmáry, Ernst Mayr, Lynn Margulis, and David Sloan Wilson. The generous assistance of Howard Bloom and his agent, Richard Curtis, is also greatly appreciated. The reviews and concrete suggestions made by Pete Richerson, Mike Ghiselin, Russ Genet, Stan Salthe, Iver Mysterud, and Connie Barlow were beyond the call of duty. My thanks also go to an anonymous reviewer. The research assistance of Zach Montz, Nayelli Gonzalez, and Pam Albert, as well as Pam's heroic efforts with the bibliography, were also greatly appreciated. Special thanks are due also to my editor at Cambridge University Press, Ward Cooper, and to the staff, whose support was indispensable and much appreciated. I was also fortunate to have had an exceptionally skillful copy editor in Anna Hodson. Finally, as always, my wife and my family have been a part of every page.

Notes

1. Prologue: The New Evolutionary Paradigm

1. There are many books about lichens. See especially V. Ahmadjian and S. Paracer (1966); D.C. Smith and A.E. Douglas (1987); V. Ahmadjian (1993); T.H. Nash III (ed.) (1996). Recent work on the phylogeny of lichen is described in F. Lutzoni *et al.* (2001).
2. A geodesic dome combines a spherical design (the strongest and most space efficient for its weight and size) with an exceedingly strong yet economical framing geometry (namely, geodesic or "great circle" tetrahedrons). What holds the whole thing together is the tensegrity in the connections between the tetrahedrons.
3. For an excellent overview of tensegrity, see R. Connelly and A. Back (1998). Tensegrity is also involved in the interconnections between our cells (the so-called "extracellular matrix"). The functional synergy between microtubules and microfilaments in the cytoskeleton is described in R.H. Gavin (1997), and in A.B. Fulton (1993). See also D.E. Ingber (1998) (2000).

2. The "Enchanted Loom"

1. Fired clay bricks date back only to the Neolithic, but the principle of building shelters out of component parts is a much older human (and animal) invention. The purist also will note that fireplaces, kilns, and other structures that are exposed to very high temperatures typically use the so-called refractory bricks that are specially made for the purpose. It is also worth noting that a number of children's toys use the synergy principle in much the same way. For instance, the old-fashioned wooden blocks, Lincoln Logs, and Lego blocks.
2. Data on the automobile industry and automobile use in the United States are from *The Statistical Abstract of the United States 2001*. Sources include K.B. Clark *et al.* (1991); J. Rinehart *et al.* (1997); Y. Monden (1998).
3. Data on the U.S. Postal Service are from *The Statistical Abstract* and the *2000 United States Postal Service Annual Report*, which is excerpted on the USPS website: www.usps.gov/history/anrpt98/.

4. A classic text on this subject is I.M. Roitt (1988). See also E. Edelson (1990); K.D. Elgert (1996). An accessible general summary can be found in G.J.V. Nossal (1993).

5. The scientific literature on this subject is vast and growing exponentially. Key references for further reading are: I. Rock (1984); J. Changeux (1985); C. Blakemore (1988); J. Jaynes (1990); R. Sekuler and R. Blake (1990); D.C. Dennett (1991); N. Humphrey (1992); J.R. Searle (1992); J.R. Searle *et al.* (1997); R. Conlan (ed.) (1999).

6. See especially R.H. Masland (1986); A. Treisman (1986); W.J. Freeman (1991); W. Singer (1999). In humans, the miracle of vision starts with those two amazingly complex organs that are designed to sense light photons and focus on visual objects. Each of the retinas in our eyes has about 125 million receptors (rods and cones) that are specialized for different aspects of the images they detect. The raw inputs to these primary receptors are then processed by three layers of cells (the so-called horizontal cells, bipolar cells, and amacrine cells) before they activate the ganglion cells that make up the optic nerve. Each ganglion is in turn specialized to transmit different aspects of the visual image – motion, size, shape, colors, shadows, etc. So a degree of processing occurs even before the image leaves the eye.

   On its way to the visual cortex – the many-layered area at the rear of the brain where the processing and analysis of various components takes place – the image is further broken down at a way-station called the lateral geniculate nucleus (LGN). We do not know how the LGN is able to perform its complex task, but we do know that the decomposed elements are then routed by the LGN to at least 20 specialized processing areas (plus several other areas that are partially involved) in the visual cortex itself. We also know that the visual cortex is hierarchically organized; a schematic of the interconnections between the various processing areas looks like the layout for a computer chip – although the resemblance ends there.

7. F. Crick (1994), p. 92. Crick's chapter on this subject provides an excellent overview, along with the discussion in H. Curtis and N.S. Barnes (1989). Other standard works on the subject include: D.H. Hubel (1988); G.M. Shepherd (ed.) (1990); J.E. Dowling (1992); J.G. Nicholls *et al.* (1992). Also highly synergistic is the way in which a neuronal signal bridges the so-called synaptic cleft, the small gap separating the junctions between many (not all) neurons. When an action potential reaches the axon terminal, or "knob," it triggers the release of various neurotransmitter chemicals (like glutamate, GABA, serotonin, norepinephrine, or acetylcholine) in the form of small packets called "vesicles" that move across the gap. When the number of neurotransmitter vesicles reaches a threshold quantity, they collectively activate enough receptors in the recipient neuron to open the gates and allow an action potential to flow through its axon. The vesicles are then rapidly diffused, or decay, or get recycled. In other words, the transmission process between any two neurons goes from electrochemical to chemical and back to electrochemical in a small fraction of a second.

8. F. Crick (1994), pp. 103–104.

9. See S. Kashima and M. Kitagawa (1997); B. Hayes (1998); F. Andrews (1999); B. Tedeschi (2000); C. Vogel (2000); G. Johnson (2002); see also "Monitor Test-Drives New Gas-Electric Hybrids from Honda and Toyota" in *Christian Science Monitor*, March 27, 2000, p.15; "Legoland to Open Next Week" (Associated Press) in *Palo Alto Daily News*, March 12, 1999, p. 6; "Here's How They Make a Cartoon" in *Christian Science Monitor*, November 9, 1999, p. 18.

10. W.C. Allee (1931) (1938).

11. J.T. Bonner (1988). See also the discussions of collective behavior in bacteria by J. Shapiro (1988) and J. Shapiro and M. Dworkin (eds.) (1997).

12. E. Hasegawa (1993).

13. T. Malthus (1789), chap. 1, p. 18.

14. E.O. Wilson (1975); A.F. Read and P.H. Harvey (1993).

15. G. Hardin (1968).

16. H. Haken (1973) (1977) (1983).

17. P. Bak and D. Chen (1991); also see P. Bak (1996).

18. F. Lara-Ochoa and A. Herrera (1992); B. Hayes (1998).

19. I. Rock and S. Palmer (1991). Also, see G. Kanizsa (1979); A. Treisman (1986). Many Gestalt effects are also found in chemistry, where the physical pattern of a molecule – its "topology" – predicts what properties it will exhibit. See D.H. Rouvray (1986).

20. S. Hamilton and C. Garber (1997).

21. C.W. Vaughan *et al.* (1997); J.T. Williams (1997). Indeed, synergies are commonplace in pharmacology. See the review in M.C. Berenbaum (1989).

22. See especially the review in P.A. Corning (2002e).

23. See the review in J. Maddox (1990); also R. Hardison (1999). The classic paper on this is J. Monod *et al.* (1965).

24. Y. Le Maho (1977).

25. B.J. Le Boeuf (1985); B.J. Le Boeuf and R.M. Laws (1994).

26. J.L. Gould and C.G. Gould (1995); T.D. Seeley (1995).

27. H.O. von Wagner (1954).

28. One report, among many in recent years, involves the red-cockaded woodpecker, where helpers at the nest indirectly benefit themselves and directly aid in the survival of the nestlings; M.Z. Khan and J.R. Walters (2002). See also the review of allomothering in S.B. Hrdy (2001).

29. G.S. Wilkinson (1984) (1988) (1990).

30. Plato, *The Republic*, Book II, p. 68.

31. A. Smith (1776), p. 5.

32. P.E. Stander (1992).

33. C. Anderson and N.R. Franks (2001); C. Anderson *et al.* (2001); N.R. Franks *et al.* (2001). See also A.W. Shingleton and W.A. Foster (2001).

34. E.O. Wilson (1975); B. Hölldobler and E.O. Wilson (1990). The 200 or so species of leaf-cutter ants are perhaps equally impressive. Famous for their fungus "gardens" (their prime source of food; they forage for leaves only to feed their fungi), the leaf-cutters engage in an elaborate division of labor for building their vast underground nests, gathering and hauling food, planting,

feeding, weeding, and tending their fungus gardens, and fighting parasites and other invaders. In addition to Hölldobler and Wilson, see L. Ariniello (1999).

35. S. Chevalier-Skolnikoff and J. Liska (1993).
36. P. Ward and A. Zahavi (1973). See the update on this work in N.J. Buckley (1997); also S.R.X. Dall (2002).
37. S. Boinski and P.A. Garber (eds.) (2000); also J.W. Bradbury and S.L. Vehrenkamp (1998).
38. M. Kawai (1965); E. Curio *et al.* (1978); R. Aisner and J. Terkel (1992).
39. R. Losick and D. Kaiser (1997); also R. Kolter and R. Losick (1998) and D.G. Davies *et al.* (1998).
40. N.R. Franks (1989). See also F. Saffre *et al.* (1999); I. Peterson (2000).
41. T.D. Seeley and R.A. Levien (1987); also see T.D. Seeley (1995); T.D. Seeley and S.C. Buhrman (1999).
42. Bingo is a game of chance in which each player has a card filled with differently numbered squares. A caller selects numbered pieces at random from a jar, a drum, or a spinner, and the first player to fill an entire row of squares wins. The tradition is that the winner calls out "Bingo" to announce his/her winning card.

3. The Magic Castle

1. T. Ferris (1997), p. 17.
2. There is a small library of books available on this subject, many of which are well written and accessible to the general reader. The principle sources used for the discussion here were J.S. Trefil (1983); P. Davies (1988); S. Weinberg (1988); B. Swimme and T. Berry (1992); N. Hathaway (1994); J. Silk (1994); T.X. Thuan (1995); P.M. Dauber and R.A. Muller (1996); C.J. Hogan (1996); B. Swimme (1996); H. Couper and N. Henbest (1997); T. Ferris (1997); C. Sagan (1997); A. Delsemme (1998); A.R. Liddle (1999); J.A. Peacock (1999); M.J. Rees (2000). In addition to the "standard model," there are a number of more radical scenarios in play these days. Lee Smolin (1997) envisions multiple universes and a process of "cosmic natural selection." Superstring Theory, on the other hand, envisions our universe as being spread out on a "membrane" in a ten-dimensional space; see N. Arkani-Hamed *et al.* (2000). There is also the recent discovery that the universe is expanding at an accelerating rate. This has led to the resurrection of what Einstein called his "greatest mistake" – the idea of a "cosmological constant." What that term refers to is a mathematical expression for an implicit cosmological force which Einstein suggested could offset the theoretical implication of an expanding universe. Einstein believed in a steady state universe. When evidence was later found that the universe is indeed expanding, Einstein abandoned the idea. The cosmological constant is now being resurrected for a very different purpose, to account for the acceleration of the universe. The energetic implications of this finding are still unclear. It may turn out to be related to the fact that some 90% or more of the predicted matter/energy in the universe has not yet been found. To quantum physicists, this and other evidence suggests the possibility of a vast,

universal field (or many interacting fields) that seethes with activity. It is reminiscent of the long-ago rejected notion of an "aether." One incarnation of this emerging new vision is what is called the "zero-point field" theory. Physicist Bernhard Haisch of the Lockheed Palo Alto Research Laboratory, along with Alfonso Rueda, an electrical engineering professor at California State University and H.E. Puthoff, director of the Institute for Advanced Studies at Austin, Texas, have proposed that the material universe is embedded in a vast electromagnetic "sea," and that some of the fundamental phenomena in physics – mass, gravity, and inertia – are all effects produced by interactions with this energy field. For an overview, see B. Haisch *et al.* (1994), pp. 26–31; also B. Haisch and A. Rueda (1996).

3. Efforts are currently under way to understand why this asymmetry occurred, making use of the strange, unstable particles called mesons, which are mixtures of matter and antimatter. See R. Irion (1999).

4. This formulation is particularly vulnerable to revision. The dream of a unified theory to account for these forces remains as strong as ever, and many physicists are hotly pursuing it. Current efforts to unify relativity and quantum theory may hold promise, and the zero-point field theory mentioned earlier represents a radically different, dark-horse alternative.

5. The name "quark" was given to these particles by their discover, the Nobel physicist Murray Gell-Mann. His inspiration was a melodic line from James Joyce's famous novel *Finnegans Wake*: "Three quarks for Muster Mark." The mass of a neutron has been variously pegged at 1836, 1839, 1842, and 1845 times that of an electron. To my knowledge, nobody has sought to reconcile those minor discrepancies. As the growing universe expanded and cooled, some of the available neutrons decayed into protons. But many more teamed up with protons to make ions of "heavy hydrogen" (so called because the addition of a neutron gives the hydrogen extra mass) and helium, which has two proton–neutron pairs. (There was also "light helium" with two protons and only one neutron.) Recent measurements by astronomers have confirmed that, at 3-plus minutes, the universe consisted of about three-quarters heavy hydrogen and one-quarter helium, along with traces (about 2%) of the third lightest element, lithium.

   Of course, this particle goulash was also laced with leptons – neutrinos and negatively charged electrons – as well as vast quantities of photons and other forms of radiation. But that was about it for the Big Whoosh. For the next several hundred thousand years, not much else happened so far as we know.

6. The mathematical calculations involve fractions of an electron volt, the basic unit of electrical charge. In the case of a neutron, the three quarks possess charges of $+2/3$, $-1/3$, and $-1/3$, which sum to zero. With protons, the fractions are $+2/3$, $+2/3$, and $-1/3$, or $+1$, which exactly offsets the negative charge of an electron.

7. Gravitons are discussed in many of the works cited above in Note 3.2. The truth is that we still don't understand what gravity is. Quantum theory cannot explain it at all. And Einstein did away with it. In relativity theory, all of the forces in nature are consequences of the geometry of space–time. It is also believed that

the influence of gravity may have been much greater in the early period after the Big Bang. In any case, we can detect the influence of gravity on a cosmic scale in the hidden "halo" effect that influences the shape of galaxies, and the "bulk motion" of galaxies, and even larger aggregations – groups, clusters, and super-clusters. A solution to the ultimate mystery of gravity may hold the key to understanding more fully the forces of nature and the dynamics of the universe.

8. S. Kauffman (1995). "The analogue in the origin-of-life theory," Kauffman explains, "will be that when a large enough number of reactions are catalyzed in a chemical reaction system, a vast web of catalyzed reactions will suddenly crystallize. Such a web, it turns out, is almost certainly autocatalytic – almost certainly self-sustaining, alive" (p. 58). Accordingly, Kauffman believes that the emergence of life should be "expected . . . almost inevitable."

9. A full understanding of water still eludes us, though progress is continuing to be made. See I. Amato (1992); S. Perkowitz (1999); F.N. Keutsch and R.J. Saykally (2001).

10. R. Burch *et al.* (1989).

11. R.J. Davey *et al.* (1993).

12. O.W. Webster (1991); A. Ravve (2000).

13. An especially important form of catalysis, in theory at least, is the "hypercycle" postulated by physicists Manfried Eigen and Peter Schuster (1979). It involves cross-catalysis, a cyclic reaction network that becomes autocatalytic. However, the number of examples identified in nature to date are limited. See D.H. Lee *et al.* (1997).

14. J.E. Lovelock (1995).

15. The literature on the anthropic principle ranges from the purely materialistic treatments of cosmologists and physicists to Judeo-Christian interpretations in terms of an intelligent God. See especially: B.J. Carr and M.J. Rees (1979); G. Gale (1981); J.D. Barrow and F.J. Tipler (1986); P. Davies (1988); G. Greenstein (1988); J. Gribbin and M.J. Rees (1989); J. Barrow (1993); J. Trefil (1997). Also useful is a bibliographic guide to the literature by Y.V. Balashov (1991). Some Design-oriented works include: J.C. Polkinghorne (1984) (1998); R. Swinburne (1990); P. Davies (1992); K.T. Gallagher (1994). Two closely related issues involve the possible "seeding" of life on Earth from elsewhere and the possibility of life elsewhere in the universe. The first, a seemingly far-fetched idea proposed by astronomer Fred Hoyle and N.C. Wickramasinghe, was based on some suggestive findings by physicist John Platt in the 1950s. The original hostility to this idea from various quarters has at least melted into an openness to the evidence as more has accumulated. See especially the "case" for this idea in F. Hoyle and N.C. Wickramasinghe (1981) (1986). The search for extraterrestrial intelligence, promoted by Carl Sagan and popularized in his novel and posthumous movie, *Contact*, has recently gained new life, as it were, by the discovery of a planetary system somewhat like our own "only" 44 light years away. For a range of views on the meaning of evolution, see C. Barlow (ed.) (1994).

16. G. Wald (1994), p. 126.

17. T. Dobzhansky (1970), p. 3. The current estimate is somewhere between 3 and 30 million species. See R.M. May (1992). P.R. Ehrlich (2000) places the number at 10 million or more (p. 44).
18. Quoted in J. Horgan (1991). His review of this issue is still quite useful.
19. Editorial (2001).
20. C. Ezzell (2002).
21. J.E. Darnell *et. al.* (1990), p.88.
22. See H.J. Morowitz (1992). The theory developed by David Deamer, Harold Morowitz, and others postulates a key role for "amphiphiles" – elongated fatty molecules that are like lipids in modern cells. Amphiphiles, which evidently were present in the prebiotic environment, have the unique ability to align themselves with respect to water and can self-assemble into "vesicles." Certain "recourse" would nonetheless have been able to be selectively transported into the vesicle to catalyze and sustain the beginnings of life according to this theory. See especially D. Deamer (ed.) (1978); H.J. Morowitz (1978b) (1981) (1992); D. Deamer and J. Oro (1980); D. Deamer and R.M. Pashley (1989).
23. Many microbiologists have played an important part, of course, but Lynn Margulis has made both theoretical contributions and, together with various collaborators, has played a leading role in arguing the case and communicating these contributions to a broader audience. See especially, L. Margulis (1970) (1981) (1993); L. Margulis and D. Sagan (1986) (1995); L. Margulis and R. Fester (eds.) (1991).
24. The assumption that ancient stromatolites were formed by the same processes that are observed in modern counterparts has recently been challenged by a mathematical analysis that suggests a purely physical origin – a "spontaneous" sprouting and aggregation process under the right conditions. See J.P. Grotzinger and D.H. Rothman (1996); also D. Pendick (1997). Moreover, many ancient stromatolites do not contain evidence of microbial activity. But others do. And there is no doubt about living stromatolites. This is an issue that needs more research.
25. A. Coghlan (1996). For a more in-depth, scientific treatment of the subject, see D.G. Allison *et al.* (eds.) (2001).
26. However, there is other evidence for the antiquity of cyanobacteria. See J.A. Shapiro (1988); J.W. Schopf (1993); E. Ben-Jacob (1997); J.A. Shapiro and M. Dworkin (eds.) (1997); E. Ben-Jacob and H. Levine (1998); E. Ben-Jacob *et al.* (1998); also S. Liebes *et al.* (1998).
27. J.A. Shapiro (1988).
28. S. Sonea (1991).
29. S. Liebes *et al.* (1998). See also L. Margulis and D. Sagan (1986) (1995).
30. C.R. Woese (2000); see also R. Jain *et al.* (1999); B.R. Levin and C.T. Bergstrom (2000); also the special supplement issue of *The American Naturalist*, vol. 154 (October, 1999).
31. L. Margulis and M. McMenamin (1990).
32. P.W. Price (1991).
33. L. Margulis and R. Fester (eds.) (1991); A. Grajal (1995); L. Margulis and D. Sagan (1995).

34. There are many experiments that show the vital importance of this symbiotic relationship. In areas where soils are depleted of *Rhizobia* and farmers rely on artificial nitrogen-rich fertilizers, efforts are currently under way to "seed" the soils with these with bacteria. See D.W. Wolfe (2001a, b).
35. J.T. Bonner (1988); C.R. Currie (2001).
36. R.D. Vetter (1991).
37. L. Margulis and M. McMenamin (eds.) (1993); L. Margulis and D. Sagan (2001); also E. Mayr (1974a).
38. L. Margulis (1993). For a still-relevant overview of our increasing appreciation of nuclear complexity, see M. Hoffman (1993).
39. T. Cavalier-Smith (1981), p. 272. Also see T. Cavalier-Smith (1991).
40. H.P. Erickson (2001); F. van den Ent *et al.* (2001).
41. J. Maynard Smith and E. Szathmáry (1999).
42. Quoted in L.N. Khakhina (1992a). See also L.N. Khakhina (1992b); L. Margulis and M. McMenamin (eds.) (1993).
43. I.E. Wallin (1927), p. 8.
44. See especially L. Margulis (1970) (1981).
45. Reviewed in L. Margulis and M. McMenamin (1990); also L. Margulis and M. Dolan (1999) and references therein.
46. See L. Margulis *et al.* (2000). A different, "three kingdom," symbiotic scenario was recently advanced by H. Hartman and A. Fedorov (2002).
47. See especially C. de Duve (1996).
48. W. Martin and M. Müller (1998). See also J. Travis (1998); G. Vogel (1998).
49. H.A. Isack and H.U. Reyer (1989).
50. D.S. Wilson and E. Sober (1989). See also P.W. Price *et al.* (1986) (1988).
51. M.J. McFall-Ngai (1994) (1998); M.J. McFall (2000).
52. G. Bell (1985).
53. The following discussion utilizes the works of A.L. Lehninger (1971); H.J. Morowitz (1978a); J.K. Hoober (1984); F.M. Harold (1986); H. Curtis and N.S. Barnes (1989); D.G. Nicholls and S.J. Ferguson (1992).
54. With thanks to Lynn Margulis and Dorion Sagan (1995), p. 158.
55. S. Conway Morris (1998a); R.A. Kerr (1998); also see the special symposium issue on the subject in the *American Zoologist*, vol. 38 (1998).
56. Needless to say, this has been a hotly debated subject. In addition to Gould's ground-breaking book, see the contrary views expressed by paleontologist Simon Conway Morris (1998). Conway Morris and Gould also held an indirect debate on the subject in a "challenge" and "reply" article, "Showdown on the Burgess Shale" (1998). There is also a lively debate at present over paleobiologist Mark McMenamin's evocatively titled *The Garden of Ediacara: Discovering the First Complex Life* (1998). McMenamin argues that many of the soft-bodied Ediacaran forms were members of a separate evolutionary lineage that did not form blastulas and went extinct during the Cambrian breakthrough. McMenamin claims that these creatures represented a possible alternative route to complex, intelligent life forms like ourselves. He believes the Ediacarans evolved in a comparatively idyllic setting in which it was not necessary to have teeth or defensive armor; they did not have

predator–prey relationships. However, in a sort of Cambrian version of the so-called "killer ape" hypothesis about human evolution, McMenamin thinks that the benign Ediacarans were ultimately done in by the predatory ancestors of modern animals – paradise lost. See also the discussion in Bennett Daviss (1998).

57. As Simon Conway Morris puts it, the Cambrian may have exhibited "an exuberance of design but appear[s] to have a fundamental similarity at a deeper genetic level" (1998a), p. 151. For a contrasting perspective see Mark McMenamin's *The Garden of Ediacara* (1998); also the discussion in B. Daviss (1998).

58. P.W. Sherman *et al.* (1992), p. 75. See also P.W. Sherman *et al.* (eds.) (1991).

59. P.W. Sherman *et al.* (1992), p. 78.

60. C.R. Darwin (1859), p. 459.

61. J. Harvey (2001), p. 463. Harvey also cites a number of recent volumes that explore this complexity in depth.

62. G. Evans (1975); M. Coe (1987); D.C. Holzman (1997).

63. K.A. Pirozynski and D.W. Malloch (1975); P.R. Atsatt (1991).

64. D. Pimentel and M. Pimentel (eds.) (1996). See also J.L. Gould and C.G. Gould (1995); S. Buchmann and G.P. Nabhan (1996); J. Hope (1998).

65. J.M. Lesinski (1996); L. Allen and C. Goodstein (1997); C.A. Kearns and D.W. Inouye (1997); O. Woodier (1998); S.A. Doebler (2000).

66. E.J. Kormondy (1969).

67. E. Curtis and N.S. Barnes (1989), p. 1120.

68. Four excellent sources were used for the following discussion: G. Vevers (1971); B. Thorne-Miller and J.G. Catena (1991); W. Gray (1993); M.M. Cerullo (1996). A point stressed in all of these writings is the delicate interdependence of the coral and its inhabitants, a set of relationships that are severely threatened by human activities. For instance, agricultural runoffs add nutrients to the water, leading to blooms in plankton algae, which in turn obscure the penetration of sunlight and photosynthesis by the coral's green algae. Global warming may also threaten the coral communities, which require a very narrow temperature range. Some may already have been stressed by rapid temperature changes; G.R. Williams (1996); D. Hinrichsen (1997). See also the special issue of the *American Zoologist* on coral reefs, vol. 39 (1999), which addresses the growing concern about environmental threats to reef ecosystems.

69. There has recently been some debate in the scientific literature about whether or not cleaner-fish and their hosts really enjoy a mutualistic relationship, or if it might be one-sided and exploitative. See R. Poulin and A.S. Grutter (1996). However, in a subsequent report, Grutter (1999) concluded that: "Cleaner-fish really do clean," and the relationship is mutualistic.

70. S. Liebes *et al.* (1998), pp. 170–171.

71. Quoted in R. Lindley (1988). The literature on Gaia reflects one of the liveliest scientific controversies of our time. Like the Selfish Gene metaphor, its ultimate value may lie not in its veracity but in its consciousness raising (to paraphrase Karl Marx) – in the thinking and research that it has spawned. Whether or not the biosphere has truly cybernetic self-adjusting properties may in the end be

less important than a deeper understanding of the many interdependencies and interactions between biological and geophysical processes. Among the many references, see especially J.E. Lovelock (1979) (1990) (1991) (1995); R. Lindley (1988); C. Barlow and T. Volk (1990) (1992); J.W. Kirchner (1990); C. Mann (1991); L. Margulis and R. Guerrero (1991); S.H. Schneider and P.J. Boston (eds.) (1991); A. Markos (1995); G.R. Williams (1996); L. Margulis and D. Sagan (1997); T. Volk (1997).

72. Lovelock and a colleague sought to defend Gaia theory with what he called his "proudest invention," a thought experiment involving a hypothetical planet, "Daisyworld." Lovelock's fictional planet was covered with two kinds of daisies, which adjusted in opposite but equal ways to changes in its Sun's output. In this way, a global environment and its life forms could be self-regulating, they claimed. See A.J. Watson and J.E. Lovelock (1983). Critics were quick to point out that this simplistic fictional planet bore no relationship to the ecology of the Earth. It did not prove anything about Gaia, which relates to our planet. Equally important, it was an experiment that actually relied on natural selection – differential survival under differing selection regimes – as D.M. Wilkinson (1999) pointed out. See also the response of J.L. Dagg (2002).

4. "Black Magic"

1. *The New York Times*, January 20, 2001, p. 1; January 25, 2001, p. 1; May 27, 2002, p. 16.
2. M.L. Sirower (1998). Sirower also offers some sound advice on how to avoid these traps. The dirty little secret of the acquisition game is that synergy is often used as a buzzword to justify a merger that mainly enriches a small number of lawyers, financiers, and top-level executives; it is not grounded in functional synergies at all. This has given the term a bad name in some quarters, understandably. Ironically, when there are real synergies to be had, the term is often not used; the synergies are self-evident. See also S. Chatterjee (1992); R. Davis and L.G. Thomas (1993); H. Geneen (1997).
3. *The New York Times*, March 18, 2000, p. B1.
4. Associated Press (David Bauder), March 1, 2000.
5. J. Graedon and T. Graedon (1999).
6. A. Silverstein *et al.* (1994); D.R. Goldmann (ed.) (1999).
7. World Bank (2002).
8. S. Milgram (1965) (1973).
9. J.A.F. Stoner (1982).
10. R.B. Zajonc (1965).
11. G. Le Bon (1896). We are so inured to collective behavior – from soccer riots to stock market panics – that we do not as a rule recognize it for what it is, a negative form of social behavior by a highly social species.
12. Among the many sources on this historic tragedy, perhaps the most comprehensive and up-to-date is the volume by D. Lynch (1992).
13. *The New York Times*, October 16, 1997, p. 1.
14. G. Regan (1993), p. 114.

15. P.B. Ryan (1985); G. Sick (1985).
16. The coverage of this non-event was extensive, before the fact. One indication is that there were some 1,300 websites devoted to the subject. Perhaps the most insightful analyses of Y2K were produced by systems scientist, Stuart Umpleby (2000a, b, c), who directed a major before-and-after research program devoted to this millennial snafu.
17. D.L. Kruse and K. Shriner (2000).
18. *The New York Times*, January 14, 1998, p. A16.
19. A. Smith (1776), pp. 10–11.
20. The term "devolution" is frowned upon by many biologists these days, for fear that it may connote the reversal of some normative definition of progress. However, the term can also be used in a more neutral, functional sense as the loss of some trait or capability for various reasons. In fact, this is a common occurrence in evolution. There are many examples: the loss of eyesight in cave-dwellers; flightless birds; the atrophied forelimbs of kangaroos; the loss in humans of the ability to synthesize ascorbic acid; the loss of mitochondria in some eukaryotic protists; the surrender by the chloroplasts in land plants of some 254 genes and, consequently, a loss of the ability to synthesize some 46 proteins that can be produced by their free-living "cousins." See L. Margulis and D. Sagan (1995); also J. Diamond (1986).
21. N. Yoffee and G. Cowgill (eds.) (1988). Also see J. Diamond (1997); P.A. Corning (2002a). Joseph Tainter's *The Collapse of Complex Societies* (1988), a formidable single-authored synthesis, deserves a special note. Tainter's objective is to develop a broad explanatory principle for socio-political devolution, and he supports his thesis with material from 20 case studies, drawn from both Old and New World settings and various historical eras. Tainter begins by noting that there have been at least 11 specific "themes" (not mutually exclusive or free of overlaps) that various authors have associated with socio-political collapses: (1) depletion or denial of a major resource, (2) the establishment of a new resource base, (3) the occurrence of an "insurmountable" catastrophe, (4) an "insufficient" response to some challenge, (5) the actions of other societies, (6) "intruders," (7) class conflicts or elite mismanagement, (8) social "dysfunction," (9) "mystical factors," (10) a chance concatenation of events, and (11) economic factors. Tainter disagrees with these theorists. He finds all of these explanations insufficient, except perhaps as contributing factors. In developing his own alternative explanation, however, Tainter truncates the scope and applicability of his theory by defining a "rise" or a "collapse" as primarily a "political process." Accordingly, he limits his theory only to reductions in "socio-political complexity." However, he does not define political "complexity" in such a way that one can measure it in real-world situations, and in his accompanying discussion he blurs the distinctions between complexity, inefficiency, bloat, the sheer number of workers, or other phenomena. Indeed, a collapse in his terms only differs from a decline in its relative suddenness and rapidity, not its concrete consequences. Nor does Tainter give us any measuring rod for devolution, or even a surrogate "indicator."

Tainter's key proposition is that the collapse of a complex socio-political system will occur when there are "declining marginal returns" – when the economic costs of additional investments in complexity outweigh the additional benefits. Not only are we left in the dark about how to measure complexity but we are not given any way of measuring the marginal value. It also begs the question: marginal value to whom? bureaucrats? a political "elite"? an underclass of slave laborers? As originally stated, Tainter's thesis could not be tested, at least with the conceptual tools provided in his volume. But even if, for the sake of argument, complexity in his terms could be defined and measured, a marginal value relationship would, at best, constitute but one "variable" – neither necessary nor sufficient to explain the many historical instances of political devolution. There is now strong evidence that, in many cases, precipitous socio-political collapses were directly attributable to such "exogenous" variables as conquests, epidemics, key resource depletions, and drastic environmental changes, independently of any discernible internal political dynamic. Conversely, there are many other cases in which political devolution has occurred when the mission was accomplished; there was no longer a need and no potential for positive synergies. It should also be noted that, in a recent, jointly authored article (Allen, Tainter, and Hoekstra 1999), the focus of Tainter's paradigm has shifted from diminishing marginal returns for the political system to a broader economic calculus associated with the marginal returns from "extracting resources" or other societal benefits. This iteration represents a major change; it is much more compatible with the thesis here, where the burden of maintaining a political system is weighed against the underlying purposes and value of the system. Even a bloated, inefficient army will continue to be publicly supported if it effectively deters potential invaders, but the converse is far less likely to be the case.

22. H. Weiss *et al.* (1993); H.N. Dalfes *et al.* (eds.) (1996); H. Weiss and R.S. Bradley (2001).
23. J. Diamond (1997); also M.B.A. Oldstone (1998).
24. B.W. Tuchman (1985), pp. 8–10.
25. This is an update of a more detailed analysis in P.A. Corning (1983) (see also accompanying endnote on p. 481, number 139, therein). More recent references include A.H.M. Jones (1986); A. Johnson and T. Earle (1987); P. Kennedy (1987); R. MacMullen (1988); J. Tainter (1988); N. Yoffee and G.L. Cowgill (eds.) (1988); C.G. Starr (1991); A. Cotterell (ed.) (1993); R. Alston (1998).
26. P. Sorokin (1937), vol. 3, pp. 414–417.
27. Some historians give more weight to a significant shift in the military balance. Contrary to the common misconception that the "barbarians" were, well, barbarian, over time the Visigoths, Huns, Vandals, and others developed an advanced civilization and technology, much of it copied from Rome. Most important, as the Teutonic warlords learned to emulate the secrets of Rome's military success, they became increasingly formidable opponents. And as they began to encroach on the periphery of the empire, they were able to replace Roman order with Gothic order, and a much lighter tax burden

to boot. In time, the word got around. For a lively discussion of the "barbarians," see R. Wright (2000). Also, see especially E.A. Thompson (1948); W.H. McNeill (1963); R. MacMullen (1988); C.G. Starr (1991); A. Cotterell (ed.) (1993).

28. Cited, with cautionary notes about the degree of uncertainty and inference involved, in R. MacMullen (1988), pp. 53–54.

29. G.W. Bowersock (1988) is representative of a revisionist school that rejects the basic thesis, going back to Gibbon (and many more before him). Rome did not decline and fall at all, Bowersock argues. Rather there was a "relocation" of power, an economic "transformation" as Rome proper declined and other areas thrived and, after 476 (when the last emperor was deposed) a political "reformation." This overlooks two fundamental points about Rome, or any other imperial power. The empire was preeminently a political system, an organized system of internal political control and an instrumentality for the defense of a bounded geographic entity. In the end it failed at both of these principal responsibilities and it was ultimately dismembered. The second point is that metropolitan Rome – the city and the surrounding countryside – was the political, administrative, economic, and cultural center, the "hub" of the empire at its height. Not only did the hub decline in the 3rd and 4th centuries AD, but it was ultimately destroyed by invaders. By any reasonable standard it was a fallen empire.

30. A.H.M. Jones (1955), p. 226.

31. Quoted in A.H. Maslow (1964), p. 156. See also A.H. Maslow and J.J. Honigman (1970); B.J. Grindal (1976).

32. The study described here was undertaken by the staff of Corning and Associates, Inc., a management consulting firm, in the spring of 1985. The results were utilized for internal purposes and a summary was published in P.A. Corning and S.P. Corning (1986).

33. M. Connor and M. Ferguson-Smith (1997); C.R. Scriver *et al.* (eds.) (2001).

34. A. Ochert (1999).

35. R.K. Herman (1993); J. Maynard Smith (1998), pp. 7–8.

36. C. Zimmer (2000); see also J.E. Donelson and M.J. Turner (1985).

37. C. Zimmer (2000); see also H. Topoff (1999). It should be noted that some biologists use "symbiosis" as a generic term that encompasses both positive and negative relationships, following the practice of the pioneering German mycologist Anton de Bary, who coined the term in the nineteenth century.

38. M.J. Daft and A.A. El-Giahmi (1978); G.J. Bethlenfalvay *et al.* (1982a, b) (1983).

39. N.G. Smith (1968).

5. The Synergism Hypothesis

1. There are many good book-length treatments of evolutionary theory, though synergy has seldom been mentioned until quite recently. Among others, see F.J. Ayala and T. Dobzhansky (eds.) (1974); D.L. Hull (1974); J. Maynard Smith (1975); E. Mayr (1976) (1982); T. Dobzhansky *et al.* (1977); E. Sober

(ed.) (1984); M. Ruse (1988) (1999) (2000); R.N. Brandon (1996); D.L. Hull and M. Ruse (1998). A number of specific issues are also addressed in R.M. Burian (1992); R. Dawkins (1992); M.J. West-Eberhard (1992).

2. E.O. Wilson (1975), p. 3.

3. Actually, the "holistic" perspective resurfaced at various times over the course of the past century. During the 1930s, embryologist Joseph Needham proposed the concept of "integrative levels" in his Herbert Spencer Lecture and in his 1937 book. The term referred to different levels of biological organization. The idea was also advanced in a 1942 symposium volume edited by the anthropologist Robert Redfield. Biologist Alex Novikoff offered a different formulation of the concept in a 1945 *Science* article. Even Julian Huxley – grandson of Darwin's chief defender, Thomas Henry Huxley, and the one who coined the term "Modern Synthesis" in his landmark 1942 book, *Evolution: The Modern Synthesis* – flirted with the idea in a later book on evolution and ethics which included the two Romanes Lectures given by himself and his grandfather. "Now and again there is a sudden and rapid passage to a totally new and more comprehensive type of order or organization, with quite new emergent properties, and involving quite new methods of further evolution" (J.S. Huxley and T.H. Huxley 1947, p. 120). Emergence reappeared once again in the 1950s as "general systems theory," under the leadership of biologist Ludwig von Bertalanffy, Anatol Rapoport, Kenneth Boulding, Ralph Gerard, James Grier Miller, and others. Today this tradition goes under the heading of the "systems sciences." See also D. Blitz (1992).

4. G.C. Williams (1966), p. 8.

5. R. Dawkins (1976), p. ix. This famous prefatory remark was only one of many provocative statements that, in retrospect, seemed designed as much as anything to rebut the playwright popularizer Robert Ardrey (1970), whose book relied, as an act of faith, on group selection theory. Here are just a few examples: "I shall argue that a predominant quality to be expected in a successful gene is ruthless selfishness" (p. 2); "Be warned that if you wish, as I do, to build a society in which individuals cooperate generously ... you can expect little help from biological nature ... we are born selfish" (p. 3); "universal love and the welfare of the species as a whole are concepts which simply do not make evolutionary sense" (pp. 2–3). His soulmate, George C. Williams, is equally sanguinary in his writings. For instance: "Natural selection maximizes shortsighted selfishness, no matter how much pain or loss it produces" (quoted in F. Roes 1998).

6. P. Little (1995). This was only a progress report on the Human Genome Project. There have been many consistent findings since. However, the recently completed preliminary "map" of the full human genome leaves many of the interrelationships and network properties of the genome still murky.

7. R. Dawkins (1986), pp. 170–171.

8. P. Kropotkin (1902), p. vi.

9. E.B. Tylor (1871), Vol. 1, p. 62. See also A.G. Keller (1915); W.G. Sumner (1934).

10. Quoted in K. Lux (1990), p. 148.

11. A. Carnegie (1889), "The Gospel of Wealth," *North American Review*, reprinted (as "Wealth") in J.F. Wall (ed.) (1992), p. 132.

12. One of the more dishonorable episodes in the history of science, it is discussed in detail in my article, "Durkheim and Spencer," in the *British Journal of Sociology* (1982). See also the new book on this subject by S. Pinker (2002).

13. E.O. Wilson (1975), pp. 4, 547. Actually, Wilson did not use the term "epigenetic rules" in *Sociobiology*, but the concept was there in spirit. It first appeared in a subsequent volume co-authored by C.J. Lumsden and E.O. Wilson (1981).

14. When my earlier book, *The Synergism Hypothesis*, was published as a full-length scholarly monograph in 1983 (by McGraw-Hill), it was mostly unnoticed or was simply dismissed without serious consideration. Reviews were also sparse. Some were ideologically motivated; others complimented the scholarship and ignored the theory.

15. See E.F. Keller (1983).

16. J. Maynard Smith (1982a, b) (1983) (1984a) (1989).

17. L.W. Buss (1987). Buss seemed unaware of *The Synergism Hypothesis*, though it had been published four years earlier. Nor has he cited it subsequently (e.g., Buss 1999). Two sophisticated, book-length treatments of the issue raised by Buss are S.A. Frank (1998) and R.E. Michod (1999).

18. D. Smillie (1993) (1995).

19. D.S. Wilson (1975) (1980); D.S. Wilson and E. Sober (1994); D.S. Wilson and L.A. Dugatkin (1997).

20. In addition to the works cited in the previous note see the edited volume by L. Keller (1999); also B.J. Bradley (1999); J. B. Wolf *et al.* (1999); S. Okasha (2001); D.S. Wilson (2001).

21. L.A. Dugatkin (1997) (1999).

22. L. Margulis and M. McMenamin (eds.) (1993). See especially L. Margulis (1970) (1981) (1990) (1993); L. Margulis and R. Fester (eds.) (1991). For the historical background, see L. N. Khakhina (1979) (1992a, b).

23. J. Maynard Smith and E. Szathmáry (1995) (1999). Nowadays, articles about synergy in evolution are routinely accepted for publication, whereas 15 years ago they were routinely rejected. Indeed, an increasing number of field studies are looking for, and finding, synergy in nature. "Emergence" has also reemerged as a hot topic in the biological sciences, from John Holland's recent book to a brand new journal with that name. A full-length article on the subject by this author is currently in press at the journal *Complexity* – Corning (2002e). It seems that synergy is an idea whose time has come.

24. J. Bronstein (2001). The general tenor of the work on cooperation is that mutualism is far more widespread and less constrained by kinship ties than had previously been supposed. A sampler of this work includes S. A. Frank (1995a); M. W. Schwartz and J. D. Hoeksema (1998); M.-A. Selosse and F. Le Tracon (1998); E. A. Herre *et al.* (1999); K. Clay (2001); J. N. Holland *et al.* (2002). On cooperative sperm, see H. Moore *et al.* (2002). Also, see the special issue of *Human Nature* (vol. 13, no. 1, 2002) that is devoted to cooperation.

25. N. Eldredge and M. Grene (1992); N. Eldredge (1995); S.A. Frank (1998); E. Sober and D.S. Wilson (1998). A number of economists have also taken an interest in this approach. See especially J. Laurent and J. Nightingale (2001); H. Ofek (2001). See also the references cited in the preceding note.
26. G. G. Simpson (1967), p. 219; E. Mayr (1976), p. 365; E.O. Wilson (1975), p. 67.
27. H. B. D. Kettlewell (1955) (1973).
28. R. Lande and S.J. Arnold (1983), p. 1210. A. Grafen (1991); see also P. Hammerstein (1996).
29. B. Clarke (1975).
30. B.R. Grant and P.R. Grant (1979) (1989) (1993). Also P.R. Grant (1986) (1991); P.R. Grant and B.R. Grant (2002). See also the Pulitzer Prize–winning account of this important work in J. Weiner (1994).
31. Readers who are familiar with the literature on "complexity" are doubtless aware that one of the more vexing issues in this field involves how to define and measure the phenomenon. A skeptical commentary by this author called "Complexity is Just a Word!" appeared in the journal *Technological Forecasting and Social Change* (1998). It is available on the ISCS website: www.complexsystems.org. Biologist Eörs Szathmáry and his colleagues (2001) have proposed a novel approach to measuring complexity that focuses on relationships and interactions in a living system, rather than an enumeration of the number of parts.
32. K. Padian and L.M. Chiappe (1998), p. 44.
33. Newly discovered fossils and new analyses (such as the controversy over the origin of the "digits" in birds) have kept this issue "in the news," but the basic scenario seems firmly established. Some recent items include: A.C. Burke and A. Feduccia (1997); R. Hinchliffe (1997); S. Chatterjee (1998); J. Qiang *et al.* (1998); C. Zimmer (1998); X. Xing *et al.* (1999); T.D. Jones *et al.* (2000); F. Zhang and Z. Zhou (2000); M.A. Norell and J. Clarke (2001).
34. The following discussion is synthesized from P.R. Bergquist (1978); J.D. George and J.J. George (1979); E.F. Ricketts *et al.* (1985); H. Curtis and N.S. Barnes (1989).
35. See especially M.A. Sleigh (1989).
36. The "selfish genome" is more than a facile metaphor. There is increasing evidence that the genome has various evolved means for exercising control over the genes. Recall the error-reduction mechanisms described in Chapter 2, under the heading of functional convergences (see R.H. Haynes, 1991). Equally important, it is becoming evident that mutation rates can also be canalized and accelerated in non-random ways. There are mutators that can modulate mutation rates, and there are "hypermutations" at certain "contingency loci" that may occur under various stress conditions. Some geneticists refer to it as an SOS response. See the reviews in B.A. Bridges (1997); E.R. Moxon and D.S. Thaler (1997); M.F. Goodman (1998). Also, see J. Cairns *et al.* (1988); J.A. Shapiro (1991) (1992); R.S. Harris *et al.* (1994); D.S. Thaler (1994). As Moxon and Thaler note: "Intuitively, it seems advantageous for organisms

to evolve mechanisms through which the environment could influence the genetic mechanisms that would alter the quantity and type of alleles in the repertoire on which selection acts." Perhaps most intriguing are the so-called "microsatellites," strange repeat DNA sequences contained in the non-coding "junk DNA" of the genome that actually perform a variety of remarkable functions and may play an important role in evolutionary change. See E.R. Moxon and C. Wills (1999). Eva Jablonka and her colleagues characterize these mechanisms as "Lamarckian" – after the pre-Darwin French naturalist who first proposed that environmental/behavioral influences were a directing agency in evolution (E. Jablonka *et al.*, 1998). We will revisit this issue in Chapter 6.

37. M. Ridley (2001). Ridley claims that what he calls "Mendel's demon" (the law of segregation) is "the key to the creation of complex life on Earth" (p. 226), because it ensures that all genes have a "fair" (equal) chance of being represented in succeeding generations (unless some "segregation distorter" is present). Calling this the "key" is a bit too strong, however. It might have been an important precondition (necessary but not sufficient). But functional synergy was the collective "key," and Mendel's demon was only one member of a large, complex team.

38. There are many articles on meerkats for general readers and on-line researchers. More technical sources used for this discussion included: D.W. Macdonald (1986); M.J. van Staaden (1994); S.P. Doolan and D.W. Macdonald (1996a, b) (1997); P.N.M. Brotherton *et al.* (2001). Although T.H. Clutton-Brock and his colleagues (1999) have shown that the meerkat sentry behavior is probably very low-risk and not altruistic, as had been supposed, nevertheless it does benefit others and is a shared cost. At very least, it is an example of by-product mutualism (see below). Indeed, Clutton-Brock's group found that smaller groups with fewer sentinels had higher rates of predation.

39. R.L. Jeanne (1986).

40. C. Campagna *et al.* (1992).

41. The term "superorganism" was coined by Herbert Spencer in the nineteenth century. It has come and gone among biologists over the years as scientific fashion and theoretical dogma have shifted. However, it has recently come back into vogue. The history of this term and some theoretical spadework on clarifying its usage can be found in Corning (2002b).

42. B. Hölldobler and E.O. Wilson (1994), p. 121.

43. C.R. Currie *et al.* (1999a, b) (2001).

44. The following discussion is drawn principally from W.D. Koenig and R.L. Mumme (1987). Also W.D. Koenig *et al.* (1995).

45. W.D. Koenig and R.L. Mumme (1987), pp. 374–375.

46. There are many good sources on this subject. One of the best remains E. O. Wilson's *Sociobiology* (1975). Other "classics" include: W.C. Allee *et al.* (1949); J.P. Scott (1958); N. Tinbergen (1965); L.D. Mech (1970); H. Kummer (1971); H. Kruuk (1972); G.B. Schaller (1972); G.F. Oster and E.O. Wilson (1978); R.L. Trivers (1985); J. Goodall (1986); J.L. Brown (1987); B. Smuts *et al.* (eds.) (1987); S.C. Strum (1987); R.I.M. Dunbar (1988). Some of the more notable

recent volumes include: A.H. Harcourt and F.B.M. de Waal (eds.) (1992); J.R. Krebs and N.B. Davies (1993); R.W. Wrangham *et al.* (eds.) (1994); F.B.M. de Waal (1996); L.A. Dugatkin (1997).

47. B.L. Partridge (1982).
48. J.M. Marzluff and B. Heinrich (1991).
49. J.L. Hoogland (1979a, b) (1981) (1995).
50. G.B. Schaller (1972). Later studies revealed that these efficiencies, and the likelihood of cooperative hunting, are very much context-dependent. See the review in C. Packer and L. Ruttan (1988).
51. J.E. Strassmann and D.C. Queller (1989).
52. C. Packer and L. Ruttan (1988).
53. S.W. Rissing and G.B. Pollack (1991).
54. S. Cahan and G.E. Julian (1999).
55. W.R. Tschinkel (1998).
56. C. Darwin (1871), pp. 115–117, 119.
57. See especially S.A. Frank (1998); also J.L. Bronstein (2001); H. Ofek (2001); J.N. Holland *et al.* (2002).
58. C. Packer and L. Ruttan (1988). See also J.R. Krebs and N.B. Davies (1993). For research related to human societies, see E.A. Smith (1985).
59. G.J. Bethlenfalvay *et al.* (1982a, b) (1983). Also M.J. Daft and A.A. El-Giahmi (1978).
60. N.G. Smith (1968).
61. W.M. Shields and J.R. Crook (1987).
62. R.I.M. Dunbar (1992).
63. L. Avilés and P. Tufiño (1998).
64. J.L. Hoogland (1979a) (1981).
65. C. Packer and A.E. Pusey (1982). Also C. Packer *et al.* (1990a, b).
66. B.J. Le Boeuf (1985); B.J. Le Boeuf and R.M. Laws (1994).
67. J.A. Raven (1992).
68. W.J.H. Peace and P.J. Grubb (1982).
69. J.B.S. Haldane (1932), p. 131.
70. W.D. Hamilton (1964a, b).
71. Among the many studies, see especially J.D. Ligon and S.H. Ligon (1982); R.L. Mumme *et al.* (1988); M.F. Clarke (1989); R.L. Mumme (1992); S.M. Haig *et al.* (1994). See also the general review in J.L. Brown (1987).
72. For examples, see A.P. Møller (1987); P.G. Parker *et al.* (1994). Also the general review in J.L. Brown (1987). Brown comments that, as a rule, helpers seem to be somewhat less conscientious where non-kin are involved. However, more recent studies suggest otherwise.
73. Reviewed in J.L. Brown (1987).
74. Reviewed in E.O. Wilson (1975); also J.L. Brown (1987).
75. J.E. Strassmann *et al.* (1997). See also D.C. Queller *et al.* (1988) (1990); P.W. Sherman *et al.* (1988); S.W. Rissing *et al.* (1989); M. Mesterton-Gibbons and L.A. Dugatkin (1992); M.D. Breed *et al.* (1994); M.P. Scott (1994); J.E. Strassmann *et al.* (1993); C.J. De Heer and K.G. Ross (1997); L. Keller (1997); G. Bernasconi and J.E. Strassmann (1999).

76. See, for instance, C. Packer and A.E. Pusey (1982); J. Moore (1984); C. Packer and L. Ruttan (1988); B. Chapais *et al.* (1991); D. Scheel and C. Packer (1991); C.B. Stanford (1992); G.S. Wilkinson (1992); T.L. Goldberg and R.W. Wrangham (1997); C. Packer *et al.* (2001).

77. See, for example, C.B. Stanford (1992); E. Avital *et al.* (1998); T.H. Clutton-Brock *et al.* (2000).

78. See the brief summary of a forum on the "Causes and Consequences of Sociality," S. Cahan *et al.* (1999); also B. Chapais (2001); A.S. Griffin and S.A. West (2002); A. Roulin (2002). Also the differing views, with respect to insects, of L. Keller and M. Chapuisat (1999).

79. S.A. West *et al.* (2002).

80. M.C. King and A.C. Wilson (1975); D. Botstein *et al.* (1997); H. Kaessmann *et al.* (1999). John Maynard Smith and I have (independently) suggested the same metaphor for what is really involved in sexual reproduction. Call it the car repair theory. The notion is that the process is rather like what happens when two non-operative older automobiles with mostly identical parts (at least the essential ones) are cannibalized to rebuild one functioning vehicle.

81. R.L. Trivers (1971). Actually, Trivers snuck in "policing" (see below) as a way to plug a large hole in his argument. To quote from Trivers's article: "Why should the rescued individual bother to reciprocate? ... Why not cheat? ... Cheating will discriminate against the cheater if cheating has later adverse affects on his life which outweigh the benefit of not reciprocating. This may happen if the altruist responds to the cheating by curtailing all future possible altruistic gestures to this individual. Assuming that the benefits of these lost altruistic acts outweigh the costs involved in reciprocating, the cheater will be selected against relative to individuals who, because neither cheats, exchange many altruistic acts." In effect, this argument presaged the tit-for-tat model of Axelrod and Hamilton (see below).

82. J. Maynard Smith (1974) (1982b) (1984a).

83. R. Axelrod and W.D. Hamilton (1981). In their original model, one point each was allowed for mutual defection; asymmetrical cooperation yielded five points for the defector and none for the cooperator; and mutual cooperation yielded a total of six points, evenly divided. Since defectors would be penalized by reciprocal defection in the next round, after two rounds the mutual cooperators would outgain a defector and the benefits would increase from there on. In other words, mutual cooperation is mutually synergistic.

84. M. Nowak and K. Sigmund (1993). For the mathematically minded, see also the much more comprehensive analysis of game theory and other evolutionary models (like the famed Lotka–Volterra equations) in J. Hofbauer and K. Sigmund (1998). See also K. Sigmund (1993); K. Binmore (1994) (1998).

85. Darwin was aware of this factor, of course. Among the more recent discussions, see R. Boyd and P.J. Richerson (1992); T.H. Clutton-Brock and G.A. Parker (1995). See also S.A. Frank (1995b) (1996) (1998) and the analyses in R.E. Michod (1996) (1999). Also see J. Hirshleifer (1999).

86. T.H. Clutton-Brock and G.A Parker (1995), p. 209. Evidence of a policing function has been documented in, among others, social insects, naked mole-rats, primates, and, needless to say, humankind. Of course, policing may sometimes fail. See S.J. Martin *et al.* (2002).

87. J. Maynard Smith and E. Szathmáry (1995), p. 261.

88. L.A. Dugatkin (1999), pp. 118–120.

89. L.A. Dugatkin (1999), p. 112. Dugatkin even quotes the originator of the concept, but he does not make clear the distinction between "accidental" or "incidental" cooperation and evolved, "intentional" adaptations (see below).

90. J.L. Brown (1987).

91. R.S.O. Harding (1973); S.C. Strum (1975a, b); R.S.O. Harding and S.C. Strum (1976); R.S.O. Harding and G. Teleki (eds.) (1981).

92. See R.D. Alexander (1987). Also R. Boyd and P.J. Richerson (1989); R.L. Mumme *et al.* (1989); M. Mesterton-Gibbons and L.A. Dugatkin (1992); M.A. Nowak and K. Sigmund (1998a, b); H. Gintis (2000a, b); T.H. Clutton-Brock *et al.* (2001); T.H. Clutton-Brock (2002).

93. Dawkins himself acknowledged as much in one of the less-frequently quoted passages of *The Selfish Gene* (1976). The genes are not really free and independent agents, he explained. "They collaborate and interact in inextricably complex ways ... Building a leg is a multi-gene, cooperative enterprise" (p. 39). To underscore the point, he too employed a metaphor from rowing. "One oarsman on his own cannot win the Oxford and Cambridge boat race. He needs eight colleagues ... Rowing the boat is a cooperative venture" (p. 40). Furthermore: "One of the qualities of a good oarsman is teamwork, the ability to fit in and cooperate with the rest of the crew" (p. 41).

94. C. Darwin (1871), pp. 146–147.

95. V.C. Wynne-Edwards (1963). See also V.C. Wynne-Edwards (1962).

96. G.C. Williams (1966), p. 211.

97. D.S. Wilson (1975) (1980); D.S. Wilson and E. Sober (1989) (1994); also L. Keller (ed.) (1999).

98. E. Sober and D.S. Wilson (1998).

99. R. Dawkins (1989), p. xxi.

100. M. Connor and M. Ferguson-Smith (1997); C.R. Scriver *et al.* (2001).

101. R. Dawkins (1986), pp. 170, 171.

102. G.C. Williams (1993).

103. Bioeconomics is in fact an emerging interdiscipline that melds economics and biology. See especially the pioneering work in this area by M. Ghiselin (1974) (1978) (1986) (1992). An effort to apply a bioeconomics paradigm to human societies in particular is undertaken in a new paper, "Fair Shares: Beyond Capitalism and Socialism, The Biological Basis of Social Justice," P.A. Corning (2002c). See also P.A. Corning (2000).

104. F. Crick (1994), p. 11. For the record, I hasten to reiterate that reductionism remains an immensely important scientific strategy; a great deal has been learned, and will be learned, by close study of the parts and their interactions, but it is only when you put the parts together that you can observe what the whole can do. Benzene, for instance, forms a colorless liquid that is lighter

than water and insoluble; it burns easily; in sufficient quantities it can be poisonous or carcinogenic; and it has many industrial uses.

105. J. H. Holland (1998), p. 2.
106. C. Lloyd Morgan (1923) (1926) (1933). Also J.C. Smuts (1926); W.M. Wheeler (1927a, b). An excellent analytical history of the concept of emergent evolution, with particular emphasis on the work of Lloyd Morgan, can be found in D. Blitz (1992). See also the review in P.A. Corning (2002e).
107. J.C. Smuts (1926), pp. 182–184.
108. Very similar ideas about the role of synergy in the evolution of complexity were advanced independently by Maynard Smith and Szathmáry in their new volume, *The Origins of Life* (1999), although they also graciously acknowledged my earlier book on the subject (Corning, 1983). One significant difference between us is that Maynard Smith and Szathmáry believe that both synergy and relatedness are preconditions for the evolution of complexity in nature. I disagree. I hold that synergy alone may be sufficient in many cases – witness symbiogenesis. Indeed, the increasing volume of evidence for horizontal gene transfers also challenges their conclusion. Even within the same species, relatedness may be a facilitator but it is clearly not a necessary prerequisite for social evolution. The convergent works of two other theorists should also be noted, though neither cited my 1983 book. These were J.T. Bonner, *The Evolution of Complexity* (1988) and L.W. Buss, *The Evolution of Individuality* (1987).
109. An oft-cited landmark is D'Arcy W. Thompson's 1917 classic *On Growth and Form*. Thompson carefully documented the many different kinds of engineering and physical principles that create imperatives for the operation of living systems. Another landmark was the study by Bernard Rensch (1947). Rensch identified 20 classes of phenomena that impose law-like constraints on the evolutionary process (like the need to base the process on carbon compounds and a water medium). Rensch also generated no less than 100 generalizations about development (morphogenesis). Other classic works on the "mechanics" of the evolutionary process include: R.M. Alexander (1968); K.S. Schmidt-Nielsen (1972); T. McMahon (1973). Philosopher of science David Hull (1974) also pointed out that there are literally thousands of "laws" relating to deterministic aspects of various biosciences – biochemistry, molecular biology, bioenergetics, neurophysiology, embryogenesis, etc.
110. S.J. Gould (1996), p. 200.

6. "The Sorcerer's Apprentice"

1. J.B. de Lamarck (1809); H.G. Cannon (1955) (1959); J.M. Savage (1977). After this chapter was completed and I began the research for Chapter 7, I discovered that biologist Jonathan Kingdon had used "The Sorcerer's Apprentice" as the title for one of the chapters in his book *Self-Made Man* (1993). Happily, he used it for a very similar purpose – as a metaphor for the active role that we have played in our own evolution.
2. J.B. de Lamarck (1809), p. 114.

3. C.R. Darwin (1859), p. 175. In truth, Lamarck also said some things that now seem like "nonsense." For instance: "In the frequent fits of anger to which the males are especially subject, the efforts of their inner feelings cause the fluids to flow more strongly towards that part of their head; in some there is hence deposited a secretion of horny matter, and in others of bony matter mixed with horny matter, which gives rise to solid protuberances: thus we have the origin of horns and antlers." Lamarck (1809), p. 122.

4. A. Weismann (1904).

5. Though Lamarck was often accused, even by Darwin, of proposing that new habits arise as a result of spontaneous "volition" or "desire," he said no such thing. This misapprehension was the apparent result of a mistranslation of the French word *besoin*; the word *volonté* was used by Lamarck only in relation to some "higher" animals. In fact, Lamarck viewed behavioral changes, by and large, as a matter of challenge and response, of externally stimulated creativity rather than a spontaneous impulse. See H.G. Cannon (1955) (1959).

6. C.R. Darwin (1859), p. 215.

7. H.M. Baldwin (1895) (1896a, b, c). See also C. Lloyd Morgan (1896b) (1900); H.F. Osborn (1896a, b) (1897).

8. G.G. Simpson (1953), p. 116.

9. C.H. Waddington (1942) (1952) (1957) (1961). To illustrate, imagine if the calluses we develop on our hands when we engage in heavy manual labor were to begin appearing spontaneously in early child development, before doing any manual labor.

10. C.H. Waddington (1975), p. 170.

11. A. Roe and G.G. Simpson (eds.) (1958).

12. Even biologist Henry Plotkin, in his important edited volume devoted to this subject, *The Role of Behavior in Evolution* (1988), argued that Mayr's claim that the pacemaker role of behavior in evolution is "self-evident" but was not developed or documented sufficiently by Mayr, or other theorists of that era. Plotkin was generally critical of the earlier work on this subject and found a relative lack of compelling evidence in the literature. However, Plotkin did not cite, and evidently did not read, the in-depth article by Mayr quoted below. Nor did he cite the 1958 volume edited by Anne Roe and George Gaylord Simpson on *Behavior and Evolution*. Indeed, Plotkin acknowledged that he had left out of his volume a number of relevant sources of evidence, namely, the research literature on habitat selection, dietary choices, migratory behaviors, predator–prey interactions, and culture (both in animals and humankind). Many of these were actually discussed in the Roe and Simpson volume. Despite this shortcoming, the contributions in Plotkin's volume by Hull, Brandon, Odling-Smee, Dunbar, and Plotkin himself were very useful additions to the literature on this subject. Mayr reiterates the point in his most recent book: "There are reasons to believe that behavioural shifts have been involved in most evolutionary innovations ..." (Mayr 2002, pp. 136–137).

13. E. Mayr (1960), pp. 371, 377–378. Although Mayr's strong views on the subject may also seem one-sided, from a modern perspective, Mayr was arguing specifically against the concept of "mutation pressure," the notion that mutations

might induce a particular evolutionary trend, independently of functionally based changes.

14. R. Byrne (1995); also J.L. Gould and C.G. Gould (1994).
15. W. Köhler (1925); R. Byrne (1995).
16. J. Goodall (1986), pp. 75–76.
17. K. von Frisch (1967). Not only have these experiments been replicated many times, but entomologists James and Carol Grant Gould have greatly extended and enriched the research in this vein. Bees seem to develop cognitive maps of their terrain and to detect tricks, like a feeder placed in the middle of a lake. See especially their 1995 volume *The Honey Bee*; also T.D. Seeley (1995).
18. J.F. Reinhardt (1952); P. Pankiw (1967).
19. M. Kawai (1965).
20. Among the many references, the volumes cited above are still relevant, namely W.H. Thorpe (1956); C.H. Waddington (1957) (1961); L.L. Whyte (1965); A.C. Hardy (1966); R.A. Hinde (1966); A. Koestler (1967). A sampler of the many more recent contributions includes P.P.G. Bateson (1988); P.P.G. Bateson *et al.* (1993); H.C. Plotkin (1988) (1994); C. Wills (1993); L. Margulis and D. Sagan (1995); R.W. Byrne (1995); E. Avital and E. Jablonka (2000); C. Heyes and L. Huber (eds.) (2000); S.I. Yoerg (2001); C.D.L. Wynne (2001).
21. See E. Avital and E. Jablonka (2000). Rather than provide even a selective listing of the relevant research reports, which would number in the thousands, the reader is referred to some of the major journals where this research is regularly reported, including: *American Naturalist, Animal Behaviour, Behavioral Ecology, Behavioral Ecology and Sociobiology, Evolutionary Ecology, Journal of Comparative Psychology, Nature, Primates, Proceedings of the National Academy of Sciences, Proceedings of the Royal Society, Series B, Quarterly Review of Biology, Science*, and *Scientific American*, among others. A "sampler" for some of the species cited here includes J.H. Thomas (1994) (nematodes); R. Dukas (1998a) (fruit flies); M. Giurfa *et al.* (1999) (honeybees); K.N. Laland and S.M. Reader (1999) (guppies); S. Pika and M. Tomasello (2001) (gorillas).
22. B. Adler (1997).
23. D.R. Griffin (2001).
24. L. Margulis and D. Sagan (1995), p. 180.
25. The research on animal "decision-making" goes back more than 30 years to the seminal work on "optimal foraging theory" by J.M. Emlen (1966); R.H. MacArthur and E.R. Pianka (1966); T.W. Schoener (1971); E.L. Charnov (1976); and others. In a recent commemoration in the journal *Evolutionary Ecology*, Oswald J. Schmitz (1997) observed that, over the years, "the field evolved into a sub-discipline of ecology in its own right that has produced a tremendous body of insight into decision making by animals. [But] it also sparked some of the most acrimonious debate ever seen in a scientific discipline." There is also the large and rapidly growing research literature in the field of "behavioral ecology," which utilizes explicit economic analyses of animal behavior. See especially the benchmark volumes edited by J.R. Krebs and N.B. Davies (1984) (1991) (1997). Also, the pioneering work of Marion Stamp

Dawkins (1983) (1988) (2000); J.L. Gould and C.G. Gould (1994); the volume on *Cognitive Ecology* (R. Dukas, ed.) (1998b); the recent papers by G. Mason *et al.* (1998a, b) on demand elasticities in animal choices; the new volume on *Economics in Nature* (Ronald Noë *et al.*, eds.) (2001).

26. R. Foley (1995).

27. S. Gilroy and A. Trewavas (2001).

28. B. Heinrich (1999). The problem-solving experiment is also reported in B. Heinrich (1995). The animal problem-solving paradigm, going back at least to the 1920s, has enjoyed a renaissance recently in comparative psychology. The overall impact of this work has been to narrow the gap between human intelligence and that of other animals and to underscore our continuity with the broader process of mental evolution in nature.

29. The work on observational learning in red squirrels was conducted by P.D. Weigl and E.V. Hanson (1980); observational learning in cats was reported by E.R. John *et al.* (1968); the social transmission of learning in pigeons was studied by B. Palameta and L. Lefebvre (1985). The research on ground squirrels was cited in R. Byrne (1995), p. 58. Among the many other works related to this subject, see especially E.O. Wilson (1975); J.T. Bonner (1980); R.L. Trivers (1985); D. McFarland (ed.) (1987); J.C. Turner *et al.* (1987); T.R. Zentall and B.G. Galef, Jr. (eds.) (1988); J. Alcock (1993); E. Avital and E. Jablonka (1994) (2000); R. Byrne (1995); J.L. Hoogland (1995); E. Jablonka and M.J. Lamb (1995); C.M. Heyes and B.G. Galef, Jr. (eds.) (1996); L.A. Dugatkin (1997) (2000).

30. E. Curio *et al.* (1978).

31. R.W. Byrne and A. Whiten (eds.) (1988); A. Whiten and R.W. Byrne (eds.) (1997). Many of the examples of animal learning described here can be found in the volumes cited in note 6.29 above. See also R. Gibson and T. Ingold (eds.) (1993).

32. J. Aisner and J. Terkel (1992).

33. E.O. Wilson (1975), p. 172. B.B. Beck (1980) and W.C. McGrew (1992) cover this subject in detail. Among the many recent reports on animal tool-use, P.W. Switzer and D.A. Cristol (1999) find that many more species of birds drop hard-shelled prey from a height to break them open; B. Bower (2000) describes Janet Mann's finding that dolphins frequently carry sponges in their mouths, perhaps for flushing prey out of their hiding places; golden lion tamarins use sticks to pry bark from trees and probe crevices, presumably to obtain invertebrate "snacks," according to T.S. Stoinski and B.B. Beck (2001). For a more general discussion of the topic, as it relates especially to primates, see C.P. van Schaik *et al.* (1999). An insightful recent analysis can be found in J.S. Turner (2000).

34. See especially the descriptions in R.W. Wrangham *et al.* (eds.) (1994); also E.O. Wilson (1975); B.B. Beck (1980); W.C. McGrew (1992).

35. See C. Boesch's chapter in K.R. Gibson and T. Ingold (eds.) (1993). However, Christophe Guinet (1991) has reported that killer whales teach hunting techniques to their young, and E.B. Ottoni and M. Mannu (2001) report having

observed free-ranging young tufted capuchin monkeys learning to use stones to break open *Syragus* nuts, though it may only involve observational learning. See also C. Guinet and J. Bouvier (1995).

36. A. Whiten *et al.* (1999). (E.g., the new report on culture in orangutans.)

37. F.B.M. de Waal (1999). Supportive articles also appeared in *The Sciences* (C.B. Stanford 2000); *Discover* (M. Small 2000); *Scientific American* (A. Whiten and C. Boesch 2001).

38. J.T. Bonner (1980). Among the works that have utilized and elaborated upon this "model" of culture, see P.A. Corning (1983); R. Boyd and P.J. Richerson (1985); R.W. Wrangham *et al.* (eds.) (1994); W.G. Runciman *et al.* (eds.) (1996); L. Cronk (1999); C. Heyes and L. Huber (eds.) (2000); F.B.M. de Waal (ed.) (2001a).

39. B.G. Galef (1992).

40. R.H. Tuttle (2001).

41. B.J. King (2001). In an effort to bring more clarity and precision to the term, W.C. McGrew (2001) has resurrected and modified the list of eight specific criteria first developed by the distinguished early twentieth-century anthropologist Alfred L. Kroeber. These criteria include: innovation, dissemination, standardization, durability, diffusion, tradition, species valid (not an artifact of human manipulation), and "transcendent."

42. There are a growing number of exceptions that, it is hoped, are indicative of a sea change. One notable exception is the philosopher of science Eva Jablonka and various colleagues, who have published a number of articles and a co-authored book on Lamarckian aspects of evolution. She stresses the multi-leveled character of "induced heritable variations" in nature, from hypermutations and other sources of environmentally influenced biases in mutational processes (a growing point in genetics) to epigenetic inheritance processes (after Waddington), social learning, and even language-based inheritance of acquired characters. See especially E. Jablonka and M.J. Lamb (1995); E. Avital and E. Jablonka (1994) (2000); Jablonka *et al.* (1998); also G. Cziko (1995) (2000). See also P.P.G. Bateson (1988); H. Plotkin (ed.) (1988); J.A. Stamps (1991); the chapter by G. Barker in P.P.G. Bateson *et al.* (1993); B.W. Robinson and R. Dukas (1999); C. ten Cate (2000). A similar argument, without the benefit of the many recent discoveries, was made in my 1983 book, *The Synergism Hypothesis*.

43. Indeed, some theorists pointedly distance themselves from Lamarck, even when they are concerned with the subject of behavior in evolution. For instance, in the edited volume by Henry Plotkin (1988) on behavior and evolution, cited above, Plotkin accuses the child psychologist Jean Piaget, who wrote a book in which he called behavior the "motor" of evolution, of being a "Lamarckian." Similarly, Matthew H. Nitecki (1990) has little to say on the subject in his book on *Evolutionary Innovations*. Gerd B. Müller and Günter P. Wagner (1991) examine the role of developmental novelties in evolution but disparage Lamarck's contribution. Perhaps most striking was the edited volume by John H. Campbell and J. William Schopf called *Creative Evolution?!* (1994),

which included an illustrious list of contributors. There was little mention of behavioral innovation, except in the final chapter by Campbell, who (briefly) cited such phenomena as sexual selection, predator–prey coevolution, and Ernst Mayr's "founder effect." His inspiration was a quote from Darwin in *The Descent of Man* (1871) regarding evolution "through the exercise of choice."

44. J.H. Campbell in J.H. Campbell and J.W. Schopf (eds.) (1994), p. 86.

45. P.A. Corning *et al.* (1977).

46. See, for example, R. Byrne (1995); also J.L. Gould and C.G. Gould (1995).

47. See especially Massimo Pigliucci's *Phenotypic Plasticity: Beyond Nature and Nurture* (2001).

48. M.L.J. Stiassny and A. Meyer (1999).

49. J.B. Losos (2001). See also the general discussion in S.A. Foster (1999).

50. T. Dobzhansky *et al.* (1977), pp. 95–96.

51. C.S. Pittendrigh (1958); E. Mayr (1974b); also see E. Mayr (1965).

52. This issue is contentious. It involves some basic misunderstandings relating to cybernetics (the science of communications and control processes in goal-oriented systems) and its relationship to other kinds of dynamic processes associated with complexity theory. This issue is dealt with in some depth in P.A. Corning and S.J. Kline (1998a, b); also P.A. Corning (2001b). In a nutshell, the two kinds of dynamics have very different properties, but many theorists treat the two as being equivalent. Two important book-length treatments of this issue are Gary Cziko's *Without Miracles* (1995) and *The Things We Do* (2000). See also the classic discussion of "information" and cybernetic control processes in living systems by biologist T.H. Waterman (1968).

53. E.O. Wilson (1975). Indeed, the concept of phenotypic plasticity is often interpreted to mean that a range of different phenotypes can be invoked without affecting the underlying genotype, which implies that it may actually have conservative effects. For instance, Laurent Keller and Kenneth G. Ross (1993), in a study of phenotypic plasticity in the fire ant (*Solenopsis invicta*) found that the two distinct forms of social organization in this species are context-dependent and are not due to genetic differences. Some colonies consist of monogamous "families" headed by a single queen, while others are composed of polygynous multifamily groups with up to several hundred queens, depending upon the genotype–environment interaction (including the social environment itself). See also the discussions in S.C. Stearns (1989), W.T. Wcislo (1989), and M.J. West-Eberhard (1989).

54. R.C. Lewontin (1978), p. 213. In fact, Lewontin has been a vehement opponent of "genitis" (as Sherwood Washburn put it) over the years and has a much more multidimensional view of the process. See for example his new book on *The Triple Helix* (2000). Lewontin's "constructionism" – as it has been called – was also celebrated in a recent special 1999 issue of the journal *Biology and Philosophy* (vol. 14, pp. 157ff).

55. This was, in fact, the title of Monod's celebrated 1971 book.

56. The term "frozen accidents" is unfortunate. Many of these primordial evolutionary "inventions" have been "procreative," important enablers for further

evolutionary developments. Moreover their preservation over the course of evolutionary history has been anything but accidental. They represent solutions to functional problems that are "conserved" because they work; they serve the functional needs of living systems. For instance, there is some research suggesting that it is possible to have more than four DNA bases. But four are sufficient, just as four wheels are enough for the functional needs of automobiles. In other cases, functional solutions are far from being "frozen" in place. Photoreceptors (eyes), for instance, have been reinvented many times (one estimate puts it at more than 40). Also, they operate on a variety of different functional principles, from pinholes to multiple tubes, movable lenses, and fixed focusing lenses like our own.

57. See the "target article" by D. Dennett (1983) and accompanying commentaries. The concept of "mind" as a factor in evolution dates back to the extensive work in comparative psychology during the latter part of the nineteenth century, much of which was rejected and all but forgotten in the twentieth century. See especially G.J. Romanes (1883); C. Lloyd Morgan (1891) (1896b) (1923); E.L. Thorndike (1911); L.T. Hobhouse (1915). In the heyday of Behaviorism, the very term was banished from the literature in psychology as "unscientific." Although it has lately been resurrected, it remains controversial. Some of the controversy involves semantics – a failure to agree on what precisely the term means. But there are also deeper theoretical differences. Major discussions relating to the "mind," and the nature of animal intelligence, can be found in D.R. Griffin (1992); J.L. Gould and C.G. Gould (1994); G. Cziko (1995) (2000); M. Bekoff and D. Jamieson (1996); L.J. Rogers (1998); M.C. Corballis and S.E.G Lea (eds.) (1999). Also see the special symposium issue of the *American Zoologist* (vol. 40, pp. 833ff). For an overview of the debate on this issue see R. Lewin (1994).

58. R.W. Sperry (1969) (1991). D.T. Campbell's coinage appeared in 1974.

59. C. Lloyd Morgan (1923) (1926) (1933).

60. This issue was discussed at length in my 1983 book. It is apparent that later writers on the subject, including Bateson, Plotkin, Jablonka, Cziko, and others were not aware of this discussion. Among the arguments presented there was the point that the "classical" Behaviorist psychologists, in their search for the universal "laws" of conditioned learning, could not explain why a particular reward or punishment (a "reinforcer") was reinforcing to an animal. Indeed, the Behaviorists and their nemeses among the biological determinists agreed on one point – the "mind" of the animal is irrelevant to an understanding of its behavior. B.F. Skinner was emphatic: "When what a person does is attributed to what is going on inside of him, investigation is brought to an end" (1974, p. 31).

61. D. Lack (1947). For a recent update on the ongoing research program in Darwin's finches, see P.R. Grant and B.R. Grant (2002).

62. T. Deacon (2003).

63. An earlier generation of evolutionists commonly used the term "pre-adaptations" to characterize how existing structures or behaviors might be applied in new ways. However, this term has been criticized on the ground

that it connotes a hidden teleology – a development designed for a future need; in fact, it is more like old dogs learning new tricks. S.J. Gould and E. Vrba (1982) were the originators of this charge, and they have a point. But their proposed alternative, the neologism "exaptation," is not crystal clear or easily grasped.

64. R. Dawkins (1976). Universal Darwinism has gained many supporters, including D.L. Hull (1982) (2000); H.C. Plotkin (1982) (1988) (1994); D. Dennett (1991) (1995); G. Cziko (1995) (2000); and others.

65. R. Dawkins (1976), pp. 206ff, p. 214.

66. D. Dennett (1991), p. 202. Following Dennett's lead, an increasing number of theorists have also adopted the meme concept to varying degrees in recent years. See especially D. Dennett (1995); R. Brodie (1996); A. Lynch (1996); S. Blackmore (1999); R. Aunger (2000) (2002); G. Cziko (1995); D.L. Hull (2000); K.N. Laland and G.R. Brown (2002).

67. S. Blackmore (1999).

68. A number of thoughtful analyses, some supportive and others quite skeptical, have appeared in the past few years. See especially L. Cronk (1999); W.C. Wimsatt (1999); R. Aunger (2000) (2002); S. Atran (2001); K.N. Laland and G.R. Brown (2002). Laland and Brown, in a balanced review of the literature, point out that "selection favouring memes with particular qualities has not been demonstrated" (p. 212). They also note that "there has been little emphasis on the possibility that humans may have evolved predispositions to prefer some memes to others ... while memetics has provided a new way of looking at the world, there is little solid evidence with which to evaluate the usefulness of the meme concept" (pp. 224–225). In what could be called a gross understatement, they also observe that "memetic evolution is sometimes directed and intentional ... at least sometimes people notice problems and try to solve them ... The stance advocated by some memeticists may be missing some of the underlying complexity to human behavior" (pp. 229, 230).

69. D. Thaler (1994).

70. W.B. Arthur (1988) (1990).

71. W.T. Powers (1973). There is a very large literature, and several journals, devoted to cybernetics and control systems. For a review that includes a new approach to information theory called "control information," see P.A. Corning (2001b). A more detailed treatment of this issue can be found in P.A. Corning (1983); also G. Cziko (1995) (2000).

72. This formulation is, of course, no better than the Behaviorist black box without some specification of what are the contents of the "mind," and exactly how Neo-Lamarckian Selection occurs. This subject was discussed in detail in my earlier book and there has been a wealth of new research and writings on the subject during the past 15 years or so. Indeed, much of the literature in behavioral ecology is focused on the "economics" of animal behavior and behavioral choices. Particularly noteworthy are the series of edited volumes by J.R. Krebs and N.B. Davies and other writings on behavioral ecology cited above. See also the path-breaking work by E. Charnov (1976) on optimal

foraging theory. Also noteworthy is the use of explicitly economic concepts and analytical techniques by Marion Stamp Dawkins (1983) (1988) (2000). See also the commentary by G. Mason *et al.* (1998b).

73. E. Mayr (1976), pp. 695–696. An especially useful discussion of the relationship between proximate and ultimate causation can be found in J.A. Stamps (1991). See also E. Jablonka *et al.* (1998). As biologist Richard D. Alexander (1990a, p. 241) points out: "Selection acts most directly on behavior, and on its underlying mechanisms only as they influence the behavior. This is as true for learned and cultural behaviors as for any others."

74. K.W. Jeon (1972) (1973). Note also that the Martin and Müller model of eukaryote origins entails evolution via symbiosis. Margulis's scenario for the origin of eukaryotes assumes predation as the initiator, but it is also behaviorally based. See Chapter 3.

75. P. Ehrlich and P. Raven (1964).

76. R. Boyd and P. J. Richerson (1985).

77. W. H. Durham (1991).

78. Some recent discussions of evolutionary "novelties" and "creativity" in evolution have stressed the importance of the sometimes complex structural changes involved in major new adaptations. For instance, the evolution of carapaces in turtles required a profound reorganization of the basic body arrangement, or *Bauplan* of four-legged creatures, as did the external pouches in pocket gophers and kangaroo rats, and the corpus callosum (the massive fiber tract that connects to two brain hemispheres) in mammals. In other words, some changes involve more than a simple modification of an existing structure. Conversely, some radical structural changes may be the result of small changes with large effects on the development of an organism. "Heterochrony," changes in the timing and rate of development, is one such mechanism. See the arguments by M.H. Nitecki (eds.) (1990); G.B. Müller and G.P. Wagner (1991); J.H. Campbell and J. W. Schopf (eds.) (1994).

79. D.J. Merrell (1949) (1953). Described also in M. Bastock (1956).

80. J. Weiner (1994), pp. 182–184.

81. K.N. Laland *et al.* (2001).

82. T. Ingold in K.R. Gibson and T. Ingold (eds.) (1993), pp. 470–471.

7. Conjuring Human Evolution: The Synergistic Ape

1. Taken from an incident described by Hans Kummer (1968), with additional details from S.C. Strum (1987) and J.A.R.A.M. van Hooff (2001). The process of collective decision making described here has also been observed in other social animals, including red deer, gorillas, and African buffalo. A theoretical model that supports such "democratic" behaviors was developed by L. Conradt and T.J. Roper (2003).

2. From an incident in Elizabeth Emanuel's *Baby Baboon* (1971).

3. Some of these conflicts have been fueled by misinterpretations, or by selective use of the available evidence, or by emphasizing a particular category of evidence. For instance, some theorists have concluded that hominids did not

start using tools until the first flaked stone tools appear in the fossil record, about 2.5 million years ago. Apart from the fact that earlier evidence of stone tools may yet be found (this is a common occurrence in paleoanthropology), a reliance on stone tools as an indicator ignores the fact that the vast majority of tools commonly used by hunter–gatherers even today do not fossilize. It also ignores one of the clichés in science to the effect that the absence of evidence is not necessarily evidence of an absence. It has also been observed that the choice of an origin point for human evolution is somewhat arbitrary. Technically, our roots could be traced back to the emergence of the primate order some 30 million years ago.

4. Quoted in M.H. Wolpoff (1999a), p. 187. See also the later versions and revisions of this male-centered scenario in R. Ardrey (1961); D. Morris (1967); L. Tiger and R. Fox (1971); A. Mann (1972); H.E. Fisher (1982); K. Hill (1982); J. Tooby and I. De Vore (1987).

5. Darwin's views on the role of warfare in human evolution are discussed in some detail in an essay by Johan van der Dennen (1999), who has written extensively on human and animal aggression. I use his citations for the quotes from Darwin's *The Descent of Man*: (1871), vol. 1, pp. 97, 199–200. Citing the works of Hobbes, Malthus, Spencer, Bagehot, Huxley, Sumner, Keith, and others, van der Dennen also points out that there have been many recurrent themes in the scholarly literature on warfare, including its role in the "progress" of human civilization, its role as a selective agency between groups, the influence of ethnocentrism, the assertion that aggression is "instinctive," the relationship between hunting and warfare, and the balance of power concept.

6. Quoted in C.B. Stanford (1999), p. 107. See R. Dart (1953) (1959).

7. S.L. Washburn and C.S. Lancaster (1968) reprinted in R.L. Ciochon and J.G. Fleagle (eds.) (1993), p. 219. The criticisms leveled at this scenario over the years provide another illustration of the tendency to misinterpret what a theorist has said. Washburn and Lancaster did not argue that the hunting mode of life initiated the process of human evolution but rather that it emerged much later, with *Homo erectus*. Although the role of hunting in human evolution is still hotly debated, the thesis that *erectus*-grade hominids were hunters, if not exclusively so, remains in contention. It is supported here.

8. Sally Linton's seminal paper appeared in 1971, but the debate was ignited by Adrienne Zihlman and Nancy Tanner in their provocative 1978 paper. Also see A.L. Zihlman (1981) and the review by L.M. Fedigan (1986).

9. G.L. Isaac (1978). See also G.L. Isaac (1981) (1983).

10. L.R. Binford (1987).

11. R. Potts and P. Shipman (1981); also R. Potts (1984) (1988); P. Shipman (1983) (1986).

12. C.O. Lovejoy (1981).

13. A. Mann (1981).

14. D. Quiatt and J. Kelso (1985).

15. J.B. Lancaster (1975) (1978).

16. P. Turke (1984).

17. M. Wolpoff (1999a), pp. 210–211.
18. See the various contributions in E. Vrba *et al.* (eds.) (1995); also J. Kingdon (1993); R. Foley (1995); J.R.M. Allen *et al.* (1999); J.F. McManus *et al.* (1999); K. Taylor (1999); R.B. Alley (2000); C.D. Keeling and T.P. Whorf (2000); S.M. Stanley (2000); P.J. Richerson (2001).
19. P.P.G. Bateson (1988). See also R. Foley (1995).
20. For example, paleoanthropologist Tim White, in a chapter for the Vrba volume, draws this significant conclusion: "The strongest pulse of speciations and extinctions in both pigs and hominids, however, occurs between ca. 1.6 and 2.0 myr [million years] ago during a period in which no dramatic global climatic oscillation signal is available. It seems likely that the end of this time window was marked by technological innovation involved with the production of the Acheulean industry" (1995, p. 378).
21. C.M. Anderson (1986); D.L. Cheney and R.W. Wrangham (1987); R.I.M. Dunbar (1988); G. Cowlishaw (1994); T. Iwamoto *et al.* (1996); R.W. Wrangham and D. Peterson (1996).
22. The evidence of predation against early hominids is discussed in C.K. Brain's writings (1981) (1985), which detail his studies at Swartkrans. There is also the hominid fossil known as OH-7, the 12–13-year-old with a gnawed foot. See also the review of predation on primates by L.A. Isbell (1995), and R. Foley (1995) and the citations in the preceding note. J. Lee-Thorp *et al.* (2000) single out leopards (*Felis*), saber-toothed cats (*Merantereon*), and hyenas (*Crocuta*) as likely to have been systematic predators on primates and australopithecines of that era.
23. See especially the analyses in P.C. Lee (1994); also R.W. Wrangham (1987).
24. M.H. Wolpoff (1999), pp. 217–219.
25. First proposed by D.J. Morton (1927). See also B. Kurtén (1984); Y. Coppens and B. Senut (1991); M.H. Wolpoff (1999a). M. Köhler and S. Moya-Solá (1997) report a recent find of a 9-million-year-old fossil specimen, which they suggest may have used bipedalism for ground foraging in their sheltered Sardinian environment. For a different point of view, see also L.A. Isbell and T.P. Young (1996).
26. R.W. Wrangham (1987).
27. As Richard Wrangham (2001) observes, the origin of bipedalism among our ancestors remains a puzzle. In addition to the contested *Oreopithecus* model, there have been suggestions that bipedalism allowed for more efficient heat dissipation; or that it allowed for more efficient carrying; or for greater mobility; or for more visibility. I would add that bipedalism also enhances the ability to utilize tools/weapons, especially in violent confrontations – a highly significant advantage. In a classic paper on human walking, John Napier (1967) pointed out that walking is "a unique activity in which the body, step by step, teeters on the edge of catastrophe." Walking also involves an exceedingly complex set of skeletal, muscular, and neurological changes – an orchestrated anatomical shift that could not have occurred overnight. Moreover, bipedalism also came at a high cost (e.g., a more complicated birthing process

and a reduction in tree-climbing ability) and, evidently, evolved long before the shift to the savanna way of life. Only the carrying hypothesis provides sufficiently important adaptive advantages to offset these costs, I would argue. But perhaps no single explanation will prevail; there may have been a suite of advantages to bipedalism in the specific context of australopithecine evolution. A recent new analysis of knuckle-walking in early hominids also weighs against the *Oreopithecus* model; see B.G. Richmond and D.S. Strait (2001).

28. Y. Haile-Selassie (2001); M. Brunet *et al.* (2002).
29. An analogue was suggested in the baboon parable, above, where unrelated males nevertheless cooperate in troop defense. For an analysis of this issue, see G. Cowlishaw (1994); also the broader review of primate social patterns in A.E. Pusey (2001).
30. On this issue, see R.W. Wrangham (2001). See also the discussion in C.B. Stanford (1999), p. 100.
31. See especially the discussions in R. Lewin (1993) and J. Kingdon (1993).
32. See especially R.G. Klein (1999); R.W. Wrangham *et al.* (1999).
33. P.A. Corning (2000).
34. L. Rose and F. Marshall (1996). Among the recent writings on intergroup aggression among primates and evolving hominids, see especially J.H. Manson and R.W. Wrangham (1991); C.B. Stanford (1999); P.A. Corning (2001a).
35. P.V. Tobias (1971) (1985); R. L. Holloway (1983a) (1996) (1997); D. Falk (1998); D. Falk *et al.* (2000).
36. See especially G.C. Conroy *et al.* (1998); L. C. Aiello and R.I.M. Dunbar (1993); R.I.M. Dunbar (1996) (1998) (2001); J.K. Rilling and T.R. Insel (1999).
37. J. Kingdon (1993).
38. M.F. Teaford and P.S. Ungar (2000). See also the suggestive parallel in *Oreopithecus bamboli*, D. Alba *et al.* (2001).
39. N. Toth (1987a, b); N. Toth, *et al.* (1993); K.D. Schick and N. Toth (1993).
40. See J. Steele (1999).
41. R. Potts (1998a, b).
42. A detailed review and analysis can be found in M.H. Wolpoff (1999a), pp. 141–142, 271–273. The research on cooperation in bonobos by Amy Randall Parish (1996), reinforces the plausibility of this scenario.
43. Reviewed in F.B.M. de Waal (1996); see also F.B.M. de Waal (ed.) (2001a).
44. Discussed in depth in E. Sober and D.S. Wilson (1998).
45. It may be argued that such a generic treatment obscures many significant differences and that it is redolent of the discredited "stages" of evolution frame of reference. This is certainly not intended, but it does emphasize the argument that a major niche shift occurred which represented a discontinuity. It also focuses our attention on the most general features of this change. It should also be pointed out that other theorists use the term "*Homo erectus* grade" in much the same way that I do with "*Homo intermedius*."
46. B. Asfaw *et al.* (1999). See also the report on meat-eating in these hominids in J. de Heinzelin *et al.* (1999).

47. Bowie knives, also called "Arkansas toothpicks," had blades ranging in length from 9 to 13 inches, along with a hand guard and a handle. They were specifically designed to be thrown as well as being wielded for various purposes. Despite the legend, there is considerable uncertainty about who actually invented the Bowie knife. Some authors believe it was Jim Bowie's brother. Today authentic Bowie knives are valuable collectors' items. See S. Latham (1973). It should also be noted that the Swiss army knife metaphor seems most often to be associated with the Acheulean axe, but it can also be applied to some of the earlier hand axes as well.

48. B. Campbell (1985), pp. 396–398.

49. R.I.M. Dunbar (1988) (1996) (1998) (2001).

50. See P.S. Rodman and H.M. McHenry (1980); H.M. McHenry (1992); R. Lewin (1993); K.L. Steudel (1994) (1996); R. Foley (1995); R.G. Klein (1999); M.H. Wolpoff (1999).

51. See S. Washburn and C.S. Lancaster (1968); R.J. Blumenshine (1987); P. Shipman and A. Walker (1989); R. W. Wrangham and D. Peterson (1996); C.B. Stanford (1999); M.H. Wolpoff (1999a); B. Wood and M. Collard (1999).

52. E. Vrba and her colleagues (1995) have proposed that climate changes were the necessary and sufficient cause for the emergence of "*Homo intermedius*." According to their "pulse hypothesis," climate changes are the primary drivers for various waves of speciation and extinction. See also S.M. Stanley (1992). However, an analysis of the extinction pattern in 510 mammal species in the Lake Turkana region during this period failed to support this theory. See A.H. Hill (1987); R. Foley (1994); R.A. Kerr (1996). The key point from our perspective is that the dramatic anatomical changes from australopithecines to *Homo erectus/Homo ergaster* did not occur until several hundred thousand years after the Pliocene climate shift occurred. In other words, behavioral adaptations came first.

53. The so-called "expensive-tissue hypothesis" is based on the evidence that the human brain represents about 2–3% of our total body mass, yet it consumes some 20% of our energy budget and even more, about 27%, in a developing infant. See R.D. Martin (1981); R.D. Martin and A.M. MacLarnon (1985); W.R. Leonard and M.L. Robertson (1994) (1997); L.C. Aiello and P. Wheeler (1995). Aiello and Wheeler point out that there may have been a trade-off – the increase in brain tissue associated with meat-eating may have been offset by a reduction in the size of our guts (also expensive tissue) in response to dietary changes. However, larger overall size, extended childhood dependency, greater longevity, and shortened birth-spacing must also be included in the bookkeeping.

54. L.C. Aiello and C. Key (2001). See also W.R. Leonard (2002).

55. In contrast with Lovejoy's food-carrying and provisioning scenario, this model relates to a much later point in human evolution, with very different causes and consequences. Lovejoy sought to explain bipedalism as a result of resource constraints at a much earlier stage in human evolution, prior to "*H. intermedius*." Like Isaac's later formulation, this scenario does not take sides on the relative importance of hunting versus scavenging but stresses that the sharing of

meat from large game animals was a major factor. Some theorists have recently called the provisioning scenario into question; e.g., K. Hawkes *et al.* (1995); C.P. van Shaik and A. Paul (1996). Their claim is that males are not motivated to raise their offspring but only to maximize their mating success. This argument overlooks the fact that natural selection is not preoccupied with sex but with "completed" reproduction. In a *K*-selected species with severe food (and other) constraints, mating success alone won't cut it. The provisioners would have had a clear adaptive advantage. See also S.B. Hrdy (2002).

56. See C.B. Stanford (1999). Also J.D. Speth (1987) (1989).

57. There is even some suggestive evidence of respiratory tract changes to accommodate a more strenuous lifestyle. See E. Trinkaus (1987).

58. R.W. Newman (1970); also P.E. Wheeler (1985) (1991). Discussions of sweating and hair loss as major adaptive changes in evolving humans can also be found in R. Foley (1995) and M.H. Wolpoff (1999a), among others. A major alternative to the hunting and hair loss hypothesis is what has been dubbed the "aquatic ape hypothesis." Originally suggested by Sir Alister Hardy many years ago, it has recently been championed by Elaine Morgan (1982) (1997). For a detailed report on a recent symposium on this issue, see M. Vaneechoutte (2000). For a critique, see J.H. Langdon (1997).

59. R. Bellomo (1994). See also B. Campbell (1985) and J. Kingdon (1993). Kingdon also points out that fire was too commonplace and too powerful in its effects to have been overlooked as a potential tool until only a few hundred thousand years ago.

60. Richard Wrangham and his colleagues call it the "cooking hypothesis," and they believe it was even more important than the addition of meat to the australopithecine diet. Cooking greatly expanded the range of edible plant materials and made them more digestible. Indeed, the staple foods of many contemporary human cultures cannot be eaten raw. In addition, cooking may have precipitated the introduction of home bases and may have been a major factor in changing the mating system toward stable pair-bonds (the nuclear family), leading in turn to such anatomical changes as the concealment of ovulation and the permanent receptivity of the females. They argue that this change was not primarily related to male provisioning, as Lovejoy supposed, but to the defense of hard-won food supplies. Wrangham calls it the "theft hypothesis." See R.W. Wrangham *et al.* (2001). In any case, the cooking hypothesis posits a synergistic cultural innovation as a major causal agency in human evolution. A parallel proposal on the early adoption of fire can be found in K. Hawkes *et al.* (1998) (2000); J.F. O'Connell *et al.* (1999).

61. See R.G. Klein (1999); M.H. Wolpoff (1999a). See also the special issue of *Science* (vol. 291, 2001) devoted to the subject of hominid migrations. Among other things, it is suggested that there were in fact several waves of migration out of Africa over time.

62. P. Richerson and R. Boyd (1992) (2002).

63. C. Boehm (1996) (1997) (1999); E. Sober and D.S Wilson (1998).

64. C. Boesch and M. Tomasello (1998); C. van Schaik (2002). For recent findings on the antiquity of various technological improvements, see O. Bar-Yosef and

S. Kuhn (1999). H. Thieme (1999) also reported the discovery of 400,000-year-old Neandertal spears. Some bone tools also date back at least 150,000 years. See B. Bower (1997). See also the broader perspective in J. Ziman (ed.) (2000).

65. See especially R.I.M. Dunbar (1988) (1996) (1998) (2001); L.C. Aiello and R.I.M. Dunbar (1993); P. J. Gannon *et al.* (1998); M.A. Nowak (2000); C.T. Snowdon (2001).

66. P. Lieberman (1998), p. 141. See also M. Donald (1993). Lieberman also holds an unusual view of how language evolved. Adopting Gordon Hewes's (1973) thesis that spoken language is derived from a more primitive mode of communication via manual gestures, Lieberman sees language as another consequence of the emergence of bipedalism and the freeing of the hands for work. Hewes based his argument on the finding that the neurological structures that are associated with the precise control of the hands are also implicated in language production. However, this scenario also has problems. On the face of it, a system of communications that relies on manual signs is relatively inefficient compared to a vocal–auditory system. It is limited to direct, face-to-face contact; it cannot be seen at a distance; and it ties up the hands precisely when, most likely, they would be needed for doing other things. Given the fact that vocal communications are widespread in the primates and especially chimpanzees, it seems rather unlikely that our hominid ancestors would have opted for a less efficient alternative that would have hampered their ability to use their hands in other ways. See also M.C. Corballis (1991) (2002); also the review of this issue in C.T. Snowdon (2001).

67. C.T. Snowdon (2001). There are many recent books on the evolution of human language. See especially W. Noble and I. Davidson (1994); D. Bickerton (1995); S. Pinker (1994); T.W. Deacon (1997); P. Lieberman (1998).

68. T.W. Deacon (1997); R.L. Holloway (1998). Holloway, Deacon, and others take the position that the use of symbols was also primeval in hominid language development, a point that is not essential to the basic functional argument advanced here.

69. The case for language as a functional imperative in evolving hominids is argued in depth by R.I.M. Dunbar (1988) (1996) (1998) (2001).

70. Indeed, T.W. Deacon (1997) cites the conclusions of endocast researchers Philip Tobias and Dean Falk that language adaptations can be discerned even in *H. habilis* (p. 342). See also R.L. Holloway (1983b).

71. P. Lieberman (1998), pp. 85–97; also I. Tattersall (1998). An alternative explanation for the disappearance of the Neandertals can be found in the profound environmental changes that occurred during the last major ice age. Beginning some 33,000 years ago, northern Europe (where Neandertal populations were centered) became increasingly inhospitable. As the glaciers advanced southward, huge areas were depopulated. However, migration to the south, an option for the Neandertals in previous ice ages, was constrained this time by a growing population of modern humans around the Mediterranean and in southern Europe. In fact, the ice age created a kind of population vice. During the early part of the cycle, sea levels declined sharply and vast areas of coastal grasslands came to be populated with an abundance of "megafauna" (as the

anthropologists would say) and, no doubt, an abundance of human hunter–gatherers. But, as the climate cycle reversed, these areas began to shrink, driving both the hunters and their prey onto higher ground. Signs of stress and intensified competition include evidence of reverse migrations from Southeast Asia and southern Europe into Africa, the occupation of more marginal environments and a sharp decline in the number of large prey animal species (a wave of extinctions), most likely caused by a combination of climate changes and "overkills" by human hunters. Where once there had been alternatives, or "escape valves" for coping with major environmental changes, the rapidly growing population and wider distribution of humankind now became a severe adaptive constraint.

72. J.M. Diamond (1995), pp. 46, 49.

73. I. Tattersall (2000), pp. 56–62.

74. T.W. Deacon (1997), p. 341.

75. It has often been assumed that the Neandertal cultural attainments were borrowed, but there is a lack of convincing evidence for this. And the presence of humans in northern Europe prior to the Upper Paleolithic is unclear and hotly debated. See M.H. Wolpoff (1999a); also F. d'Errico *et al.* (1998); P. Mellars (1999) and replies.

76. J.M. Diamond (1995), pp. 50–51. Diamond is not alone in this view, of course.

77. This is, of course, a very controversial subject. No doubt the relationship between mental imaging, thinking, and language is complex. However, it is also increasingly evident that the three functions are not isomorphic. Indeed, the bulk of our thinking, decision-making, and purposeful actions (inclusive of feedback responses) go on at the preconscious level. To cite an obvious illustration, it would be impossible to drive a car, much less talk on a cell phone at the same time, if driving were predominantly a language-based skill in which every action had to be thought about consciously and verbalized.

78. Although many theorists adhere to an adaptationist view of language, others do not. D. Bickerton (1995), for instance, argues that language is an outgrowth of the development of "mind" – the underlying mental capacities for making representations (or mental pictures). But this view ignores the context; sophisticated communications skills were an absolute requisite for evolving hominid societies. More likely, the capacity for communications and the capacity for forming mental representations of experience evolved both independently and interactively.

79. On this point, see especially the discussion in F.B.M. de Waal (1996); also C.T. Snowdon (2001).

80. Anthropologist Richard G. Klein (2000) is the strongest proponent of a "punctuational" biological cause. "The shift to a fully modern behavioral mode and the geographic expansion of modern humans were coproducts of a selectively advantageous genetic mutation" (pp. 17–18). Klein finds the human ability to innovate otherwise puzzling and speaks of the "sudden origin of the modern capacity for culture" (p. 24). I do not find it so inexplicable but instead view cultural evolution as a cumulative learning process. Would anyone say that the radical shift from the horse and buggy to a complex modern

driving machine required a genetic mutation? I do not think the distance between the Mousterian industry of the lower Paleolithic period and the more
advanced Aurignacian techniques of the upper Paleolithic, taken as a whole,
is comparably greater.

81. In keeping with their radically different scenarios for the final stages of human
evolution, Milford Wolpoff and Richard Klein interpret the available fossils
and other artifacts in very different ways. Wolpoff sees a multiregional pattern
of progressive changes. Klein sees a more unitary process with periods of relative stasis interrupted by punctuational changes. Yet both acknowledge that
progressive changes were occurring.

82. M.H. Wolpoff (1999a), p. 554. Wolpoff's views have been supported by recent
discoveries of more sophisticated artifacts with much earlier dates and in diverse regions. See the overview by J.N. Wilford (2002); also A.S. Brooks *et al.*
(1995); J.E. Yellen *et al.* (1995); S. McBrearty and A.S. Brooks (2000); S.L.
Kuhn *et al.* (2001).

83. R.G. Klein (2000), p. 26.

84. M.H. Wolpoff (1984) (1999a, b).

85. See especially R.L. Cann *et al.* (1987); L.L. Cavalli-Sforza *et al.* (1988) (1994);
M. Stoneking (1993); M.F. Hammer (1995); M. Nei (1995); S. Pääbo (1999)
(2001); J.H. Relethford (1995); L. Jin *et al.* (1999); R.G. Klein (1999) (2000); B.
Su *et al.* (1999); P.R. Ehrlich (2000); O. Semino *et al.* (2000). There is some variation in the estimates for inclusive dates. Between 150,000 and 250,000 years
ago is also mentioned.

86. Actually, it is possible that some initial, tentative migrations may have occurred
earlier. This would account for the human remains found at Skhul, Israel,
that date to about 100,000 years and the recent finding by Alan Thorne that
an Australian fossil human, known as Mungo 3, now appears to be about
60,000 years old.

87. M.H. Wolpoff *et al.* (2001) present the most compelling contrary evidence
to date in fossils from Australia and Central Europe. Their conclusion is
that a middle-ground position might be justified. Replacement occurred in
some/many areas but not everywhere. For a critique of both scenarios, see
P.R. Ehrlich (2000). The *FOXP2* gene discovery was reported in W. Enard
et al. (2002). See also R.G. Klein (2000), p. 17.

88. See especially C. Wills (1993) (1998); J.M. Diamond (1997); R.G. Klein (1999);
L.L. Cavalli-Sforza *et al.* (1994); P.R. Ehrlich (2000); I. Tattersall (2002). Actually, the Aurignacian was not a single, unified industry. The label encompasses
a diversified range of regional variants. Nevertheless, it may be that certain key
technologies were associated with the humans that migrated out of Africa.

89. The history of this theory, from Darwin to the present day, is briefly reviewed
in J.M.G. van der Dennen (1999). Among the many more in-depth works, see
especially R.D. Alexander (1979); J. Keegan (1993); P. Crook (1994); J.M.G.
van der Dennen (1995); R.W. Wrangham and D. Peterson (1996); P.A. Corning
(2001a).

90. A starting point for the extensive literature on ethnocentrism and xenophobia is V. Reynolds *et al.* (1987); also R.P. Shaw and Y. Wong (1989);

J.M.G. van der Dennen and V.S.E. Falger (eds.) (1990); I. Eibl-Eibesfeldt and F.K. Salter (eds.) (1998). In case it needs to be said, the assertion that we may have a psychological predisposition is not to condone it, or to accept it fatalistically. There is also good evidence that cultural means can be used to contain and even rise above our innate urges, but this is more likely to occur if we recognize them for what they are. Indeed, human populations often do live in peace with one another and even form alliances against common enemies.

91. J.D. Clark (1992). These axes had about ten times as much cutting edge per pound of material as the more primitive hand axes, according to Ian Tattersall. The manufacture of these advanced tools/weapons apparently involved a highly specialized process that could not readily be imitated. A form of glue – bitumen or a similar substance – heated to a very high temperature was used for attaching the stone blades to wooden shafts. It was not easy to do, and the technique for making hafted stone axes may have been among the earliest military secrets. However, this does not mean that these implements were invented only for the purpose of making war. As we noted earlier, things that are devised for one purpose (both in biological and cultural evolution) may subsequently be adapted for quite different purposes. See also M. Farmer (1994); R.G. Klein (1999).

92. J.M. Diamond (1997), pp. 53–57.

93. The following derived from M. Gluckman (1940) (1969) and D.R. Morris (1965). Other examples can be found in J. Keegan (1993).

94. There is, of course, a vast literature on this subject, both historical and contemporary. Recent writings on the subject from Darwinian/evolutionary perspective include L. Bramson and G. Goethals (eds.) (1964); R. Bigelow (1969) (1975); L. Tiger and R. Fox (1971); R.D. Alexander (1979) (1990b); P. Slurink (1993); J.M.G. van der Dennen (1995); P.A. Corning (2001a).

95. This literature is reviewed in P.A. Corning (2001a). See also the works cited in the previous note. Also, K. Lorenz (1966); E.O. Wilson (1975); R.W. Wrangham and D. Peterson (1996); C.B. Stanford (1999).

96. Many of the classics on human evolution point to the obvious relationship between group hunting and group aggression. These range from Darwin to Dart, Washburn, Keith, Bigelow, Tiger and Fox, Alexander, van der Dennen, and many others. In addition to the works cited above in note 7.94, see Robert Ardrey's *The Hunting Hypothesis* (1976).

97. See especially P.S. Martin (1967); P.S. Martin and R.G. Klein (eds.) (1984); R. Klein (1992); R.D.E. MacPhee (1999); C. Barlow (2000). The latter espouses an alternative to the overkill hypothesis, what has been called the disease hypothesis. For a review, see M. Miller (2001).

98. The extensive evidence on this issue is reviewed in Jared Diamond's *Guns, Germs and Steel* (1997), especially Chapter 11. Diamond also provides an indepth annotated bibliography. Among other things, he notes that the documentation goes back to Thucydides in *The Peloponnesian War*.

99. While it is true that the Tasmanians lost the ability to make and use fire, their special situation (and marginal existence) is an exception that proves the rule. The provocative theory of Paul M. Bingham (2000) should also be

mentioned. Bingham believes there is a "simple, unitary, inexorable logic to the entire human story" (p. 255). What he proposes is, in essence, a synergistic nexus – cooperation by non-kin coalitions that were reinforced by policing and ethics, novel technologies that allowed for "killing at a distance" (which greatly reduced individual risks), and competition between groups. Perhaps so, but the logic of it seems "inexorable" only in retrospect.

100. The evidence for micro-evolutionary change as an ongoing process in human evolution was reviewed by Christopher Wills in his recent book, *Children of Prometheus* (1998).

8. Conjuring History: Does Cultural Evolution Have an "Arrow"?

1. Aristotle, *The Politics*; also see G.H. Sabine (1961).
2. Quoted in G.H. Sabine (1961), p. 571.
3. J.-A.-N. Condorcet (1795), p. 210. Condorcet and his writings helped to spark the French Revolution, but he was ultimately one of its victims.
4. Quoted in G.H. Sabine (1961), p. 717.
5. R.W. Burkhardt (1995).
6. J.B. de Lamarck (1809). See also H.G. Cannon (1955).
7. See C.R. Darwin (1868). On Lamarck, see S.J. Gould (1999a, b). Gould quotes the following key passage from Lamarck's 1820 book: "Let us consider the most influential cause for everything done by nature, the only cause that can lead to an understanding of everything that nature produces ... This cause resides in the power that circumstances have to modify all operations of nature, to force nature to change continually the laws that she would have followed without [the intervention of] these circumstances, and to determine the character of each of her products. The extreme diversity of nature's productions must also be attributed to this cause."
8. H. Spencer (1852a), p. 10. Also see P.A. Corning (1982).
9. H. Spencer (1852a).
10. P.A. Corning (1982), p. 362.
11. L.A. White (1959), p. 56.
12. L.A. White (1949), pp. 39, 330, 335.
13. L.A. White (1975), p. 46.
14. E. Boserup (1965); D.E. Dumond (1965); M.H. Fried (1967); M.J. Harner (1970).
15. M. N. Cohen (1977), p. 285.
16. R. Bigelow (1969).
17. R.D. Alexander (1979), pp. 222, 223.
18. R.L. Carneiro (1970a).
19. R.L. Carneiro (1981).
20. K.F. Ottebein (1970) (1985) (1994); H.J.M. Claessen and P. Skalník (eds.) (1978); R.P. Shaw and Y. Wong (1989); J. Haas (ed.) (1990); J.M.G. van der Dennen and V.S.E. Falger (eds.) (1990); J. Keegan (1993); S. Brown (1994); J.M.G. van der Dennen (1995); D. Dawson (1996).

21. See the review of this issue in P.A. Corning (2001a).
22. R.A. Hackenberg (1962); E. Brumfiel (1976); R. Cohen (1978a, b, c).
23. See the discussion and references in P.A. Corning (1983), p. 371.
24. R.B. Lee (1968), plus M. Harris (1975); see also E. Cook (1971).
25. See especially the review by R.F. Salisbury (1973).
26. A. Smith (1776), p. 5.
27. Data obtained from displays at the Gold Mining Museum, Angels Camp, California.
28. M. Rothschild (1990). Rothschild also provides a detailed description and analysis (pp. 167–176).
29. R.L. Heilbroner (1994), p. 77. Heilbroner's original essay is also reprinted in this volume. See also M.R. Smith and L. Marx (eds.) (1994).
30. M. Rothschild (1990), p. 185, passim.
31. J. Cohn (1997).
32. B.H. Beard (1981). See also the data provided by the U.S. Department of Agriculture, National Agricultural Statistics Service, and the National Sunflower Association: www.sunflowers.com
33. Among the many historical sources on the automobile revolution, see especially J.J. Flink (1970); T.R. Nicholson (1970); also the *Encyclopedia Britannica*.
34. R. Taagepera (1968) (1978a, b) 1979).
35. E.R. Service (1971), p. 25.
36. J. Schumpeter (1911), p. 58.
37. C.B. Drucker (1978).
38. See, for example, J.H. Steward (1955); M.D. Coe and K.V. Flannery (1967); R. McC. Adams (1966) (1972); E.R. Service (1971) (1975); K.V. Flannery (ed.) (1976) (1982); R. Cohen and E.R. Service (eds.) (1978).
39. B.M. Fagan (1998), p. 364.
40. R.L. Carneiro (1973), p. 108.
41. G.P. Murdock (1967); also see T.J. O'Leary and D. Levinson (eds.) (1991). The correlation analysis can be found in R.L. Carneiro (1970b).
42. R.L. Carneiro (1967). A more comprehensive survey and analysis of this issue can be found in P.A. Corning (1983), pp. 345–362.
43. R.A. Hackenberg (1962).
44. Discussed in P.A. Corning (1983), p. 301. See also J.E. Pfeiffer (1977).
45. J.H. Steward (1938) (1941) (1943). Also, see A.W. Johnson and T. Earle (1987) (2000).
46. See the classic study by A. Balikci (1970).
47. S.C. Oliver (1962).
48. A.W. Johnson and T. Earle (1987), pp. 84–85. See also A.W. Johnson and T. Earle (2000).
49. The main source for the Tereumiut is R. Spencer's (1959) classic study. It is summarized in A.W. Johnson and T. Earle (1987), pp. 132–138.
50. Among others, see especially A.W. Johnson and T. Earle (1987) (2000); also T.J. O'Leary and D. Levinson (eds.) (1991).
51. Political evolution is discussed at length in P.A. Corning (1983), Chapter VI. On "workarounds," see Richerson and Boyd (1999).

52. This example is taken from B.M. Fagan (1998), pp. 255–258.

53. My interpretation relies on the long-term studies of archaeologist Andrew Moore (1985) (1992). See also A.M.T. Moore and G.C. Hillman (1992).

54. See R. Kunzig (1999); see also J. Mellaart (1970) (1972) and I. Hodder (1996) (2000).

55. Some insight might be gained from comparing Çatalhöyük to Jericho, the legendary walled community that arose at an abundant natural spring in the Jordan Valley about 8500 BC. Not only was Jericho one of the longest-lived of the agricultural pioneers but it produced the most impressive public works – a model of collective synergy that foreshadowed the future pattern of urban civilizations. Its massive walls and cylindrical towers (10 feet thick and 13 feet high), with a rock-cut "moat" that was 9 feet deep and 10 feet wide (it may have served as the stone quarry for the walls) were most likely designed to protect Jericho's bounty from its jealous, or desperate, competitors. However, Çatalhöyük had no such defensive fortification, perhaps for lack of a readily available stone quarry. B.M. Fagan (1998).

56. See especially the review in T.K. Earle (1987) (1989); also R.L. Carneiro (1981).

57. On "superorganisms," see P.A. Corning (2002b); the disparities in wealth are amply documented in the Marxist literature, both in economics and anthropology. The causes and consequences are more contentious.

58. H. Weiss (1996) (2000); H. Weiss and R.S. Bradley (2001).

59. J. Diamond (1997), pp. 188–189; the quote is found on p. 157.

60. J. Diamond (1997).

61. Although this theory has long been debated, the accumulating weight of evidence seems increasingly to favor it. See especially P.S. Martin (1967); P.S. Martin and R.G. Klein (eds.) (1984); R. Klein (1992). Among the most recent findings is a study by R.P. Duncan et al. (2002) which showed that the bird species preferentially hunted by the Maori in New Zealand were much more likely to have gone extinct than birds that were not on their menu.

62. M.E. Basgall (1987).

9. The Science of History

1. Among the references to Pythagoras and Pythagoreanism, see especially R.W. Bernard (1958); W. Burkert (1972); K.S. Guthrie et al. (1987). Other theorists point to the philosopher Thales of Miletus, also in the sixth century BC, as the founding father of Western science, a claim that dates back to Aristotle. Thales was supposedly the first Western thinker to seek the underlying basis of the natural world through the use of reason. (He singled out water.) However, Pythagoras, a younger contemporary of Thales, was reputed to be the first to deploy mathematics as a systematic scientific tool. There may also have been some Greek chauvinism in Aristotle's claim. Samos is close to the coast of Turkey, and Pythagoras later migrated to Crotona, Italy. He was not a part of the Ionian circle. Novelist Arthur Koestler (1959), a student of that era, believed that Pythagoras was perhaps the most original and important

thinker of his age. But as Jamie James (1993) notes, Pythagoras's achievements languish in obscurity because none of his writings survive.

2. J. James (1993), p. 31.

3. J. Godwin (1987), p.145. There are a number of other good biographies and biographical histories that recount Kepler's life and works. See especially A. Koestler (1960); also A. Van Helden (1985). Also relevant is N. Kollerstrom (1989).

4. S. Weinberg (1992), p. 18. See also the lively exchange between S. Weinberg (1987) and E. Mayr (1988). Many other theorists seem to share Weinberg's vision. For instance, G. Cziko (1995) declares that the universe is the "mechanical consequence of the principles of physics and chemistry" (p. 3).

5. E.O. Wilson (1998), p. 55.

6. E.O. Wilson (1998), p. 266. Of course, Wilson also leaves some room for historical contingencies. The Marquis de Laplace in the eighteenth century still holds prize for scientific arrogance. With sufficient information about the operation of the laws of nature, he asserted, the future could be predicted.

7. Quoted in F. Miele (1998), p. 79.

8. Clausius is quoted in F.M. Harold (1986). This issue is treated in some depth in P.A. Corning and S.J. Kline (1998a, b). The issue is also discussed briefly in Chapter 10, especially note 10.29.

9. J.B. de Lamarck (1809); H. Spencer (1852a); H. Bergson (1907); H.A.E. Dreisch (1909); P. Teilhard de Chardin (1959); P.P. Grassé (1973); J. Piaget (1978).

10. E. Schrödinger (1945), p. 72.

11. H.J. Morowitz (1968), pp. 2, 120, 146.

12. H. Spencer (1852a), p. 10. See also H. Spencer (1862), p. 216.

13. I. Prigogine *et al.* (1972a, b); (1977), p. 18.

14. Per Bak is the scientist who is principally responsible for identifying and analyzing the phenomenon of "self-organized criticality." In his 1996 book, *How Nature Works: The Science of Self-Organized Criticality*, Bak boldly proposes that criticality explains complexity, both in the physical world of sand piles and in biological evolution – not to mention black holes, pulsars, and earthquakes. "I will argue that complex behavior in nature reflects the tendency of large systems with many components to evolve to a poised, 'critical' state, way out of balance, where minor disturbances may lead to events, called avalanches, of all sizes ... The state is established solely because of the dynamical interactions among individual elements of the system: the critical state is *self-organized* [his emphasis]. Self-organized criticality is so far the only known general mechanism to generate complexity" (pp. 2–3). The problem with Bak's reasoning is that he's got things backwards. Criticality does not create the sand pile; it does not cause complexity. Nor does some hidden universal causal force produce self-organization and a march to criticality. An unstable state may arise in various complex physical and biological processes for various reasons, and these may well produce catastrophic, avalanche-like phase transitions. But the underlying causal dynamics are not homogeneous; they are certainly not somehow embedded in the mathematics. Indeed, many biological and social

"catastrophes" are produced by the synergy-minus-one phenomenon – the loss of a vital part (as we have already noted).

15. I. Prigogine *et al.* (1972a), pp. 27–28; (1977), pp. 32, 34.

16. S. Kauffman (1995), p. 8.

17. S. Kauffman (1995), p. 25.

18. S. Kauffman (1995), pp. 16, 60.

19. S. Kauffman (1995), pp. 58, 61.

20. S. Kauffman (1995), p. 19.

21. S. Kauffman (2000), p. 5.

22. S. Camazine *et al.* (2001). See also F.E. Yates *et al.* (1987); G.M. Whitesides and B. Grzybowski (2002).

23. B. Goodwin (1994). For an review devoted to this remarkable unicellular organism, see D.F. Mandoli (1998).

24. J. Maynard Smith (1998). The importance of environmental conditions, even where self-organizing processes are concerned, was illustrated in an experiment reported by C. Papaseit and his colleagues (2000). When microtubules, a major constituent of the cytoskeleton in eukaryotic cells, are grown in the zero-gravity conditions of space flight, they show almost no self-organization.

25. R.E. Page and S.D. Mitchell (1998).

26. S. Camazine *et al.* (2001), p. 89. See also R.C. Richardson (2001).

27. C.R. Darwin (1871), p. 150.

28. J. Monod (1971).

29. F. Jacob (1977).

30. A major change in this worldview may be under way, however. There seems to be a growing movement within the scientific community to reaffirm humanistic and even religious belief systems. See especially the cover story by Gregg Easterbrook (1998a), and his book *Beside Still Waters* (1998b). See also such recent works by K. Miller (1999); S.J. Gould (1999c); J. Haught (ed.) (2000); M. Ruse (2000); and D.S. Wilson (2002). Scientists also seem to have an unlikely ally in the Pope, whose most recent encyclical, *Fides et Ratio* ("Faith and Reason"), defended reason and the search for truth against the relativism and cynicism embodied in the philosophical schools of "postmodernism" and "constructivism."

31. S.J. Gould (1999); also (1994). Gould's world view is sharply critiqued by R. Wright (1999). It should also be noted that Gould's metaphor begs the question. Is he playing fair? Will all of the determinants – the causal relationships at all levels (including the environment) – also be replayed? If so, then the scope for truly stochastic phenomena is greatly circumscribed. A betting man might be willing to take Gould's odds.

32. C.E. Cleland (2001). A concerted effort to resuscitate an historical perspective in economics, another science that went astray, was undertaken by Geoffrey M. Hodgson in his new book, *How Economics Forgot History* (2001).

33. C.R. Darwin (1859), pp. 485–486.

34. For a full-length discussion of the issue of "progress" in evolution, see M.H. Nitecki (ed.) (1988). There have been many books devoted to "bioengineering" in the natural world. Three recent examples are the works

of R.M. Alexander (1968) (1989) (1992) (1996); S. Vogel (1988) (1998); M.J. French (1994).

35. J.H. Holland (1998), p. 2.
36. The common practice of proposing a "thought experiment" to advance an argument is often abused. As the late Stephen Jay Kline, founder of Stanford University's renowned "Science, Technology and Society Program," pointed out in his last book, it is bad science to posit some hypothetical state or action that is "wildly infeasible" – that has no realistic possibility of ever occurring. See S.J. Kline (1995) (1997). An example is the nineteenth-century physicist James Clerk Maxwell's famous "demon" – a thought experiment in which an imaginary "creature" is supposed to be able to sort out fast- and slow-moving molecules in a large volume of gas. This implies capabilities for perception, detection, data collection, and actions that are totally impracticable, not to mention totally uneconomical. Nevertheless, generations of physicists have argued over the merits of this paradigm and have dreamed up many variations in an effort to make the benefits outweigh the costs. See especially the collection of articles in H.S. Leff and A.F. Rex (1990).
37. S. Wolfram (2002), pp. 1, 2.
38. S. Brenner, cited in Lewontin (2000a), p. 10.
39. R. Lewontin (2000a), p. 75 passim. Lewontin (2000b) provides another example in a recent article, where he describes the famous experiment by Jens Clausen, David Keck, and William Hiesey. Three identical clones from each of seven *Achillea millefolium* plants were grown at different altitudes, 100 feet, 4,600 feet, and 10,000 feet, above sea level. Not surprisingly, the differences in the mature offspring were striking, even to a layman.
40. S. Wright (1964), p. 122.
41. D.M. Kennedy, personal communication.
42. M. Polanyi (1968). For the record, it should be noted that many other theorists, before and since, have appreciated the hierarchical organization of nature and its theoretical significance. This is discussed in some detail in P.A. Corning (1996b) (2002a). Also, see especially such landmarks as L. von Bertalanffy (1968); P.A. Weiss (ed.) (1971); H.H. Pattee (ed.) (1973); J.G. Miller (1978). Especially notable among the more recent contributions are S.N. Salthe (1985); L.W. Buss (1987); J. Maynard Smith and E. Szathmáry (1995); M.T. Ghiselin (1997); R.E. Michod (1999). Many others have written on the subject, including Rapoport, Boulding, Sperry, Wimsatt, Eldredge, Vrba, S.J. Gould, and Jablonka, to name a few.
43. P.W. Anderson (1972).
44. R.B. Laughlin and D. Pines (2000). See also R.B. Laughlin *et al.* (2000); and N. Goldenfeld and L.P. Kadanoff (1999).

10. Conjuring the Future: What Can We Predict?

1. R. Naroll (1967).
2. R.L. Carneiro (1978), p. 219.

3. C.J. Lumsden and E.O. Wilson (1981), pp. 554, 356–357, 360.

4. In fact, Wright uses the term synergy at least 12 times in ways that are not conceptually different from the idea of non-zero-sumness.

5. R. Wright (2000), p. xiii.

6. R. Wright (2000), pp. xi, 103.

7. This point is discussed in some detail in P.A. Corning (1983), pp. 164–165. Also, see P.A. Corning (1996). Wright, in a lengthy appendix, says he cannot think of a better term than non-zero-sumness, even though he uses the term synergy at various places in the book. It became evident that he has a very truncated view of the concept of synergy.

8. R. Wright (2000), p. 74.

9. R. Wright (2000), ca. p. 169.

10. R. Wright (2000), pp. 74, 124, 169.

11. J. Diamond (1997), p. 312. Anthropologists are well aware of such evolutionary convergences, which amount to cultural analogues of the similar phenomenon of convergent evolution in nature. Alfred Kroeber (1948) provided what may still be the most thoroughgoing and well-reasoned analysis of cultural convergences. They are likely to occur only when there is a parallel convergence of many other factors, and preconditions, and needs. See also the discussion in P.A. Corning (1983), Chapter V.

12. R. Wright (2000), p. xiii.

13. R. Wright (2000), p. xvi. An interesting counterpoint to Wright's thesis was a special symposium in the journal *Politics and the Life Sciences*, vol. 18(2) (1999) on the topic "Is Humanity Destined to Self-Destruct?" Opinions were mixed, but nobody displayed unbounded optimism.

14. D. Dörner (1996).

15. B.M. Fagan (1998), p. 373.

16. W. Fey and A.C.W. Lam (1998) (1999) (2000).

17. P. Ward and A. Rockman (2002).

18. J. Malone (1997).

19. G.H. May (1996).

20. Editorial (2001b).

21. B. Holland (1999).

22. P.A. Corning (2000). This lengthy article includes a more extensive discussion of the basic needs framework as an analytical construct for economic and political analysis, including measuring-rods called "Survival Indicators."

23. Among the many references on this issue, see especially W.K. Stevens (1999a, b, c) (2000a); B. McKibben (1999); A.C. Revkin (2000); W.S. Broecker (2001); Committee on the Science of Climate Change (2001).

24. W.S. Broecker (1994) (1995) (1996); J. Overpeck and R. Webb (2000); Committee on Abrupt Climate Change (2002).

25. H. Weiss *et al.* (1993); H. Weiss (1996) (2000); H. Weiss and R.S. Bradley (2001); also H.N. Dalfes *et al.* (eds.) (1996) and W.K. Stevens (2000b).

26. U.S. Department of Agriculture (1999) (2000).

27. *The New York Times*, Editorial, March 20, 2000. See especially D. Pimentel *et al.* (1997); P.H. Gleick (2000) (2001); J.R. McNeill (2000); S. Postel (2000).

28. D.G. McNeil, Jr. (1998); D. Pimentel *et al.* (1998); W.J. Broad *et al.* (1999); P. Farmer (1999).

29. C. Perrow (1984), pp. 3, 4.

30. J. Markoff (1999).

31. From a U.S. Department of Energy Report cited in an Associated Press wire June 9, 2000.

32. Among others, see P.K. Dayton (1998); D. Pauly *et al.* (1998).

33. T. Halliday (1998).

34. J. Gowdy (1997).

35. *The New York Times*, Editorial, March 20, 2000.

36. J.R. McNeill (2000). For a sampler of the voluminous literature on this subject, see especially G.C. Daily and P.R. Ehrlich (1992); J. Harte (1996); G.C. Daily *et al.* (1998); G.W. Barrett and E.P. Odum (2000); R. Costanza *et al.* (2000).

37. T.F. Homer-Dixon *et al.* (1993).

38. J.W. Toigo (2000).

39. A. McLaughlin (1999).

40. Quoted in J. Markoff (2000).

41. A. Smith (1776), vol. 1, p. 12.

42. K. Kelly (1994).

43. For an in-depth discussion of this issue, see P.A. Corning (2002c), and the references cited therein.

44. H. Brandon (1969).

45. G. Hardin (1971).

46. L.M. Krauss and G.D. Starkman (1999).

47. This issue is treated in some depth in P.A. Corning and S.J. Kline (1998a). Clausius's prediction about the "heat death" of the universe is increasingly dubious, and lies 10 to 100 billion years in the future in any case. In the meantime, much of the known energy in the universe has been "frozen." It is "stored" in the many billions of galaxies and trillions of stars. But more to the point, what is frequently overlooked by the adherents to this gloomy scenario is the fact that the universe does not have a fixed, ever-depleting stock of available energy, like a box filled with hot gas molecules. Gravitational forces are continually generating new sources of available energy in the myriad of evolving stars and galaxies. On this important but underappreciated point, see especially F.J. Dyson (1971).

48. See especially M.S. Turner (2000). Also, see the zero-point field theory of B. Haisch *et al.* (1994); B. Haisch and A. Rueda (1996). L.M. Krauss and G.D. Starkman (1999) claim that it would be "impossible" to extract usable energy from any dark energy source. However, their conclusion hangs on the assumption of a quantum vacuum – an extremely low energy state in the universe. An alternative view of dark energy assumes that it is powerful enough to influence the expansion of the universe and the behavior of entire galaxies. Stay tuned.

Afterword

1. M.T. Ghiselin (1974) (1978) (1986); J.R. Krebs and N. B. Davies (1984) (1991) (1993) (1997); N. Eldredge and M. Grene (1992); N. Eldredge (1995); R.E. Ricklefs (1996).

2. S.G. Winter (1964); N. Georgescu-Roegen (1971) (1975a, b); G. Tullock (1971) (1979) (1994); J. Hirschleifer (1977) (1978a, b) (1999); K. Boulding (1981); R. Costanza (1989) (1991); U. Witt (1991a, b); G. Hodgson (1993); J.M. Gowdy (1994).

3. See H. Gintis (2000b); E. Fehr and S. Gächter (2000a, b) (2002); R. Sethi and E. Somanathan (2001); E. Fehr *et al.* (2002).

4. E.g., M. Rabin (1993); E. Fehr and S. Gächter (2000a, b) (2002); E. Fehr and K.M. Schmidt (1999); J. Henrich and R. Boyd (2001); J. Henrich *et al.* (2001); M.E. Price *et al.* (2002).

5. R. Boyd and P. Richerson (2002).

6. C. Boesch (2002); S.F. Brosnan and F.B.M. de Waal (2002).

7. K. Hill (2002); also, see M. Alvard (2001); M. Gurven *et al.* (2000); R. Sosis (2000).

8. Among many others, see especially D.E. Brown (1991); J.Q. Wilson (1993); F.B.M. de Waal (1996) (2001); M. Ridley (1997); E. Sober and D.S. Wilson (1998); P.R. Ehrlich (2000); S. Pinker (2002).

9. P.A. Corning (2000).

10. This subject is reviewed in some detail in P.A. Corning (2002d).

11. S. Weinberg (2002).

12. N. Goldenfeld and L.P. Kadanoff (1999), p. 89.

13. P.A. Corning (1996b) (1997b).

References

Adams, R.McC. (1966) *The Evolution of Urban Society: Early Mesopotamia and Prehispanic Mexico*. Chicago, IL: Aldine Press.

Adams, R.McC. (1972) "Patterns of Urbanization in Early Southern Mesopotamia." In *Man, Settlement and Urbanism*, eds. P.J. Ucko *et al.*, pp. 735–750. London: Duckworth.

Adler, B. (1997) *Outwitting Critters: A Human Guide for Confronting Devious Animals and Winning*. Guilford, CT: Lyons Press.

Ahmadjian, V. (1993) *The Lichen Symbiosis*. New York: John Wiley & Sons.

Ahmadjian, V., and S. Paracer (1966) *Symbiosis: An Introduction to Biological Associations*. Lebanon, NH: University Press of New England (for Clark University).

Aiello, L.C., and R.I.M Dunbar (1993) "Neocortex Size, Group Size, and the Evolution of Language." *Current Anthropology*, 34(2): 184–192.

Aiello, L.C., and C. Key (2001) "The Energetic Consequences of Being a *Homo erectus* Female." Prepared for the Human Behaviour and Evolution Society Meeting, University College London, June 13–17.

Aiello, L.C., and P. Wheeler (1995) "The Expensive-Tissue Hypothesis." *Current Anthropology*, 36: 199–221.

Aisner, R., and J. Terkel (1992) "Ontogeny of Pine Cone Opening Behaviour in the Black Rat, *Rattus rattus*." *Animal Behaviour*, 44: 327–336.

Alba, D., *et al.* (2001) "Canine Reduction in the Miocene Hominid *Oreopithecus bambolii*: Behavioural and Evolutionary Implications." *Journal of Human Evolution*, 40: 1–16.

Alcock, J. (1993) *Animal Behaviour: An Evolutionary Approach*, 5th edn. Sunderland, MA: Sinauer Associates.

Alexander, R.D. (1979) *Darwinism and Human Affairs*. Seattle: University of Washington Press.

Alexander, R.D. (1987) *The Biology of Moral Systems*. New York: Aldine de Gruyter.

Alexander, R.D. (1990a) "Epigenetic Rules and Darwinian Algorithms: The Adaptive Study of Learning and Development." *Ethology and Sociobiology*, 11: 241–303.

Alexander, R.D. (1990b) *How Did Humans Evolve? Reflections on the Uniquely*

Unique Species. Special Publication No. 1. Ann Arbor: Museum of Zoology, University of Michigan.

Alexander, R.M. (1968) *Animal Mechanics*. Seattle: University of Washington Press.

Alexander, R.M. (1989) *Dynamics of Dinosaurs and Other Extinct Giants*. New York: Columbia University Press.

Alexander, R.M. (1992) *Exploring Biomechanics: Animals in Motion*. New York: Scientific American.

Alexander, R.M. (1996) *Optima for Animals*. Princeton, NJ: Princeton University Press.

Allee, W.C. (1931) *Animal Aggregations: A Study in General Sociology*. Chicago, IL: University of Chicago Press.

Allee, W.C. (1938) *Cooperation among Animals: With Human Implications*. 1951 edn. New York: Henry Schuman.

Allee, W.C., *et al.* (1949) *Principles of Animal Ecology*. Philadelphia, PA: W.B. Saunders.

Allen, J.R.M., *et al.* (1999) "Rapid Environmental Changes in Southern Europe During the Last Glacial Period." *Nature*, 400: 740–743.

Allen, L., and C. Goodstein (1997) "Where Have All the Pollinators Gone?" *Science World*, 53(12): 12–16.

Allen, P. (1983) *Who Sank the Boat?* New York: Coward-McCann.

Allen, T.F.H., *et al.* (1999) "Supply-side Sustainability." *Systems Research and Behavioral Science*, 16(5): 403–427.

Alley, R.B. (2000) "Ice-core Evidence of Abrupt Climate Changes." *Proceedings of the National Academy of Sciences*, 97(4): 1331–1334.

Allison, D.G., *et al.* (2001) *Community Structure and Cooperation in Biofilms*. London: Pergamon Press.

Alston, R. (1998) *Aspects of Roman History AD 14–117*. London: Routledge.

Alvard, M. (2001) "Mutualistic Hunting." In *The Early Human Diet: The Role of Meat,* eds. C. Stanford and H. Bunn, pp. 261–278. Oxford, U.K.: Oxford University Press.

Amato, I. (1992) "A New Blueprint for Water's Architecture." *Science*, 256: 1764.

Anderson, C., and N.R. Franks (2001) "Teams in Animal Societies." *Behavioral Ecology*, 12(5): 534–540.

Anderson, C., *et al.* (2001) "The Complexity and Hierarchical Structure of Tasks in Insect Societies." *Animal Behaviour*, 62: 643–651.

Anderson, C.M. (1986) "Predation and Primate Evolution." *Primates*, 27: 15–39.

Anderson, P.W. (1972) " 'More is Different': Broken Symmetry and the Nature of the Hierarchical Structure of Science." *Science*, 177: 393–396.

Andrews, F. (1999) "It's Not the Product That's Different, It's the Process." *The New York Times*, December 15, 1999, p. C14.

Ardrey, R. (1961) *African Genesis*. New York: Dell.

Ardrey, R. (1970) *The Social Contract*. New York: Atheneum.

Ardrey, R. (1976) *The Hunting Hypothesis*. New York: Atheneum.

Ariniello, L. (1999) "Protecting Paradise." *BioScience*, 49(10): 760–763.

Aristotle. *The Metaphysics* (trans. H. Tredennick). Cambridge, MA: Harvard University Press.

Aristotle. *The Politics.* 1946 edn., trans. E. Barker. Oxford, U.K.: Oxford University Press.

Arkani-Hamed, N., *et al.* (2000) "The Universe's Unseen Dimensions." *Scientific American*, 283(2): 62–69.

Arthur, W.B. (1988) "Self-Reinforcing Mechanisms in Economics." In *The Economy as an Evolving Complex System*, eds. P.W. Anderson, *et al.*, pp. 9–31. Reading, MA: Addison-Wesley.

Arthur, W.B. (1990) "Positive Feedbacks in the Economy." *Scientific American*, 266: 92–99.

Asfaw, B., *et al.* (1999) "*Australopithecus garhi*: A New Species of Early Hominid from Ethiopia." *Science*, 284: 629–635.

Atran, S. (2001) "The Trouble with Memes: Inference versus Imitation in Cultural Creation." *Human Nature*, 12(4): 351–381.

Atsatt, P.R. (1991) "Fungi and the Origin of Land Plants." In *Symbiosis as a Source of Evolutionary Innovation*, eds. L. Margulis and R. Fester, pp. 301–305. Cambridge, MA: MIT Press.

Aunger, R. (2000) *Darwinizing Culture: The Status of Memetics as a Science.* Oxford, U.K.: Oxford University Press.

Aunger, R. (ed.) (2002) *The Electric Meme: A New Theory of How We Think and Communicate.* New York: Free Press.

Avilés, L., and P. Tufiño (1998) "Colony Size and Individual Fitness in the Social Spider *Anelosimus eximius*." *American Naturalist*, 152(3): 403–418.

Avital, E., and E. Jablonka (1994) "Social Learning and the Evolution of Behaviour." *Animal Behaviour*, 48: 1195–1199.

Avital, E. and E. Jablonka (2000) *Animal Traditions: Behavioural Inheritance in Evolution.* Cambridge, U.K.: Cambridge University Press.

Avital, E., *et al.* (1998) "Adopting Adoption." *Animal Behaviour*, 55: 1451–1459.

Axelrod, R., and W. Hamilton (1981) "The Evolution of Cooperation." *Science*, 211: 1390.

Ayala, F.J., and T. Dobzhansky (eds.) (1974) *Studies in Philosophy of Biology.* Berkeley: University of California Press.

Bak, P. (1996) *How Nature Works: The Science of Self-Organized Criticality.* New York: Copernicus.

Bak, P., and K. Chen (1991) "Self-Organized Criticality." *Scientific American*, 261(1): 46–53.

Balashov, Y.V. (1991) "Resource Letter AP-1: The Anthropic Principle." *American Journal of Physics*, 59(12): 1069–1076.

Baldwin, H.M. (1895) *Mental Development in the Child and the Race: Methods and Processes.* New York: Macmillan.

Baldwin, H.M. (1896a) "Heredity and Instinct: Discussions (Revised) following Professor C. Lloyd Morgan before the New York Academy of Sciences, January 31, 1896." *Science*, 3: 438–441, 558–561.

Baldwin, H.M. (1896b) "On Criticisms of Organic Selection." *Science*, 4: 724–727.

Baldwin, H.M. (1896c) "A New Factor in Evolution." *American Naturalist*, 30: 441–451, 536–553.

Balikci, A. (1970) *The Netsilik Eskimo.* Garden City, NY: Natural History Press.

Bar-Yusef, O. and S. Kuhn (1999). "The Big Deal about Blades: Laminar Technologies and Human Evolution." *American Anthropologist* 101(2): 322–338.

Barker, G. (1993) "Models of Biological Change: Implications of Three Studies of Lamarckian Change." In *Perspectives in Ethology: Behavior and Evolution*, eds. P.P.G. Bateson *et al.*, pp. 229–246. New York: Plenum Press.

Barlow, C. (ed.) (1994) *Evolution Extended: Biological Debates on the Meaning of Life*. Cambridge, MA: MIT Press.

Barlow, C. (2000) *The Ghosts of Evolution: Nonsensical Fruit, Missing Partners, and Other Ecological Anachronisms*. New York: Basic Books.

Barlow, C., and T. Volk (1990) "Open Systems Living in a Closed Biosphere: A New Paradox for the Gaia Debate." *BioSystems*, 23: 371–384.

Barlow, C., and T. Volk (1992) "Gaia and Evolutionary Biology." *BioScience*, 42(9): 686–693.

Barrett, G.W., and E.P. Odum (2000) "The Twenty-First Century: The World at Carrying Capacity." *BioScience*, 50: 363–368.

Barrow, J.D. (1993) "Anthropic Principles in Cosmology." *Vistas in Astronomy*, 37: 409–427.

Barrow, J.D., and F.J. Tipler (1986) *The Anthropic Cosmological Principle*. New York: Oxford University Press.

Basgall, M.E. (1987) "Resource Intensification among Hunter–Gatherers: Acorn Economies in Prehistoric California." *Research in Economic Anthropology*, 9: 21–52.

Bastock, M. (1956) "A Gene Mutation which Changes a Behavior Pattern." *Evolution*, 10: 421–439.

Bateson, P.P.G. (1988) "The Active Role of Behavior in Evolution." In *Evolutionary Processes and Metaphors*, eds. M.W. Ho and S.W. Fox, pp. 191–207. New York: John Wiley & Sons.

Bateson, P.P.G., *et al.* (1993) *Perspectives in Ethology*, vol. 10, *Behavior and Evolution*. New York: Plenum Press.

Beard, B.H. (1981) "Sunflower Crop." *Scientific American*, 244(5): 150–161.

Beck, B.B. (1980) *Animal Tool Behavior*. New York: Garland Press.

Bekoff, M., and D. Jamieson (1996) *Readings in Animal Cognition*. Cambridge, MA: MIT Press.

Bell, G. (1985) "Origin and Early Evolution of Germ Cells as Illustrated by the Volvocales." In *Origin and Evolution of Sex*, eds. H.O. Halvorson and A. Monroy, pp. 221–256. New York: Alan R. Liss, Inc.

Bellomo, R. (1994) "Methods of Determining Early Hominid Behavioral Activities Associated with the Controlled Use of Fire at FxJj Main, Koobi Fora, Kenya." *Journal of Human Evolution*, 27(1–3): 173–195.

Ben-Jacob, E. (1997) "Cooperative Formation of Complex Patterns (From Snowflake Formation to Growth of Bacterial Colonies)." *Contemporary Physics*, 38(3): 205–241.

Ben-Jacob, E., and H. Levine (1998) "The Artistry of Microorganisms: Colonies of Bacteria or Amoebas Form Complex Patterns that Blur the Boundary Between Life and Non-Life." *Scientific American*, 279(4): 82–87.

Ben-Jacob, E., *et al.* (1998) "Cooperative Organization of Bacterial Colonies: From Genotype to Morphotype." *Annual Review of Microbiology*, 52: 779–806.

Berenbaum, M.C. (1989) "What is Synergy?" *Pharmacological Reviews,* 1989(41): 93–141.

Bergquist, P.R. (1978) *Sponges.* Berkeley: University of California Press.

Bergson, H. (1907) *L'Évolution Créatrice.* Paris: F. Alcan.

Bernard, R.W. (1958) *Pythagoras, the Immortal Sage.* Mokelumne Hill, CA: Health Research.

Bernasconi, G., and J.E. Strassmann (1999) "Cooperation among Unrelated Individuals: The Ant Foundress Case." *Trends in Ecology and Evolution,* 14(12): 477–482.

Bertalanffy, L. von (1968) *General System Theory: Foundations, Development, Applications.* New York: George Braziller.

Bethlenfalvay, G.J., *et al.* (1982a) "Parasitic and Mutualistic Associations between a Mycorrhizal Fungus and Soybean: Development of the Host Plant." *Phytopathology,* 72: 889–893.

Bethlenfalvay, G.J., *et al.* (1982b) "Parasitic and Mutualistic Associations between a Mycorrhizal Fungus and Soybean: Development of the Endophyte." *Phytopathology,* 72: 894–897.

Bethlenfalvay, G.J., *et al.* (1983) "Parasitic and Mutualistic Associations between a Mycorrhizal Fungus and Soybean: The Effect of Phosphorus on Host Plant–Endophyte Interactions." *Physiologia Plantarum,* 57: 543–48.

Bickerton, D. (1995) *Language and Human Behavior.* Seattle: University of Washington Press.

Bigelow, R. (1969) *The Dawn Warriors: Man's Evolution Towards Peace.* Boston, MA: Little, Brown.

Bigelow, R. (1975) "The Role of Competition and Cooperation in Human Aggression." In *War, Its Causes and Correlates,* eds. M.A. Nettleship, R.D. Givens, and A. Nettleship, pp. 235–261. The Hague: Mouton.

Binford, L.R. (1987) "The Hunting Hypothesis, Archaeological Methods, and the Past." *Yearbook of Physical Anthropology,* 30: 1–9.

Bingham, P.M. (2000) "Human Evolution and Human History: A Complete Theory." *Evolutionary Anthropology,* 9(6): 248–257.

Binmore, K. (1994) *Game Theory and the Social Contract,* vol. 1, *Playing Fair.* Cambridge, MA: MIT Press.

Binmore, K. (1998) *Game Theory and the Social Contract,* vol. 2, *Just Playing.* Cambridge, MA: MIT Press.

Blackmore, S. (1999) *The Meme Machine.* Oxford, U.K.: Oxford University Press.

Blackmore, S. (2000) "The Power of Memes." *Scientific American,* 283(4): 64–73.

Blakemore, C. (1988) *The Mind Machine.* London: BBC Books.

Blitz, D. (1992) *Emergent Evolution: Qualitative Novelty and the Levels of Reality.* Dordrecht, The Netherlands: Kluwer.

Bloom, F.E., and A. Lazerson (1988) *Brain, Mind, and Behavior,* 2nd edn. New York: W.H. Freeman and Co.

Blumenshine, R.J. (1987) "Characteristics of an Early Hominid Scavenging Niche." *Current Anthropology,* 28(4): 383–407.

Boehm, C. (1996) "Emergency Decisions, Cultural-Selection Mechanics, and Group Selection." *Current Anthropology,* 37: 763–793.

Boehm, C. (1997) "Impact of the Human Egalitarian Syndrome on Darwinian Selection Mechanics." *American Naturalist*, 150: 5100–5121.

Boehm, C. (1999) *Hierarchy in the Forest: The Evolution of Egalitarian Behavior.* Cambridge, MA: Harvard University Press.

Boesch, C. (1988) "Teaching among Wild Chimpanzees." *Animal Behaviour*, 41: 530–532.

Boesch, C. (1993) "Towards a New Image of Culture in Wild Chimpanzees." *Behavioral and Brain Sciences*, 16: 514–515.

Boesch, C. (2002). "Cooperative Hunting Roles Among Taï Chimpanzees." *Human Nature*, 13: 27–46.

Boesch, C., and Tomasello, M. (1998) "Chimpanzee and Human Cultures." *Current Anthropology*, 39(5): 591–614.

Boinski, S., and P.A. Garber (eds.) (2000) *On the Move: How and Why Animals Travel in Groups.* Chicago, IL: University of Chicago Press.

Bonner, J.T. (1980) *The Evolution of Culture in Animals.* Princeton, NJ: Princeton University Press.

Bonner, J.T. (1988) *The Evolution of Complexity by Means of Natural Selection.* Princeton, NJ: Princeton University Press.

Boserup, E. (1965) *The Conditions of Agricultural Growth: The Economies of Agrarian Change under Population Pressure.* Chicago, IL: Aldine Press.

Botstein, D., *et al.* (1997) "Yeast as a Model Organism." *Science*, 277: 1259–1260.

Boulding, K.E. (1981) *Evolutionary Economics.* Beverly Hills, CA: Sage Publications.

Bower, B. (1997) "Cave Finds Make Point about Early Humans." *Science News*, 152: 342.

Bower, B. (2000) "Culture of the Sea: Whales and Dolphins Strut their Social Stuff for Scientists." *Science News*, 158(18): 284–286.

Bowersock, G.W. (1988) "The Dissolution of the Roman Empire." In *The Collapse of Ancient States and Civilizations*, eds. N. Yoffee and G.L. Cowgill, pp. 165–175. Tucson: University of Arizona Press.

Boyd, R., and P.J. Richerson (1985) *Culture and Evolutionary Process.* Chicago, IL: University of Chicago Press.

Boyd, R., and P.J. Richerson (1989) "The Evolution of Indirect Reciprocity." *Social Networks*, 11: 213–236.

Boyd, R., and P.J. Richerson (1992) "Punishment Allows the Evolution of Cooperation (or Anything Else) in Sizable Groups." *Ethology and Sociobiology*, 13: 171–195.

Boyd, R., and P. J. Richerson (2002) "Group Beneficial Norms Can Spread Rapidly in a Structured Population." *Journal of Theoretical Biology*, 210: 287–296.

Bradbury, J.W., and S.L. Vehrencamp (1998) *Principles of Animal Communication.* Sunderland, MA: Sinauer Associates.

Bradley, B.J. (1999) "Levels of Selection, Altruism, and Primate Behavior." *Quarterly Review of Biology*, 74(2): 171–194.

Brain, C.K. (1981) *The Hunters or the Hunted?* Chicago, IL: University of Chicago Press.

Brain, C.K. (1985) "Interpreting Early Hominid Death Assemblies: The Use of Taphonomy Since 1925." In *Hominid Evolution: Past, Present and Future,*

Proceedings of the Taung Diamond Jubilee International Symposium, ed. P.V. Tobias, pp. 41–46. New York: Alan R. Liss, Inc.

Bramson, L., and G. Goethals (eds.) (1964) *War: Studies from Psychology, Sociology and Anthropology*. New York: Basic Books.

Brandon, H. (1969) "A Talk with Walter Lippmann, at 80, About this 'Minor Dark Age'." *The New York Times Magazine*, September 14, 1969, pp. 25ff.

Brandon, R.N. (1996) *Concepts and Methods in Evolutionary Biology*. New York: Cambridge University Press.

Breed, M.D., *et al.* (1994) "Kin Discrimination within Honey Bee (*Apis mellifera*) Colonies: An Analysis of the Evidence." *Behavioral Process*, 33: 25–40.

Bridges, B.A. (1997) "Hypermutation under Stress." *Nature*, 387: 557–558.

Broad, W.J., *et al.* (1999) "Small Pox: The Once and Future Scourge?" *The New York Times*, June 15, 1999, p. D1ff.

Brodie, R. (1996) *Virus of the Mind: The New Science of the Meme*. Seattle, WA: Integral Press.

Broecker, W.S. (1994) "Massive Iceberg Discharges as Trigger for Global Climate Change." *Nature*, 372: 421–424.

Broecker, W.S. (1995) "Chaotic Climate." *Scientific American*, 273(5): 62–68.

Broecker, W.S. (1996) "The Once and Future Climate." *Natural History*, 105(9): 31–38.

Broecker, W.S. (2001) "Glaciers that Speak in Tongues and Other Tales of Global Warming." *Natural History*, 110(8): 60.

Bronstein, J.L. (2001) "The Costs of Mutualism." *American Zoologist*, 41: 825–839.

Brooks, A.S., *et al.* (1995) "Dating and Context of Three Middle Stone Age Sites with Bone Points in the Upper Semliki Valley, Zaire." *Science*, 268: 548.

Brosnan, S.F., and F.B.M. de Waal (2002) "A Proximate Perspective on Reciprocal Altruism." *Human Nature*, 13: 129–152.

Brotherton, P.N.M., *et al.* (2001) "Offspring Food Allocation by Parents and Helpers in a Cooperative Mammal." *Behavioral Ecology*, 12: 590–599.

Brown, D.E. (1991) *Human Universals*. Philadelphia, PA: Temple University Press.

Brown, J.L. (1987) *Helping and Communal Breeding in Birds*. Princeton, NJ: Princeton University Press.

Brown, S. (1994) *The Causes and Prevention of War*, 2nd edn. New York: St. Martin's Press.

Browne, M.W. (1997) "Signal Travels Farther and Faster than Light in Tracking the Paths of 'Twin Photons'." *The New York Times*, July 22, 1997, pp. C1, C2.

Brumfiel, E. (1976) "Regional Growth in the Eastern Valley of Mexico: A Test of the 'Population Pressure' Hypothesis." In *The Early Mesoamerican Village*, ed. K.V. Flannery, pp. 234–248. New York: Academic Press.

Brunet, M., *et al.* (2002) "A New Hominid from the Upper Miocene of Chad, Central Africa." *Nature*, 418: 145–151.

Buchmann, S., and G.P. Nabhan (1996) *Forgotten Pollinators*. Washington, D.C.: Island Press.

Buckley, N.J. (1997) "Experimental Tests of the Information-Center Hypothesis with Black Vultures (*Coragyps atratus*) and Turkey Vultures (*Cathartes aura*)." *Behavioral Ecology and Sociobiology*, 41: 267–279.

Burch, R., *et al.* (1989) "Synergy between Copper and Zinc Oxide during Methanol Synthesis." *Journal of the Chemical Society, Faraday Transactions*, 85(10): 3569–3578.

Burian, R.M. (1992) "Adaptation: Historical Perspectives." In *Keywords in Evolutionary Biology*, eds. E.F. Keller and E.A. Lloyd, Cambridge, MA: Harvard University Press.

Burke, A.C., and A. Feduccia (1997) "Developmental Patterns and the Identification of Homologies in the Avian Hand." *Science*, 278: 666–668.

Burkert, W. (1972) *Lore and Science in Ancient Pythagoreanism*, trans. L. Minar, Jr. Cambridge, MA: Harvard University Press.

Burkhardt, R.W. (1995) *The Spirit of System: Lamarck and Evolutionary Biology.* Cambridge, MA: Harvard University Press.

Buss, L.W. (1987) *The Evolution of Individuality.* Princeton, NJ: Princeton University Press.

Buss, L.W. (1999) "Slime Molds, Ascidians, and the Utility of Evolutionary Theory." *Proceedings of the National Academy of Sciences*, 96: 8801–8803.

Byrne, R.W. (1995) *The Thinking Ape: Evolutionary Origins of Intelligence.* Oxford, U.K.: Oxford University Press.

Byrne, R.W., and A. Whiten (eds.) (1988) *Machiavellian Intelligence: Social Expertise and the Evolution of Intellect in Monkeys, Apes and Humans.* Oxford, U.K.: Clarendon Press.

Cahan, S., and G.E. Julian (1999) "Fitness Consequences of Cooperative Colony Founding in the Desert Leaf-cutter Ant." *Behavioral Ecology*, 10(5): 585–591.

Cahan, S., *et al.* (1999) "Causes and Consequences of Sociality." *Ethology, Ecology and Evolution*, 11(1): 85–87.

Cairns, J., *et al.* (1988) "The Origin of Mutants." *Nature*, 335: 142–145.

Camazine, S., *et al.* (2001) *Self-Organization in Biological Systems.* Princeton, NJ: Princeton University Press.

Campagna, C., *et al.* (1992) "Group Breeding in Sea Lions: Pups Survive Better in Colonies." *Animal Behaviour*, 43: 541–548.

Campbell, B. (1985) *Human Evolution: An Introduction to Man's Adaptations*, 3rd edn. New York: Aldine Press.

Campbell, D.T. (1974) "Downward Causation in Hierarchically Organized Biological Systems." In *Studies in the Philosophy of Biology*, eds. T. Dobzhansky and F.J. Ayala, pp. 85–90. Berkeley: University of California Press.

Campbell, J.H. (1994) "Organisms Create Evolution." In *Creative Evolution?!* eds. J.H. Campbell and J.W. Schopf, pp. 85–102. Boston, MA: Jones & Bartlett.

Campbell, J.H., and J.W. Schopf (eds.) (1994) *Creative Evolution?!* Boston, MA: Jones & Bartlett.

Cann, R.L. (2002) "Tangled Genetic Roots." *Nature*, 416: 32–33.

Cann, R.L., *et al.* (1987) "Mitochondrial DNA and Human Evolution." *Nature*, 325: 31–36.

Cannon, H.G. (1955) "What Lamarck Really Said." *Proceedings of the Linnean Society of London*, 168: 70–87.

Cannon, H.G. (1959) *Lamarck and Modern Genetics.* Manchester, U.K.: University of Manchester Press.

Carnegie, A. (1889) "Wealth." *North American Review*, June, 653–664.

Carneiro, R.L. (1967) "On the Relationship between Size of Population and Complexity of Social Organization." *Southwestern Journal of Anthropology*, 23: 234–243.

Carneiro, R.L. (1970a) "A Theory of the Origin of the State." *Science*, 169: 733–738.

Carneiro, R.L. (1970b) "Scale Analysis, Evolutionary Sequences and the Rating of Cultures." In *A Handbook of Method in Cultural Anthropology*, eds. R. Naroll and R. Cohen, pp. 834–871. Garden City, NY: Natural History Press.

Carneiro, R.L. (1973) "The Four Faces of Evolution." In *Handbook of Social and Cultural Anthropology*, ed. J.H. Honigmann, pp. 89–110. Chicago, IL: Rand McNally.

Carneiro, R.L. (1978) "Political Expansion as an Expression of the Principle of Competitive Exclusion." In *Origins of the State: The Anthropology of Political Evolution*, eds. R.N. Cohen and E.R. Service, pp. 205–223. Philadelphia, PA: I.S.H.I.

Carneiro, R.L. (1981) "The Chiefdom: Precursor of the State." In *The Transition to Statehood in the New World*, eds. G.D. Jones and R.R. Kautz, pp. 37–79. Cambridge, U.K.: Cambridge University Press.

Carr, B.J., and M.J. Rees (1979) "The Anthropic Principle and the Structure of the Physical World." *Nature*, 278: 610.

Cartwright, N. (1993) *How the Laws of Physics Lie*. New York: Oxford University Press.

Cavalier-Smith, T. (1981) "The Origin and Early Evolution of the Eukaryote Cell." *Symposia of the Society of General Microbiology*, 36: 33–84.

Cavalier-Smith, T. (1991) "The Evolution of Cells." In *Evolution of Life: Fossils, Molecules and Culture,* eds. S. Osawa and T. Honjo, pp. 271–304. Tokyo: Springer-Verlag.

Cavalli-Sforza, L.L., *et al.* (1988) "Reconstruction of Human Evolution: Bringing Together Genetic, Archaeological, and Linguistic Data." Proceedings of the National Academy of Sciences, 85: 6002–6006.

Cavalli-Sforza, L.L., *et al.* (1994) *The History and Geography of Human Genes*. Princeton, NJ: Princeton University Press.

Cerullo, M.M. (1996) *Coral Reef*. New York: Cobblehill Books.

Changeux, J. (1985) *Neuronal Man: The Biology of Mind,* trans. L. Garey. New York: Pantheon Books.

Chapais, B. (2001) "Primate Nepotism: What Is the Explanatory Value of Kin Selection?" *International Journal of Primatology*, 22(2): 203–230.

Chapais, B., *et al.* (1991) "Non-kin Alliances, and the Stability of Matrilineal Dominance Relations in Japanese Macaques." *Animal Behaviour*, 41: 481–491.

Charnov, E.L. (1976) "Optimal Foraging, the Marginal Value Theorem." *Theoretical Population Biology*, 9: 129–136.

Chatterjee, S. (1992) "Sources of Value in Takeovers: Synergy or Restructuring – Implications for Target and Bidder Firms." *Strategic Management Journal*, 13: 267–286.

Chatterjee, S. (1998) "Counting the Fingers of Birds and Dinosaurs." *Science*, 280: 355a.

Cheney, D.L. and R.W. Wrangham (1987) "Predation." In *Primate Societies*, eds. B.B. Smuts *et al.*, pp. 227–239. Chicago, IL: University of Chicago Press.

Chevalier-Skolnikoff, S., and J. Liska (1993) "Tool Use by Wild and Captive Elephants." *Animal Behavior*, 46: 209–219.

Ciochon, R.L., and J.G. Fleagle (eds.) (1993) *The Human Evolution Source Book*. Englewood Cliffs, NJ: Prentice-Hall.

Claessen, H.J.M., and P. Skalník (eds.) (1978) *The Early State*. The Hague: Mouton.

Clark, J.D. (1992) "African and Asian Perspectives on the Origins of Modern Humans." *Philosophical Transactions of the Royal Society of London, Series B*, 337: 201–215.

Clark, K.B., *et al.* (1991) *Product Development Performance: Strategy, Organization, and Management in the World Auto Industry*. Cambridge, MA: Harvard University Press.

Clarke, B. (1975) "The Causes of Biological Diversity." *Scientific American*, 232(2): 50–60.

Clarke, M.F. (1989) "The Pattern of Helping in the Bell Miner." *Ethology*, 80: 292–306.

Clay, K. (2001) "Symbiosis and Regulation of Communities." *American Zoologist*, 41: 810–824.

Cleland, C.E. (2001) "Historical Science, Experimental Science, and the Scientific Methodology." *Geology*, 29: 987–990.

Clutton-Brock, T.H. (2002) "Breeding Together: Kin Selection and Mutualism in Cooperative Vertebrates." *Science*, 296: 69–72.

Clutton-Brock, T.H., and G.A. Parker (1995) "Punishment in Animal Societies." *Nature*, 373: 209–216.

Clutton-Brock, T.H., *et al.* (1999) "Selfish Sentinels in Cooperative Mammals." *Science*, 284: 1640–1644.

Clutton-Brock, T.H., *et al.* (2000) "Individual Contributions to Babysitting in a Cooperative Mongoose, *Suricata suricatta*." *Proceedings of the Royal Society of London B*, 267: 301–305.

Clutton-Brock, T.H., *et al.* (2001) "Contributions to Cooperative Rearing in Meerkats." *Animal Behaviour*, 61(4): 705–710.

Coe, M. (1987) "Unforeseen Effects of Control." *Nature*, 327: 367.

Coe, M.D., and K.V. Flannery (1967) *Early Cultures and Human Ecology in South Coastal Guatemala*. Washington, D.C.: Smithsonian Institution Press.

Coghlan, A. (1996) "Slime City." *New Scientist*, 151: 31–36.

Cohen, M.N. (1977) *The Food Crisis in Prehistory: Overpopulation and the Origins of Agriculture*. New Haven, CT: Yale University Press.

Cohen, R.N. (1978a) "Introduction." In *Origins of the State: The Anthropology of Political Evolution*, eds. R.N. Cohen and E.R. Service, pp. 1–20. Philadelphia, PA: I.S.H.I.

Cohen, R.N. (1978b). "State Foundations: A Controlled Comparison." In *Origins of the State: The Anthropology of Political Evolution*, eds. R.N. Cohen and E.R. Service, pp. 141–160. Philadelphia, PA: I.S.H.I.

Cohen, R. (1978c) "State Origins: A Reappraisal." In *The Early State*, eds. H.J.M. Claessen and P.S. Skalník, pp. 31–75. The Hague: Mouton.

Cohen, R., and E.R. Service (eds.) (1978) *Origins of the State: The Anthropology of Political Evolution*. Philadelphia, PA: I.S.H.I.

Cohn, J. (1997) "How Wild Wolves Became Domestic Dogs." *BioScience*, 47: 725–728.

Committee on Abrupt Climate Change, National Research Council (2002) *Abrupt Climate Changes: Inevitable Surprises*. Washington, D.C.: National Academy Press.

Committee on the Science of Climate Change, National Research Council (2001) *Climate Change Science: An Analysis of Key Questions*. Washington, D.C.: National Academy Press.

Condorcet, J.-A.-N. (1795) *Esquisse d'un Tableau Historique des Progrès de l'Esprit Humain*, 1970 edn: ed. O.H. Prior. Paris: Librairie Philosophique.

Conlan, R. (ed.) (1999) *States of Mind: New Discoveries about How our Brains Make Us Who We Are*. New York: John Wiley & Sons.

Connelly, R., and A. Back (1998) "Mathematics and Tensegrity." *American Scientist*, March–April: 142–151.

Connor, M., and M. Ferguson-Smith (1997) *Essential Medical Genetics*, 5th edn. Cambridge, MA: Blackwell Science.

Conrad, L., and T.J. Roper (2003) "Group Decision-Making in Animals." *Nature*, 421: 155–158.

Conroy, G.C., *et al.* (1998) "Endocranial Capacity in an Early Hominid Cranium from Sterkfontein, South Africa." *Science*, 280: 1730–1731.

Conway Morris, S. (1998a) *The Crucible of Creation: The Burgess Shale and the Rise of Animals*. Oxford, U.K.: Oxford University Press.

Conway Morris, S. (1998b) "Early Metazoan Evolution: Reconciling Paleontology and Molecular Biology." *American Zoologist*, 38: 867–877.

Conway Morris, S., and S.J. Gould (1998) "Showdown on the Burgess Shale." *Natural History*, 107(10): 48–55.

Cook, E. (1971) "The Flow of Energy in a Hunting Society." *Scientific American*, 224(3): 134–147.

Coppens, Y., and B. Senut (1991) *Origine(s) de la Bipédie les Hominidés*. Paris: Cahiers de Paléanthropologie, Centre National de la Recherche Scientifique.

Corballis, M.C. (1991) *The Lopsided Ape: Evolution of the Generative Mind*. New York: Oxford University Press.

Corballis, M.C. (2002) *From Hand to Mouth: The Origins of Language*. Princeton, NJ: Princeton University Press.

Corballis, M.C., and S.E.G. Lea (eds.) (1999) *The Descent of Mind: Psychological Perspectives on Hominid Evolution*. New York: Oxford University Press.

Corning, P.A. (1982) "Durkheim and Spencer." *British Journal of Sociology*, 33(3): 359–382.

Corning, P.A. (1983) *The Synergism Hypothesis: A Theory of Progressive Evolution*. New York: McGraw-Hill.

Corning, P.A. (1996a) "Evolutionary Economics: Metaphor or Unifying Paradigm." *Journal of Social and Evolutionary Systems*, 18: 421–435.

Corning, P.A. (1996b) "Evolution and Ethics . . . An Idea Whose Time Has Come? (Part 1)." *Journal of Social and Evolutionary Systems*, 19: 277–285.

Corning, P.A. (1997a) "Holistic Darwinism: 'Synergistic Selection' and the Evolutionary Process." *Journal of Social and Evolutionary Systems*, 20(4): 363–400.

Corning, P.A. (1997b) "Evolution and Ethics ... An Idea Whose Time Has Come? (Part 2)." *Journal of Social and Evolutionary Systems*, 20(3): 323–333.

Corning, P.A. (1998) "Complexity is Just a Word!" *Technological Forecasting and Social Change*, 58: 1–4.

Corning, P.A. (2000) "Biological Adaptation in Human Societies: A 'Basic Needs' Approach." *Journal of Bioeconomics*, 2: 41–86.

Corning, P.A. (2001a) "Synergy Goes to War: An Evolutionary Theory of Collective Violence." Prepared for the annual meeting, *Association for Politics and the Life Sciences*, Charleston, SC, October 18–21, 2001.

Corning, P.A. (2001b) "Control Information." *Kybernetes*, 30 (9/10): 1272–1288.

Corning, P.A. (2002a) " 'Devolution' as an Opportunity to Test the 'Synergism Hypothesis' and a Cybernetic Theory of Political Systems." *Systems Research and Behavioral Science*, 19: 3–26.

Corning, P.A. (2002b) "Synergy and the Evolution of 'Superorganisms', Past, Present, and Future." Presented at the annual meeting, Association for Politics and Life Sciences, Montreal, Canada, August 11–14, 2002.

Corning, P.A. (2002c) "'Fair Shares': Beyond Capitalism and Socialism; The Biological Basis of Social Justice." Presented at the annual meeting, Association for Politics and the Life Sciences, Montreal, Canada, August 11–14, 2002.

Corning, P.A. (2002d) "Thermoeconomics: Beyond the Second Law." *Journal of Bioeconomics*, 4: 57–88.

Corning, P.A. (2002e) "The Re-Emergence of 'Emergence': A Venerable Concept in Search of a Theory." *Complexity*, 7(6): 18–30.

Corning, P.A., and S.P. Corning (1986) *Winning with Synergy: How America Can Regain the Competitive Edge.* San Francisco, CA: Harper & Row.

Corning, P.A., and S.J. Kline (1998a) "Thermodynamics, Information and Life Revisited, Part I: To Be or Entropy." *Systems Research and Behavioral Science*, 15: 273–295.

Corning, P.A., and S.J. Kline (1998b) "Thermodynamics, Information and Life Revisited, Part II: Thermoeconomics and Control Information." *Systems Research and Behavioral Science*, 15: 453–482.

Corning, P.A., et al. (1977) "Three Kinds of Aggressive Behavior in Laboratory Mice." *Behavior Genetics*, 7(1): 51–52.

Corruccini, R.S., and H.M. McHenry (2001) "Knuckle Walking Hominid Ancestors." *Journal of Human Evolution*, 40: 507–511.

Costanza, R. (1989) "What Is Ecological Economics?" *Ecological Economics*, 1: 1–7.

Costanza, R. (1991) *Ecological Economics: The Science and Management of Sustainability.* New York: Columbia University Press.

Costanza, R., et al. (2000) "Managing Our Environmental Portfolio." *BioScience*, 50: 149–155.

Cotterell, A. (ed.) (1993) *The Penguin Encyclopedia of Classical Civilizations.* New York: Viking.

Couper, H., and N. Henbest (1997) *Big Bang.* New York: Dorling Kindersley.

Cowlishaw, G. (1994) "Vulnerability to Predation in Baboon Populations." *Behaviour,* 131(3–4): 293–304.

Crick, F. (1994) *The Astonishing Hypothesis: The Scientific Search for the Soul.* New York: Charles Scribner's Sons.

Cronk, L. (1999) *That Complex Whole: Culture and the Evolution of Human Behavior.* Boulder, CO: Westview Press.

Crook, P. (1994) *Darwinism, War and History: The Debate over the Biology of War from the "Origin of Species" to the First World War.* New York: Cambridge University Press.

Curio, E., *et al.* (1978) "Cultural Transmission of Enemy Recognition: One Function of Mobbing." *Science,* 202: 899–901.

Currie, C.R. (2001) "A Community of Ants, Fungi, and Bacteria: A Multilateral Approach to Studying Symbiosis." *Annual Reviews of Microbiology,* 55: 357–380.

Currie, C.R., *et al.* (1999a) "Fungus-Growing Ants Use Antibiotic-Producing Bacteria to Control Garden Parasites." *Nature,* 398: 701–703.

Currie, C.R. *et al.* (1999b) "The Agricultural Pathology of Ant Fungus Gardens." *Proceedings of the National Academy of Sciences,* 96: 7998–8002.

Curtis, H., and N.S. Barnes (1989) *Biology,* 5th edn. New York: Worth.

Cziko, G. (1995) *Without Miracles: Universal Selection Theory and the Second Darwinian Revolution.* Cambridge, MA: MIT Press.

Cziko, G. (2000) *The Things We Do: Using the Lessons of Bernard and Darwin to Understand the What, How, and Why of Our Behavior.* Cambridge, MA: MIT Press.

Daft, M.J., and A.A. El-Giahmi (1978) "Effect of Arbuscular Mycorrhiza on Plant Growth, VIII. Effects of Defoliation and Light on Selected Hosts." *New Phytology,* 80: 365–372.

Dagg, J.A. (2002) "Unconventional Bed Mates: Gaia and the Selfish Gene." *Oikos,* 96: 182–186.

Dahlberg, F. (ed.) (1981) *Woman the Gatherer.* New Haven, CT: Yale University Press.

Daily, G.C., and P.R. Ehrlich (1992) "Population, Sustainability, and Earth's Carrying Capacity." *BioScience,* 42: 761–771.

Daily, G.C., *et al.* (1998) "Food Production, Population Growth, and the Environment." *Science,* 281: 1291–1292.

Dalfes, H.N., *et al.* (eds.) (1996) *Third Millennium BC Climate Change and Old World Collapse.* Berlin: Springer-Verlag.

Dall, S.R.X. (2002) "Can Information Sharing Explain Recruitment to Food from Communal Roosts?" *Behavioral Ecology,* 13(1): 42–51.

Darnell, J., *et al.* (1990) *Molecular Cell Biology,* 2nd edn. New York: Scientific American Books.

Dart, R. (1953) "The Predatory Transition from Ape to Man." *International Anthropological and Linguistic Review,* 1: 201–219.

Dart, R. (1959) *Adventures with the Missing Link.* New York: Harper & Brothers.

Darwin, C.R. (1859) *On the Origin of Species by Means of Natural Selection, or the Preservation of Favoured Races in the Struggle for Life.* London: John Murray.

Darwin, C.R. (1868) *The Variations of Animals and Plants under Domestication.* London: John Murray.

Darwin, C.R. (1871) *The Descent of Man, and Selection in Relation to Sex.* London: John Murray.

Dauber, P.M., and R.A. Muller (1996) *The Three Big Bangs.* Reading, MA: Addison-Wesley.

Davey, R.J., *et al.* (1993) "Stabilization of a Metastable Crystalline Phase by Twinning." *Nature,* 366: 248–250.

Davies, D.G., *et al.* (1998) "The Involvement of Cell-to-cell Signals in the Development of a Bacterial Biofilm." *Science,* 280: 295–298.

Davies, P. (1988) *The Cosmic Blueprint: New Discoveries in Nature's Creative Ability to Order the Universe.* New York: Simon & Schuster.

Davies, P. (1992) *The Mind of God: The Scientific Basis for a Rational World.* New York: Simon & Schuster.

Davis, R., and L.G. Thomas (1993) "Direct Estimation of Synergy: A New Approach to the Diversity-Performance Debate." *Management Science,* 39: 1334–1346.

Daviss, B. (1998) "Cast Out of Eden." *New Scientist,* 158: 26–30.

Dawkins, M.S. (1983) "Battery Hens Name their Price: Consumer Demand Theory and the Measurement of Ethological Needs." *Animal Behaviour,* 31: 1195–1205.

Dawkins, M.S. (1988) *Through Our Eyes Only? The Search for Animal Consciousness.* Oxford, U.K.: W.H. Freeman/Spektrum.

Dawkins, M.S. (2000) "Animal Minds and Animal Emotions." *American Zoologist,* 40: 883–888.

Dawkins, R. (1976) *The Selfish Gene.* Oxford, U.K.: Oxford University Press.

Dawkins, R. (1992) "Progress." In *Keywords in Evolutionary Biology,* eds. E.F. Keller and E.A. Lloyd. Cambridge, MA: Harvard University Press.

Dawkins, R. (1986) *The Blind Watchmaker.* Oxford, U.K.: Oxford University Press.

Dawkins, R. (1989) *The Selfish Gene,* 2nd edn. Oxford, U.K.: Oxford University Press.

Dawson, D. (1996) "The Origins of War: Biological and Anthropological Theories." *History and Theory,* 35(1): 1–28.

Dayton, P.K. (1998) "Reversal of the Burden of Proof in Fisheries Management." *Science,* 279: 821–822.

de Duve, C. (1996) "The Birth of Complex Cells." *Scientific American,* 274(4): 56–63.

de Heinzelin, J., *et al.* (1999) "Environment and Behavior of 2.5-Million Year-Old Bouri Hominids." *Science,* 284: 625–629.

De Heer, C.J., and K.G. Ross (1997) "Lack of Detectable Nepotism in Multiple Queen Colonies of the Fire Ant *Solenopsis invicta* (Hymenoptera: Formicidae)." *Behavioral Ecology and Sociobiology,* 40: 27–33.

de Waal, F.B.M. (1996) *Good Natured: The Origins of Right and Wrong in Humans and Other Animals.* Cambridge, MA: Harvard University Press.

de Waal, F.B.M. (1997) *Bonobo: The Forgotten Ape.* Berkeley: University of California Press.

de Waal, F.B.M. (1999) "Cultural Primatology Comes of Age." *Nature*, 399: 635–636.

de Waal, F.B.M. (2000) "Reading Nature's Tea Leaves." *Natural History*, 109(10): 66–71.

de Waal, F.B.M. (ed.) (2001a) *Tree of Origin: What Primate Behavior Can Tell Us about Human Social Evolution.* Cambridge, MA: Harvard University Press.

de Waal, F.B.M. (2001b) *The Ape and the Sushi Master: Cultural Reflections by a Primatologist.* New York: Basic Books.

Deacon, T.W. (1997) *The Symbolic Species: The Co-Evolution of Language and the Brain.* New York: W.W. Norton.

Deacon, T.W. (2003) *Homunculus: Evolution, Self-Organization, and Consciousness.* New York: W.W. Norton.

Deamer, D.W. (ed.) (1978) *Light Transcending Membranes: Structure, Function and Evolution.* New York: Academic Press.

Deamer, D.W., and J. Oro (1980) "Role of Lipids in Prebiotic Structures." *BioSystems*, 12: 167–175.

Deamer, D.W., and R.M. Pashley (1989) "Amphiphilic Components of the Murchison Carbonaceous Chondrite: Surface Properties and Membrane Formation." *Origins of Life*, 74: 319–327.

Delsemme, A. (1998) *Our Cosmic Origins: From the Big Bang to the Emergence of Life and Intelligence.* New York: Cambridge University Press.

Dennett, D.C. (1983) "Intentional Systems in Cognitive Ethology." *Behavioral and Brain Sciences*, 6: 343–390.

Dennett, D.C. (1991) *Consciousness Explained.* Boston, MA: Little, Brown.

Dennett, D.C. (1995) *Darwin's Dangerous Idea: Evolution and the Meaning of Life.* New York: Simon & Schuster.

Deutsch, K.W. (1979) *Tides among Nations.* New York: Free Press.

Diamond, J. (2000) "Threescore and Ten." *Natural History*, 109(10): 24–35.

Diamond, J.M. (1986) "Why Do Disused Proteins Become Genetically Lost or Repressed?" *Nature*, 321: 565–566.

Diamond, J.M. (1992) *The Third Chimpanzee: The Evolution and Future of the Human Animal.* New York: HarperCollins.

Diamond, J.M. (1995) "The Evolution of Human Inventiveness." In *What Is Life? The Next Fifty Years*, eds. M.P. Murphy and L.A.J. O'Neill, pp. 41–45. Cambridge, U.K.: Cambridge University Press.

Diamond, J.M. (1997) *Guns, Germs and Steel: The Fates of Human Societies.* New York: W.W. Norton.

Dobzhansky, T. (1970) *Genetics of the Evolutionary Process.* New York: Columbia University Press.

Dobzhansky, T., *et al.* (1977) *Evolution.* San Francisco, CA: W.H. Freeman.

Doebler, S.A. (2000) "The Rise and Fall of the Honeybee." *BioScience*, 50: 738.

Donald, M. (1993) "Précis of Origins of the Modern Mind: Three Stages in the Evolution of Culture and Cognition." *Behavioral and Human Sciences*, 16: 737–791.

Donelson, J.E., and M.J. Turner (1985) "How the Trypanosome Changes its Coat." *Scientific American*, 252(2): 44–51.

Doolan, S.P., and D.W. Macdonald (1996a) "Diet and Foraging Behaviour of Group-Living Meerkats, *Suricata suricatta*, in the Southern Kalahari." *Journal of the Zoological Society of London*, 239: 697–716.

Doolan, S.P., and D.W. Macdonald (1996b) "Dispersal and Extra-territorial Prospecting by Slender-tailed Meerkats (*Suricata suricatta*) in the South-western Kalahari." *Journal of the Zoological Society of London*, 240: 59–73.

Doolan, S.P., and D.W. Macdonald (1997) "Breeding and Juvenile Survival among Slender-tailed Meerkats (*Suricata suricatta*) in the South-western Kalahari: Ecological and Social Influences." *Journal of the Zoological Society of London*, 242: 309–327.

Dopfer, K. (ed.) (2001) *Evolutionary Economics: Program and Scope*. Boston, MA: Kluwer.

Dörner, D. (1996) *The Logic of Failure*. Reading, MA: Addison-Wesley.

Dowling, J.E. (1992) *Neurons and Networks: An Introduction to Neuroscience*. Cambridge, MA: Belknap Press of Harvard University Press.

Dreisch, H.A.E. (1909) *Philosophie des Organischen*. Leipzig: Engelmann.

Drucker, C.B. (1978) "The Price of Progress in the Philippines." *Sierra*, 63: 22–26.

Dugatkin, L.A. (1997) *Cooperation among Animals: An Evolutionary Perspective*. New York: Oxford University Press.

Dugatkin, L.A. (1999) *Cheating Monkeys and Citizen Bees*. New York: Free Press.

Dugatkin, L.A. (2000) *The Imitation Factor: Evolution beyond the Gene*. New York: Free Press.

Dukas, R. (1998a) "Ecological Relevance of Associative Learning in Fruit Fly Larvae." *Behavioral Ecology and Sociobiology*, 19: 195–200.

Dukas, R. (ed.) (1998b) *Cognitive Ecology: The Evolutionary Ecology of Information Processing and Decision Making*. Chicago, IL: University of Chicago Press.

Dumond, D.E. (1965) "Population Growth and Cultural Change." *Southwestern Journal of Anthropology*, 21: 302–324.

Dunbar, R.I.M. (1988) *Primate Social Systems*. London: Croom Helm.

Dunbar, R.I.M. (1992) "Time: A Hidden Constraint on the Behavioural Ecology of Baboons." *Behavioral Ecology and Sociobiology*, 31: 35–49.

Dunbar, R.I.M. (1996) "Determinants of Group Size in Primates: A General Model." In *Evolution of Social Behaviour Patterns in Primates and Man*, eds. W.G. Runciman *et al.*, pp. 33–58. Oxford, U.K.: Oxford University Press.

Dunbar, R.I.M. (1998) "The Social Brain Hypothesis." *Evolutionary Anthropology*, 6: 178–190.

Dunbar, R.I.M. (2001) "Brain on Two Legs: Group Size and the Evolution of Intelligence." In *Tree of Origin: What Primate Behaviour Can Tell Us about Human Social Evolution*, ed. F.B.M. de Waal, pp. 173–191. Cambridge, MA: Harvard University Press.

Duncan, R.P., *et al.* (2002) "Prehistoric Bird Extinctions and Human Hunting." *Proceedings of the Royal Society of London B*, 269: 517–521.

Durham, W.H. (1991) *Coevolution: Genes, Culture, and Human Diversity*. Stanford, CA: Stanford University Press.

Dyson, F.J. (1971) "Energy in the Universe." In *Energy and Power* (A *Scientific American* Book), pp. 19–27. San Francisco, CA: W.H. Freeman.

Earle, T.K. (1987) "Chiefdoms in Archaeological and Ethnohistorical Perspective." *Annual Review of Anthropology*, 16: 279–308.

Earle, T.K. (1989) "The Evolution of Chiefdoms." *Current Anthropology*, 30(1): 84–88.

Easterbrook, G. (1998a) "Science Sees the Light." *The New Republic*, 219(15): 24–29.

Easterbrook, G. (1998b) *Beside Still Waters: Searching for Meaning in an Age of Doubt*. New York: William Morrow.

Edelson, E. (1990) *The Immune System (Encyclopedia of Health/The Healthy Body)*. Philadelphia, PA: Chelsea House Publishing.

Editorial (2000) "Losing Ground against Microbes." *The New York Times*, June 18, 2000, p. 14.

Editorial (2001a) "Pursuing Arrogant Simplicities." *Nature*, 416: 247.

Editorial (2001b) "Economics Focus: A Global Euro." *The Economist*, July 28, 2001, p. 70.

Ehrlich, P.R. (1968) *The Population Bomb*. New York: Ballantine.

Ehrlich, P.R. (2000) *Human Natures: Genes, Cultures, and the Human Prospect*. Washington, D.C.: Island Press.

Ehrlich, P.R., and P.H. Raven (1964) "Butterflies and Plants: A Study in Coevolution." *Evolution*, 18: 586–608.

Eibl-Eibesfeldt, I., and F.K. Salter (eds.) (1998) *Indoctrinability, Ideology, and Warfare: Evolutionary Perspectives*. New York: Berghahn Books.

Eldredge, N. (1995) *Reinventing Darwin*. New York: John Wiley & Sons.

Eldredge, N., and M. Grene (1992) *Interactions: The Biological Context of Social Systems*. New York: Columbia University Press.

Elgert, K.D. (1996) *Immunology: Understanding the Immune System*. New York: Wiley–Liss.

Emanuel, E. (1971) *Baby Baboon*. San Francisco, CA: Golden Gate Junior Books.

Ember, C.R. (1978) "Myths about Hunter–Gatherers." *Ethnology*, 27: 239–448.

Emlen, J.M. (1966) "The Role of Time and Energy in Food Preference." *American Naturalist*, 100: 611–617.

Enard, W, *et al.* (2002) "Molecular Evolution of *FOXP2*, a Gene Involved in Speech and Language." *Nature*, 418: 869–872.

Erickson, H.P. (2001) "Evolution in Bacteria." *Nature*, 413: 30.

d'Errico, F., *et al.* (1998) "Neanderthal Acculturation in Western Europe?" *Current Anthropology*, 39: S1-42.

Evans, G. (1975) *The Life of Beetles*. New York: Hafner Press.

Ezzell, C. (2002) "Proteins Rule." *Scientific American*, 286(4): 41–49.

Fagan, B.M. (1998) *People of the Earth: An Introduction to World Prehistory*, 9th edn. New York: Addison Wesley Longman, Inc.

Falk, D. (1998) "Hominid Brain Evolution: Looks Can be Deceiving." *Science*, 280: 1714.

Falk, D., *et al.* (2000) "Early Hominid Brain Evolution: A New Look at Old Endocasts." *Journal of Human Evolution*, 38: 695–717.

Farmer, M. (1994) "The Origins of Weapon Systems." *Current Anthropology*, 35(5): 679–681.

Farmer, P. (1999) "TB Superbugs: The Coming Plague on All Our Houses." *Natural History*, 108(3): 46–53.

Fedigan, L.M. (1986) "The Changing Role of Women in Models of Human Evolution." *Annual Review of Anthropology*, 15: 25–66.

Fehr, E., and S. Gächter (2000a) "Cooperation and Punishment in Public Goods Experiments." *American Economic Review*, 90: 980–994.

Fehr, E., and S. Gächter (2000b) "Fairness and Retaliation: The Economics of Reciprocity." *Journal of Economic Perspectives*, 14: 159–181.

Fehr, E., and S. Gächter (2002) "Altruistic Punishment in Humans." *Nature*, 415: 137–140.

Fehr, E., *et al.* (2002) "Strong Reciprocity, Human Cooperation, and the Enforcement of Social Norms." *Human Evolution*, 13: 1–25.

Fehr. E., and K.M. Schmidt (1999) "A Theory of Fairness, Competition and Cooperation." *Quarterly Journal of Economics*, 114: 817–868.

Ferris, T. (1997) *The Whole Shebang: A State-of-the-Universe(s) Report*. New York: Simon & Schuster.

Fey, W., and A.C.W. Lam. (1998) "Pie in the Sky: A System Dynamics Perspective of Sustainability." In *Proceedings of the 42nd Annual Conference of the International Society for the Systems Sciences (Atlanta, GA)*, eds. J.K. Allen and J. Wilby. (CD-ROM)

Fey, W., and A.C.W. Lam (1999) "The Ecocosm Paradox: Humanity's Tragic Dilemmas Arising from Earth's Environmental Crisis." http://www.ecocosmdynamics. org/ED/catalog.asp

Fey, W., and A.C.W. Lam (2000) "A Bridge to Humanity's Future." http://www. ecocosmdynamics.org/ED/catalog.asp

Fisher, H.E. (1982) *The Sex Contract*. New York: William Morrow.

Flannery, K.V. (ed.) (1976) *The Early Mesoamerican Village*. New York: Academic Press.

Flannery, K.V. (ed.) (1982) *Maya Subsistence: Studies in Memory of Dennis E. Puleston*. New York: Academic Press.

Flink, J.J. (1970) *America Adopts the Automobile, 1895–1910*. Cambridge, MA: MIT Press.

Foley, R. (1994) "Speciation, Extinction and Climatic Change in Hominid Evolution." *Journal of Human Evolution*, 26(4): 275–289.

Foley, R. (1995). *Humans before Humanity: An Evolutionary Perspective*. Oxford, U.K.: Blackwell Scientific Publications.

Foster, S.A. (1999) "The Geography of Behaviour: An Evolutionary Perspective." *Trends in Ecology and Evolution*, 14(5): 190–195.

Frank, S.A. (1995a) "The Origin of Synergistic Symbiosis." *Journal of Theoretical Biology*, 176: 403–410.

Frank, S.A. (1995b) "Mutual Policing and Repression of Competition in the Evolution of Cooperative Groups." *Nature*, 377: 520–522.

Frank, S.A. (1996) "Policing and Group Cohesion when Resources Vary." *Animal Behaviour*, 52: 1163–1169.

Frank, S.A. (1998) *Foundations of Social Evolution*. Princeton, NJ: Princeton University Press.

Franks, N.R. (1989) "Army Ants: A Collective Intelligence." *American Scientist,* 77(2): 139–145.

Franks, N.R., *et al.* (2001) "Division of Labour within Teams of New World and Old World Army Ants." *Animal Behaviour,* 62: 635–642.

Freeman, W.J. (1991) "The Physiology of Perception." *Scientific American,* 264(2): 34–41.

French, M.J. (1994) *Invention and Evolution: Design in Nature and Engineering,* 2nd edn. New York: Cambridge University Press.

Fried, M.H. (1967) *The Evolution of Political Society.* New York: Random House.

Fulton, A.B. (1993) "Spatial Organization of the Synthesis of Cytoskeletal Proteins." *Journal of Cellular Biochemistry,* 52(2): 148–152.

Gale, G. (1981) "The Anthropic Principle." *Scientific American,* 245(6): 154–171.

Galef, B.G., Jr. (1992) "The Question of Animal Culture." *Human Nature,* 3(2): 157–178.

Gallagher, K.T. (1994) "Remarks on the Argument from Design." *Review of Metaphysics,* 48(1): 30.

Gannon, P.J., *et al.* (1998) "Asymmetry of Chimpanzee Planum Temporale: Humanlike Pattern of Wernicke's Brain Language Area Homolog." *Science,* 279: 220–222.

Gavin, R.H. (1997) "Microtubule–Microfilament Synergy in the Cytoskeleton." *International Review of Cytology,* 173: 207–242.

Geneen, H., with B. Bowers (1997) *The Synergy Myth and Other Ailments of Business Today.* New York: St. Martin's Press.

George, J.D., and J.J. George (1979) *Marine Life.* New York: John Wiley & Sons.

Ghiselin, M.T. (1974) *The Economy of Nature and the Evolution of Sex.* Berkeley: University of California Press.

Ghiselin, M.T. (1978) "The Economy of the Body." *American Economic Review,* 68: 233–237.

Ghiselin, M.T. (1986) "Principles and Prospects for General Economy." In *Economic Imperialism: The Economic Approach Applied Outside the Field of Economics,* eds. G. Radnitzky and P. Bernholz, pp. 21–31. New York: Paragon.

Ghiselin, M.T. (1992) "Biology, Economics and Bioeconomics." In *Universal Economics: Assessing the Achievements of the Economic Approach,* ed. G. Radnitzky, pp. 71–118. New York: Paragon.

Ghiselin, M.T. (1997) *Metaphysics and the Origin of Species.* Albany: State University of New York Press.

Georgescu-Roegen, N. (1971) *The Entropy Law and Economic Process.* Cambridge, MA: Harvard University Press.

Georgescu-Roegen, N. (1976a) "Bioeconomics: A New Look at the Nature of Economic Activity." In *The Political Economy of Food and Energy,* ed. L. Junker, pp. 105–134. Ann Arbor: University of Michigan Press.

Georgescu-Roegen, N. (1976b) *Energy and Economic Myths.* New York: Pergamon Press.

Gibson, R., and T. Ingold (eds.) (1993) *Tools, Language, and Cognition in Human Evolution.* Cambridge, U.K.: Cambridge University Press.

Gilroy, S., and A. Trewavas (2001) "Signal Processing and Transduction in Plant Cells: The End of the Beginning." *Nature Reviews*, 2: 307–314.

Gintis, H. (2000a) *Game Theory Evoloving: A Problem-Centered Introduction to Modeling Strategic Behavior*. Princeton, NJ: Princeton University Press.

Gintis, H. (2000b) "Strong Reciprocity and Human Sociality." *Journal of Theoretical Biology*, 206: 169–179.

Giurfa, M., *et al.* (1999) "Pattern Learning by Honeybees: Conditioning Procedure and Recognition Strategy." *Animal Behaviour*, 57: 315–324.

Gleick, P.H. (2000) *The World's Water 2000–2001: The Biennial Report of Freshwater Resources*. Washington, D.C.: Island Press.

Gleick, P.H. (2001) "Safeguarding Our Water." *Scientific American*, 284(2): 41–45.

Gluckman, M. (1940) "The Kingdom of the Zulu of South Africa." In *African Political Systems*, eds. M. Fortes and E.E. Evans-Pritchard, pp. 25–55. London: Oxford University Press.

Gluckman, M. (1969) "The Rise of a Zulu Empire." *Scientific American*, 202(4): 157–168.

Godwin, J. (1987) *Harmonies of Heaven and Earth*. London: Thames & Hudson.

Goldberg, T.L., and R.W. Wrangham (1997) "Genetic Correlates of Social Behaviour in Wild Chimpanzees: Evidence from Mitrochondrial DNA." *Animal Behaviour*, 54(3): 559–570.

Goldenfeld, N., and L.P. Kadanoff (1999) "Simple Lessons from Complexity." *Science*, 284: 87–89.

Goldmann, D.R. (ed.) (1999) *American College of Physicians Complete Home Medical Guide*. New York: Dorling Kindersley.

Goodall, J. (1986) *The Chimpanzees of Gombe: Patterns of Behavior*. Cambridge, MA: Harvard University Press.

Goodman, M.F. (1998) "Purposeful Mutations." *Nature*, 395: 221–222.

Goodwin, B.C. (1994) *How the Leopard Changed its Spots: The Evolution of Complexity*. London: Wiedenfeld & Nicolson.

Gould, J.L., and C.G. Gould (1994) *The Animal Mind*. New York: Scientific American Library.

Gould, J.L., and C.G. Gould (1995) *The Honey Bee*. New York: Scientific American Library.

Gould, S.J. (1989) *Wonderful Life: The Burgess Shale and the Nature of History*. New York: W.W. Norton.

Gould, S.J. (1994) "Tempo and Mode in the Macroevolutionary Reconstruction of Darwinism." *Proceedings of the National Academy of Sciences*, 91: 6764–6771.

Gould, S.J. (1996) *Full House: The Spread of Excellence from Plato to Darwin*. New York: Harmony Books.

Gould, S.J. (1997) "A Tale of Two Worksites." *Natural History*, 106(9): 18–28.

Gould, S.J. (1999a) "A Division of Worms." *Natural History*, 108(2): 18.

Gould, S.J. (1999b) "Branching through a Wormhole: Lamarck's Ladder Collapses." *Natural History*, 108(2): 24–27.

Gould, S.J. (1999c) *Rock of Ages: Science and Religion in the Fullness of Life*. New York: Ballantine.

Gould, S.J., and E.S. Vrba (1982) "Exaptation – A Missing Term in the Science of Form." *Paleobiology*, 8: 4–15.

Gowdy, J. (1994) *Coevolutionary Economics: The Economy, Society and the Environment*. Boston, MA: Kluwer.

Gowdy, J. (1997) "Introduction: Biology and Economics." *Structural Change and Economic Dynamics*, 8: 377–383.

Graedon, J., and T. Graedon (1999) *Dangerous Drug Interactions*. New York: St. Martin's Press.

Grafen, A. (1991) "Modelling in Behavioral Ecology." In *Behavioural Ecology: An Evolutionary Approach*, 3rd edn.: eds. J.R. Krebs and N.B. Davies, pp. 5–31. Oxford, U.K.: Blackwell Scientific Publications.

Grajal, A. (1995) "Digestive Efficiency of the Hoatzin, *Opisthocomus hoazin*: A Folivorous Bird with Foregut Fermentation." *IBIS*, 137: 383–388.

Grant, B.R., and P.R. Grant (1979) "Darwin's Finches: Population Variation and Sympatric Speciation." *Proceedings of the National Academy of Sciences*, 76: 2359–2363.

Grant, B.R., and P.R. Grant (1989) "Natural Selection in a Population of Darwin's Finches." *American Naturalist*, 133(3): 377–393.

Grant, B.R., and P.R. Grant (1993) "Evolution of Darwin's Finches Caused by a Rare Climatic Event." *Proceedings of the Royal Society of London B*, 251: 111–117.

Grant, P.R. (1986) *Ecology and Evolution of Darwin's Finches*. Princeton, NJ: Princeton University Press.

Grant, P.R. (1991) "Natural Selection and Darwin's Finches." *Scientific American*, 265(4): 82–87.

Grant, P.R., and B.R. Grant (2002) "Adaptive Radiation of Darwin's Finches." *American Scientist*, 90(2): 130–139.

Grassé, P.P. (1973) *Evolution of Living Organisms: Evidence for a New Theory of Transformation*, 1977 edn. New York: Academic Press.

Gray, W. (1993) *Coral Reefs and Islands*. Newton Abbot, U.K.: David & Charles.

Greenstein, G. (1988) *The Symbiotic Universe*. New York: William Morrow.

Gribbin, J., and M.J. Rees (1989) *Cosmic Coincidences*. New York: Bantam Books.

Griffin, A.S., and S.A. West (2002) "Kin Selection: Fact or Fiction." *Trends in Ecology and Evolution*, 17(1): 15–21.

Griffin, D.R. (1992) *Animal Minds*. Chicago, IL: University of Chicago Press.

Griffin, D.R. (2001) "Animals Know More than We Used to Think." *Proceedings of the National Academy of Sciences*, 98(9): 4833–4834.

Grindal, B.J. (1976) "The Idea of Synergy and its Bearing on Anthropological Humanism." *Anthropology and Humanism Quarterly*, 1: 4–6.

Grinnell, J., *et al.* (1995) "Cooperation in Male Lions: Kinship, Reciprocity or Mutualism?" *Animal Behaviour*, 49: 95–105.

Grotzinger, J.P., and D.H. Rothman (1996) "An Abiotic Model for Stromatolite Morphogenesis." *Nature*, 383: 423–425.

Grutter, A.S. (1999) "Cleaner Fish Really Do Clean." *Nature*, 398: 672–673.

Guinet, C. (1991) "Intentional Stranding Apprenticeship and Social Play in Killer Whales (*Orcinus orca*)." *Canadian Journal of Zoology*, 69: 2712–2716.

Guinet, C., and J. Bouvier (1995) "Development of Intentional Stranding Hunting Techniques in Killer Whale (*Orcinus orca*) Calves at Crozet Archipelago." *Canadian Journal of Zoology*, 73:27–33.

Gurven, M. *et al.* (2000) "Food Transfers Among Hiwi Foragers of Venezuela." *Journal of Human Ecology,* 28: 171–218.

Guthrie, K.S., *et al.* (1987) *The Pythagorean Sourcebook and Library: An Anthology of Ancient Writings which Relate to Pythagoras and Pythagorean Philosophy.* Grand Rapids, MI: Phanes Press.

Hass, J. (ed.) (1990) *The Anthropology of War.* Cambridge, U.K.: Cambridge University Press.

Hackenberg, R.A. (1962) "Economic Alternatives in Arid Lands: A Case Study of the Pima and Papago Indians." *Ethnology,* 1: 186–95.

Haig, S.M., *et al.* (1994) "Genetic Evidence for Monogamy in the Cooperatively Breeding Red-cockaded Woodpecker." *Behavioral Ecology and Sociobiology,* 34: 295–303.

Haile-Selassie, Y. (2001) "Late Miocene Hominids from the Middle Awash, Ethiopia." *Nature,* 412: 178–180.

Haisch, B., and A. Rueda (1996) "A Quantum Broom Sweeps Clean." *Mercury,* 25 (2): 12–15.

Haisch, B., *et al.* (1994) "Beyond $E = mc^2$." *The Sciences,* 34(6): 26–31.

Haken, H. (1973) *Cooperative Phenomena.* New York: Springer-Verlag.

Haken, H. (1977) *Synergetics.* Berlin: Springer-Verlag.

Haken, H. (1983) *Advanced Synergetics.* Berlin: Springer-Verlag.

Haldane, J.B.S. (1932) *The Causes of Evolution.* New York: Harper & Row.

Halliday, T. (1998) "A Declining Amphibian Conundrum." *Nature,* 394: 418–419.

Hamilton, S., and L. Garber (1997) "Deep Blue's Hardware–Software Synergy." *Computer,* 30(10): 29–35.

Hamilton, W.D. (1964a) "The Genetical Evolution of Social Behavior, I." *Journal of Theoretical Biology,* 7: 1–16.

Hamilton, W.D. (1964b) "The Genetical Evolution of Social Behavior, II." *Journal of Theoretical Biology,* 7: 17–52.

Hammer, M.F. (1995) "A Recent Common Ancestry for Human Y Chromosomes." *Nature,* 378: 376–378.

Hammerstein, P. (1996) "Darwinian Adaptation, Population Genetics and the Streetcar Theory of Evolution." *Journal of Mathematical Biology,* 34: 511–532.

Harcourt, A.H., and F.B.M. de Waal (eds.) (1992) *Coalitions and Alliances in Humans and Other Animals.* Oxford, U.K.: Oxford University Press.

Hardin, G. (1968) "The Tragedy of the Commons." *Science,* 162: 1243–1248.

Hardin, G. (1971) "Genetic Consequences of Cultural Decisions in the Realm of Population." Paper prepared for the syposium on *Human Evolution: Past, Present, and Future,* American Museum of Natural History, New York City, November 18–20, 1971.

Harding, R.S.O. (1973) "Predation by a Troop of Olive Baboons (*Papio anubis*)." *Symposia of the Zoological Society of London,* 10: 49–56.

Harding, R.S.O., and S.C. Strum (1976) "Predatory Baboons of Kepopey." *Natural History,* 85: 46–53.

Harding, R.S.O., and G. Teleki (eds.) (1981) *Omnivorous Primates: Gathering and Hunting in Human Evolution.* New York: Columbia University Press.

Hardison, R. (1999) "The Evolution of Hemoglobin." *American Scientist*, 87: 126–137.

Hardy, A.C. (1965) *The Living Stream: A Restatement of Evolution Theory and its Relation to the Spirit of Man.* London: Collins.

Harner, M.J. (1970) "Population Pressure and the Social Evolution of Agriculturalists." *Southwestern Journal of Anthropology*, 26: 67–86.

Harold, F.M. (1986) *The Vital Force: A Study of Bioenergetics.* New York: W.H. Freeman.

Harris, M. (1975) *Culture, People, Nature: An Introduction to General Anthropology*, 2nd edn. New York: Crowell.

Harris, R.S., *et al.* (1994) "Recombination in Adaptive Mutation." *Science*, 264: 258–260.

Harte, J. (1996) "Feedbacks, Thresholds and Synergies in Global Change: Population as a Dynamic Factor." *Biodiversity and Conservation*, 5: 1069–1083.

Hartman, H., and A. Fedorov (2002) "The Origin of the Eukaryotic Cell: A Genomic Investigation." *Proceedings of the National Academy of Sciences*, 99: 1420–1425.

Harvey, J. (2001) "The Natural Economy." *Nature*, 413: 463.

Hasegawa, E. (1993) "Nest Defense and Early Production of the Major Workers in the Dimorphic Ant *Colobopsis nipponicus* (Wheeler) (Hymenoptera: Formicidae)." *Behavioral Ecology and Sociobiology*, 33: 73–77.

Hathaway, N. (1994) *The Friendly Guide to the Universe.* New York: Viking.

Haught, J. (ed.) (2000) *Science and Religion in Search of Cosmic Purpose.* Washington, D.C.: Georgetown University Press.

Hawkes, K., *et al.* (1995) "The Male's Dilemma: Increased Offspring Production Is More Paternity to Steal." *Evolutionary Ecology*, 9: 662–677.

Hawkes, K., *et al.* (1998) "Grandmothering, Menopause, and the Evolution of Human Life Histories." *Proceedings of the National Academy of Sciences*, 95: 1336–1340.

Hawkes, K., *et al.* (2000) "The Grandmother Hypothesis and Human Evolution." In *Adaptation and Human Behaviour: An Anthropological Perspective*, eds. L. Cronk *et al.*, pp. 237–258. New York: Aldine de Gruyter.

Hayes, B. (1998) "Collective Wisdom." *American Scientist*, 86: 118–122.

Haynes, R.H. (1991) "Modes of Mutation and Repair in Evolutionary Rhythms." In *Symbiosis as a Source of Evolutionary Innovation*, eds. L. Margulis and R. Fester, pp. 40–56. Cambridge, MA: MIT Press.

Heilbroner, R. (1994) "Technological Determinism Revisited." In *Does Technology Drive History?* eds. M.R. Smith and L. Marx, pp. 67–78. Cambridge, MA: MIT Press.

Heinrich, B. (1995) "An Experimental Investigation of Insight in Common Ravens (*Corvus corax*)." *Auk*, 112: 994–1003.

Heinrich, B. (1999) *Mind of the Raven: Investigations and Adventures with Wolf-Birds.* New York: HarperCollins.

Henrich, J., and R. Boyd (2001) "Why People Punish Defectors: Weak Conformist Transmission Can Stabilize Costly Enforcement of Norms in Cooperative Dilemmas." *Journal of Theoretical Biology*, 208: 79–89.

Henrich, J., *et al.* (2001) "In Search of *Homo economicus*: Behavioral Experiments in 15 Small-Scale Societies." *American Economic Review*, 91: 73–79.

Herman, R.K. (1993) "Genes Make Worms Behave." *Nature*, 364: 282–283.

Herre, E.A., *et al.* (1999) "The Evolution of Mutualisms: Exploring the Paths between Conflict and Cooperation." *Trends in Ecology and Evolution*, 14(2): 49–52.

Hewes, G.W. (1973) "Primate Communication and the Gestural Origins of Language." *Current Anthropology*, 14: 5–24.

Heyes, C.M., and B.G. Galef, Jr. (eds.) (1996) *Social Learning in Animals: The Roots of Culture*. San Diego, CA: Academic Press.

Heyes, C.M., and L. Huber (eds.) (2000) *The Evolution of Cognition*. Cambridge, MA: MIT Press.

Hill, A.H. (1987) "Causes of Perceived Faunal Change in the Later Neogene of East Africa." *Journal of Human Evolution*, 16(7/8): 583–596.

Hill, K. (1982) "Hunting and Human Evolution." *Journal of Human Evolution*, 11: 521–544.

Hill, K. (2002) "Altruistic Cooperation During Foraging by the Ache, and the Evolved Human Predisposition to Cooperate." *Human Nature*, 13: 105–128.

Hinchliffe, R. (1997) "The Forward March of the Bird–Dinosaurs Halted?" *Science*, 278: 597–598.

Hinde, R.A. (1966) *Animal Behavior: A Synthesis of Ethology and Comparative Psychology*. New York: McGraw-Hill.

Hinrichsen, D. (1997) "Coral Reefs in Crisis." *BioScience*, 47: 554–558.

Hirschleifer, J. (1977) "Economics from a Biological Viewpoint." *Journal of Law and Economics*, 20: 1–52.

Hirschleifer, J. (1978a) "Competition, Cooperation, and Conflict in Economics and Biology." *American Economic Review*, 68: 238–243.

Hirschleifer, J. (1978b) "Natural Economy versus Political Economy." *Journal of Social and Biological Structures*, 1: 319–337.

Hirshleifer, J. (1999) "There Are Many Evolutionary Pathways to Cooperation." *Journal of Bioeconomics*, 1: 73–93.

Hobhouse, L.T. (1915) *Mind in Evolution*. New York: Macmillan.

Hobsbawm, E.J. (1996) *The Age of Extremes: A History of the World, 1914–1991*. New York: Pantheon Books.

Hodder, I. (ed.) (1996) *On the Surface: Çatalhöyük, 1993–1995*. Cambridge, U.K.: McDonald Institute for Archaeological Research.

Hodder, I. (2000) *Towards Reflexive Method in Archaeology: The Example at Çatalhöyük*. Cambridge, U.K.: McDonald Institute for Archaeological Research.

Hodgson, G.M. (1993) *Economics and Evolution: Bringing Life back into Economics*. Cambridge, U.K.: Polity Press.

Hodgson, G.M. (2001) *How Economics Forgot History: The Problems of Historical Specificity in Social Science*. London: Routledge.

Hofbauer, J., and K. Sigmund (1998) *Evolutionary Games and Population Dynamics*. Cambridge, U.K.: Cambridge University Press.

Hoffman, M. (1993) "The Cell's Nucleus Shapes Up." *Science*, 259: 1257–1259.

Hogan, C.J. (1996) "Primordial Deuterium and the Big Bang." *Scientific American,* 275(6): 68–73.

Holland, B. (1999) "You Can't Keep a Good Prophet Down." *Smithsonian,* April: 69–80.

Holland, J.H. (1998) *Emergence: From Chaos to Order.* Reading, MA: Addison-Wesley.

Holland, J.N., et al. (2002) "Population Dynamics and Mutualism: Functional Responses of Benefits and Costs." *American Naturalist,* 159(3): 231–244.

Hölldobler, B., and E.O. Wilson (1990) *The Ants.* Cambridge, MA: Harvard University Press.

Hölldobler, B., and E.O. Wilson (1994) *Journey to the Ants.* Cambridge, MA: Belknap Press of Harvard University Press.

Holloway, R.L. (1975) "Early Hominid Endocasts: Volumes, Morphology, and Significance for Hominid Evolution." In *Primate Functional Morphology and Evolution,* ed. R. Tuttle, pp. 393–415. The Hague: Mouton.

Holloway, R.L. (1983a) "Cerebral Brain Endocast Pattern of *Australopithecus afarensis* Hominid." *Nature,* 303: 420–422.

Holloway, R.L. (1983b) "Human Paleontological Evidence Relevant to Language Behavior." *Human Neurobiology,* 2: 105–114.

Holloway, R.L. (1996) "Evolution of the Human Brain." In *Handbook of Human Symbolic Evolution,* eds. A. Lock and C.R. Peters, pp. 74–125. Oxford, U.K.: Blackwell Scientific Publications.

Holloway, R.L. (1997) "Brain Evolution." In *Encyclopedia of Human Biology,* vol. 2, 2nd edn., ed. R. Dulbecco, pp. 189–200. New York: Academic Press.

Holloway, R.L. (1998) "Book Review: *The Symbolic Species: The Co-Evolution of Language and the Brain.*" *American Scientist,* 86(2): 184–187.

Holzman, D.C. (1997) "Nature's Own Pooper-Scoopers Keep Earth Livable for All of Us." *Smithsonian,* 28(3): 116–124.

Homer-Dixon, T.F., et al. (1993) "Environmental Change and Violent Conflict." *Scientific American,* 268(2): 38–44.

Hoober, J.K. (1984) *Chloroplasts.* New York: Plenum Press.

Hoogland, J.L. (1979a) "Effect of Colony Size on Individual Alertness of Prairie Dogs." *Animal Behaviour,* 27: 394–407.

Hoogland, J.L. (1979b) "Aggression, Ectoparasitism, and Other Possible Costs of Prairie Dog (Sciuridae, *Cynomys* Spp.) Coloniality." *Behaviour,* 69(1–2): 1–34.

Hoogland, J.L. (1981) "The Evolution of Coloniality in White-tailed and Black-tailed Prairie Dogs (Sciuridae: *Cynomys leucurus* and *C. ludovicianus).*" *Ecology,* 62: 252–272.

Hoogland, J.L. (1995) *Black-Tailed Prairie Dog: Social Life of a Burrowing Mammal.* Chicago, IL: University of Chicago Press.

Hope, J. (1998) "In Search of a Bee Tree." *Natural History,* 107(9): 26–30.

Horgan, J. (1991) "In the Beginning." *Scientific American,* 264(2): 116–125.

Hoyle, F., and N.C. Wickramasinghe (1981) *Evolution from Space: A Theory of Cosmic Creationism.* New York: Simon & Schuster.

Hoyle, F., and N.C. Wickramasinghe (1986) "The Case for Life as a Cosmic Phenomenon." *Nature,* 322: 509–511.

Hrdy, S.B. (2001) "Mothers and Others." *Natural History,* 110(4): 50–62.

Hrdy, S.B. (2002) "Maternal Love and Ambivalence in the Pleistocene, the Eighteenth Century, and Right Now." Paper presented at the Plenary Session of the 16th Biennial Meeting, International Society of Human Ethology, Montreal, Canada, August 7–10, 2002.

Hubel, D.H. (1988) *Eye, Brain and Vision.* New York: W.H. Freeman.

Hull, D.L. (1974) *Philosophy of Biological Science.* Englewood Cliffs, NJ: Prentice-Hall.

Hull, D.L. (1982) "The Naked Meme." In *Learning, Development, and Culture: Essays in Evolutionary Epistemology,* ed. H.C. Plotkin, pp. 273–327. Chichester, U.K.: John Wiley & Sons.

Hull, D.L. (2000) "Taking Memetics Seriously: Memetics Will Be What We Make It." In *Darwinizing Culture: The Status of Memetics as a Science,* ed. R. Aunger, pp. 43–67. Oxford, U.K.: Oxford University Press.

Hull, D.L., and M. Ruse (1998) *The Philosophy of Biology.* New York: Oxford University Press.

Humphrey, N. (1992) *A History of the Mind: Evolution and the Birth of Consciousness.* New York: Simon & Schuster.

Huxley, J.S. (1942) *Evolution: The Modern Synthesis.* New York: Harper & Row.

Huxley, J.S., and T.H. Huxley (1947) *Evolution and Ethics: 1893–1943.* London: Pilot Press.

Ingber, D.E. (1998) "The Architecture of Life." *Scientific American,* 278(1): 48–57.

Ingber, D.E. (2000) "The Origin of Cellular Life." *BioEssays,* 22: 1160–1170.

Ingold, T. (1993) "Epilogue." In *Tools, Language and Cognition in Human Evolution,* eds. R. Gibson and T. Ingold, pp. 447–472. Cambridge, U.K.: Cambridge University Press.

Irion, R. (1999) "The Lopsided Universe." *New Scientist,* 161: 26–29.

Isaac, G.L. (1978) "The Food Sharing Behavior of Proto Human Hominids." *Scientific American,* 238(4): 90–108.

Isaac, G.L. (1981) "Emergence of Human Behaviour Patterns." *Philosophical Transactions of the Royal Society of London, Series B,* 292: 177–188.

Isaac, G.L. (1983) "Aspects of Human Evolution." In *Evolution from Molecules to Men,* ed. D. Bendell, pp. 509–543. New York: Cambridge University Press.

Isack, H.A., and H.U. Reyer (1989) "Honeyguides and Honey Gatherers: Interspecific Communication in a Symbiotic Relationship." *Science,* 243: 1343–1346.

Isbell, L.A. (1995) "Predation on Primates: Ecological Patterns and Evolutionary Consequences." *Evolutionary Anthropology,* 3(2): 61–71.

Isbell, L.A., and T.P. Young (1996) "The Evolution of Bipedalism in Hominids and Reduced Group Size in Chimpanzees: Alternative Responses to Decreasing Resource Availability." *Journal of Human Evolution,* 30: 389–397.

Iwamoto, T., *et al.* (1996) "Anti-predator Behavior of Gelada Baboons." *Primates,* 37(4): 389–397.

Jablonka, E., and M.J. Lamb (1995) *Epigenetic Inheritance and Evolution: The Lamarckian Dimension.* New York: Oxford University Press.

Jablonka, E., *et al.* (1998) "Lamarckian Mechanisms in Darwinian Evolution." *Trends in Ecology and Evolution*, 13(5): 206–210.

Jacob, F. (1977) "Evolution and Tinkering." *Science*, 196: 1161–1166.

Jain, R., *et al.* (1999) "Horizontal Gene Transfer among Genomes: The Complexity Hypothesis." *Proceedings of the National Academy of Sciences*, 96: 3801–3806.

James, J. (1993) *The Music of the Spheres: Music, Science and the Natural Order of the Universe*. New York: Grove.

Jaynes, J. (1990) *The Origin of Consciousness in the Breakdown of the Bicameral Mind*. New York: Houghton Mifflin.

Jeanne, R.L. (1986) "The Organization of Work in *Polybia occidentalis*: Costs and Benefits of Specialization in a Social Wasp." *Behavioral Ecology and Sociobiology*, 19: 333–341.

Jeon, K.W. (1972) "Development of Cellular Dependence in Infective Organisms: Micurgical Studies in Amoebas." *Science*, 176: 1122–1123.

Jeon, K.W. (ed.) (1973) *The Biology of Amoeba*. New York: Academic Press.

Jin, L., *et al.* (1999) "Distribution of Haplotypes from a Chromosome 21 Region Distinguishes Multiple Prehistoric Human Migrations." *Proceedings of the National Academy of Sciences*, 96: 3796.

John, E.R., *et al.* (1968) "Observation Learning in Cats." *Science*, 159: 1489–1491.

Johnson, A.W., and T. Earle (1987) *The Evolution of Human Societies: From Foraging Group to Agrarian State*. Stanford, CA: Stanford University Press.

Johnson, A.W., and T. Earle (2000) *The Evolution of Human Societies: From Foraging Group to Agrarian State*, 2nd edn. Stanford, CA: Stanford University Press.

Johnson, G. (1997) "Researchers on Complexity Ponder What It's All About." *The New York Times*, May 5, 1997, pp. B9, B13.

Johnson, G. (2002) "Supercomputing '@Home' Is Paying Off." *The New York Times*, April 23, 2002, pp. D1, D6.

Jones, A.H.M. (1955) "The Decline and Fall of the Roman Empire." *History*, 40: 209–226.

Jones, A.H.M. (1986) *The Later Roman Empire, 284–602: A Social Economic and Administrative Survey*. Oxford, U.K.: Basil Blackwell.

Jones, T.D., *et al.* (2000) "Nonavian Feathers in a Late Triassic Archosaur." *Science*, 288: 2202–2205.

Kaessmann, H., V. Wiebe, and S. Pääbo (1999) "Extensive Nuclear DNA Sequence Diversity Among Chimpanzees." *Science*, 286: 1159.

Kanizsa, G. (1979) *Organization in Vision: Essays on Gestalt Perception*. New York: Praeger.

Kano, T. (1992) *The Last Ape: Pygmy Chimpanzee Behavior and Ecology*, trans. E.O. Vineburg. Stanford, CA: Stanford University Press.

Karl, P. (1991) *Animal and Human Aggression*. New York: Oxford University Press.

Kashima, S., and M. Kitagawa (1997) "The Longest Suspension Bridge." *Scientific American*, 277(6): 89–93.

Kauffman, S.A. (1995) *At Home in the Universe: The Search for the Laws of Self-Organization and Complexity*. New York: Oxford University Press.

Kauffman, S.A. (2000) *Investigations*. New York: Oxford University Press.

Kawai, M. (1965) "Newly Acquired Pre-cultural Behavior of a Natural Troop of Japanese Monkeys on Koshima Island." *Primates*, 6: 1–30.

Kearns, C.A., and Inouye, D.W (1997) "Pollinators, Flowering Plants, and Conservation Biology: Much Remains to be Learned About Pollinators and Plants." *BioScience*, 47(5): 297–308.

Keegan, J. (1993) *A History of Warfare*. New York: Alfred A. Knopf.

Keeley, L.H. (1996) *War Before Civilization: The Myth of the Peaceful Savage*. Oxford, U.K.: Oxford University Press.

Keeling, C.D., and T.P. Whorf (2000) "The 1,800-year Oceanic Tidal Cycle: A Possible Cause of Rapid Climate Change." *Proceedings of the National Academy of Sciences*, 97(8): 3814–3819.

Keller, A.G. (1915) *Societal Evolution*, 1931 edn. New Haven, CT: Yale University Press.

Keller, E.F. (1983) *A Feeling for the Organism*. New York: W.H. Freeman.

Keller, L. (1997) "Indiscriminate Altruism: Unduly Nice Parents and Siblings." *Trends in Ecology and Evolution*, 12: 99–103.

Keller, L. (ed.) (1999) *Levels of Selection in Evolution*. Princeton, NJ: Princeton University Press.

Keller, L., and K.G. Ross (1993) "Phenotypic Plasticity and 'Cultural Transmission' of Alternative Social Organizations in the Fire Ant *Solenopsis invicta*." *Behavioral Ecology and Sociobiology*, 33: 121–129.

Keller, L., and M. Chapuisat (1999) "Cooperation among Selfish Individuals in Insect Societies." *BioScience*, 49(11): 899–909.

Kelly, K. (1994) *Out of Control: The Rise of Neo-Biological Civilization*. Reading, MA: Addison-Wesley.

Kennedy, P. (1987) *The Rise and Fall of Great Powers*. New York: Random House.

Kerr, R.A. (1996) "New Mammalian Data Challenge Evolutionary Pulse Theory." *Science*, 273: 431–432.

Kerr, R.A. (1998) "Pushing Back the Origins of Animals." *Science*, 279: 803–804.

Kettlewell, H.B.D. (1955) "Selection Experiments on Industrial Melanism in the Lepidoptera." *Heredity*, 9: 323–342.

Kettlewell, H.B.D. (1973) *The Evolution of Melanism: The Study of a Recurring Necessity*. Oxford, U.K.: Clarendon Press.

Keutsch, F.N., and R.J. Saykally (2001) "Water Clusters: Untangling the Mysteries of the Liquid, One Molecule at a Time." *Proceedings of the National Academy of Sciences*, 98: 10533–10540.

Khakhina, L.N. (1979) *Concepts of Symbiogenesis*. Leningrad, USSR: Akademie NAUK (in Russian).

Khakhina, L.N. (1992a) "Evolutionary Significance of Symbiosis: Development of the Symbiogenesis Concept." *Symbiosis*, 14: 217–228.

Khakhina, L.N. (1992b) "Contemporary Concepts of Symbiogenesis." In *Concepts of Symbiogenesis: A Historical and Critical Study of Russian Botanists*, eds. L. Margulis and M. McMenamin, pp. 95–118. New Haven, CT: Yale University Press.

Khan, M.Z., and J.R. Walters (2002) "Effects of Helpers on Breeder Survival in the Red-cockaded Woodpecker (*Picoides borealis*)." *Behavioral Ecology and Sociobiology*, 51: 336–344.

King, B.J. (2001) "Debating Culture." *Current Anthropology*, 42(3): 441–443.

King, M.C., and A.C. Wilson (1975) "Evolution at Two Levels in Humans and Chimpanzees." *Science*, 188: 107–116.

Kingdon, J. (1993) *Self-Made Man: Human Evolution from Eden to Extinction?* New York: John Wiley & Sons.

Kirchner, J.W. (1990) "Gaia Metaphor Unfalsifiable." *Nature*, 345: 470.

Klein, R. (1992) "The Impact of Early People on the Environment: The Case of Large Mammal Extinctions." In *Human Impact on the Environment*, eds. J.E. Jacobsen and J. Firor, pp. 13–34. Boulder, CO: Westview Press.

Klein, R.G. (1999) *The Human Career: Human Biological and Cultural Origins*, 2nd edn. Chicago, IL: University of Chicago Press.

Klein, R.G. (2000) "Archeology and the Evolution of Human Behavior." *Evolutionary Anthropology*, 9(1): 17–36.

Klein, R.G., with B. Edgar (2002) *The Dawn of Human Culture.* New York: John Wiley & Sons.

Kline, S.J. (1995) *Conceptual Foundations for Multidisciplinary Thinking.* Stanford, CA: Stanford University Press.

Kline, S.J. (1997) *The Semantics and Meaning of the Entropies.* Report CB-1, Department of Mechanical Engineering, Stanford University, Stanford, CA.

Knauft, B.M. (1991) "Violence and Sociality in Human Evolution." *Current Anthropology*, 32(4): 391–409.

Koenig, W.D., and R.L. Mumme (1987) *Population Ecology of the Cooperatively Breeding Acorn Woodpecker.* Princeton, NJ: Princeton University Press.

Koenig, W.D., *et al.* (1995) "Acorn Woodpecker." In *Birds of North America (No. 194)*, eds. A. Poole and F. Gill, pp. 1–23. Philadelphia, PA: The Academy of Natural Sciences of Philadelphia.

Koestler, A. (1959) *The Sleep Walkers.* New York: Macmillan.

Koestler, A. (1960) *The Watershed: A Biography of Johannes Kepler.* Garden City, NY: Anchor Books.

Koestler, A. (1967) *The Ghost in the Machine.* New York: Macmillan.

Koestler, A., and J.R. Smythies (eds.) (1969) *Beyond Reductionism: New Perspectives in the Life Sciences*, London: Hutchinson.

Köhler, M. (1925) *The Mentality of Apes.* London: Routledge & Kegan Paul.

Köhler, M., and S. Moya-Solá (1997) "Ape-like or Hominid-like? The Positional Behavior of *Oreopithecus bambolii* Reconsidered." *Proceedings of the National Academy of Sciences*, 94: 11747–11750.

Kollerstrom, N. (1989) "Kepler's Belief in Astrology." In *History and Astrology*, ed. A. Kitson, pp. 47–48. London: Unwin.

Kolter, R., and R. Losick (1998) "One for All and All for One" *Science*, 280: 226.

Kormondy, E.J. (1969) *Concepts of Ecology.* Englewood Cliffs, NJ: Prentice-Hall.

Krauss, L.M., and G.D. Starkman (1999) "The Fate of Life in the Universe." *Scientific American,* 281(5): 58–65.

Krebs, J.R., and N.B. Davies (eds.) (1984) *Behavioural Ecology: An Evolutionary Approach,* 2nd edn. Oxford, U.K.: Blackwell Scientific Publications.

Krebs, J.R., and N.B. Davies (eds.) (1991) *Behavioural Ecology: An Evolutionary Approach,* 3rd edn. Oxford, U.K.: Blackwell Scientific Publications.

Krebs, J.R., and N.B. Davies (1993) *An Introduction to Behavioural Ecology,* 3rd edn. Oxford, U.K.: Blackwell Scientific Publications.

Krebs, J.R., and N.B. Davies (eds.) (1997) *Behavioural Ecology: An Evolutionary Approach,* 4th edn. Oxford, U.K.: Blackwell Scientific Publications.

Kroeber, A. (1948) *Anthropology: Race, Language, Culture, Psychology, Prehistory.* New York: Harcourt, Brace.

Kropotkin, P. (1902) *Mutual Aid: A Factor of Evolution.* New York: McClure Phillips & Co.

Kruse, D.L., and K. Shiner (2000) "A Study of the Political Behavior of People with Disabilities." Final Report to the Disability Research Consortium, Bureau of Economic Research, Rutgers University and New Jersey Developmental Disabilities Council.

Kruuk, H. (1972) *The Spotted Hyena: A Study of Predation and Social Behaviour.* Chicago, IL: University of Chicago Press.

Kuhn, S.L., *et al.* (2001) "Ornaments of the Earliest Upper Paleolithic: New Insights from the Levant." *Proceedings of the National Academy of Sciences,* 98: 7641.

Kuhn, T. (1962) *The Structure of Scientific Revolutions.* Chicago, IL: University of Chicago Press.

Kummer, H. (1968) *Social Organization of Hamadryas Baboons: A Field Study.* Chicago, IL: University of Chicago Press.

Kummer, H. (1971) *Primate Societies: Group Techniques of Ecological Adaptation.* Chicago: Aldine-Atherton.

Kunzig, R. (1999) "A Tale of Two Obsessed Archeologists, One Ancient City, and Nagging Doubts about Whether Science Can Even Hope to Reveal the Past." *Discover,* 20(5): 84–92.

Kurtén, B. (1984) *Not from the Apes.* New York: Columbia University Press.

Lack, D.L. (1947) *Darwin's Finches,* 1961 edn. New York: Harper & Row.

Laland, K.N., and G.R. Brown (2002) *Sense and Nonsense: Evolutionary Perspectives on Human Behavior.* Oxford, U.K.: Oxford University Press.

Laland, K.N., and S.M. Reader (1999) "Foraging Innovation in the Guppy." *Animal Behaviour,* 57: 331–340.

Laland, K.N., *et al.* (2001) "Cultural Niche Construction and Human Evolution." *Journal of Evolutionary Biology,* 14: 22–33.

Lamarck, J.B. de (1809) *Zoological Philosophy,* 1963 edn., trans. H. Elliot, New York: Hafner; 1984 edn: trans. H. Elliot, Chicago: University of Chicago Press.

Lancaster, J.B. (1975) *Primate Behavior and the Emergence of Human Culture.* New York: Holt, Rinehart & Winston.

Lancaster, J.B. (1978) "Carrying and Sharing in Human Evolution." *Human Nature,* 1: 83–89.

Lancaster, J.B., and C.S. Lancaster (1983) "Parental Investment: The Hominid Adaptation." In *How Humans Adapt,* ed. D.J. Ortner, pp. 33–65. Washington, D.C.: Smithsonian Institution Press.

Lande, R., and S.J. Arnold (1983) "The Measurement of Selection on Correlated Characters." *Evolution,* 37: 1210–1226.

Langdon, J.H. (1997) "Umbrella Hypothesis and Parsimony in Human Evolution: A Critique of the Aquatic Ape Hypothesis." *Journal of Human Evolution,* 33: 479–494.

Lara-Ochoa, F., and A. Herrera (1992) "A Dynamical Model for a Co-operative Enzyme." *Journal of Theoretical Biology,* 154: 189–204.

Latham, S. (1973) *Knives and Knifemakers.* New York: Winchester Press.

Laughlin, R.B., and D. Pines (2000) "The Theory of Everything." *Proceedings of the National Academy of Sciences,* 97: 28–31.

Laughlin, R.B., *et al.* (2000) "The Middle Way." *Proceedings of the National Academy of Sciences,* 97: 32–37.

Laurent, J., and J. Nightingale (eds.) (2001) *Darwinism and Evolutionary Economics.* Cheltenham, U.K.: Edward Elgar.

Le Boeuf, B.J. (1985) "Elephant Seals." In *The Natural History of Año Nuevo,* eds. B.J. Le Boeuf and S. Kaza, pp. 326–374. Pacific Grove, CA: Boxwood Press.

Le Boeuf, B.J., and R.M. Laws (1994) *Elephant Seals: Population Ecology, Behavior and Physiology.* Berkeley: University of California Press.

Le Bon, G. (1896) *The Crowd: A Study of the Popular Mind.* New York: Macmillan.

Le Maho, Y. (1977) "The Emperor Penguin: A Strategy to Live and Breed in the Cold." *American Scientist,* 65: 680–693.

Lee, D.H., *et al.* (1997) "Emergence of Symbiosis in Peptide Self-Replication through a Hypercyclic Network." *Nature,* 390: 591–594.

Lee, P.C. (1994) "The New Evolutionary Paradigm." In *Behaviour and Evolution,* eds. P. Slater and T. Halliday, pp. 266–300. Cambridge, U.K.: Cambridge University Press.

Lee, R.B. (1968) "What Hunters Do for a Living, or, How to Make Out on Scarce Resource." In *Man the Hunter,* eds. R.B. Lee and I. DeVore, pp. 30–48. Chicago, IL: Aldine Press.

Lee, R.B., and I. DeVore (eds.) (1968) *Man the Hunter.* Chicago, IL: Aldine Press.

Lee-Thorp, J., *et al.* (2000) "The Hunters and the Hunted Revisited." *Journal of Human Evolution,* 39(6): 565–576.

Leff, H.S., and A.F. Rex (1990) *Maxwell's Demon, Entropy, Information, Computing.* Princeton, NJ: Princeton University Press.

Lehninger, A.L. (1971) *Bioenergetics: The Molecular Basis of Biological Energy Transformations.* Menlo Park, CA: Benjamin/Cummings.

Leonard, W.R. (2002) "Food for Thought: Dietary Change Was a Driving Force in Human Evolution." *Scientific American,* 287(6): 106–115.

Leonard, W.R., and M.L. Robertson (1994) "Evolutionary Perspectives on Human Nutrition: The Influence of Brain and Body Size on Diet and Metabolism." *American Journal of Human Biology,* 6: 77–88.

Leonard, W.R., and M.L. Robertson (1997) "Comparative Primate Energetics and Hominid Evolution." *American Journal of Physical Anthropology*, 102(2): 265–281.

Lesinski, J.M. (1996) *Exotic Invaders*. New York: Walker & Co.

Levin, B.R., and C.T. Bergstrom (2000) "Bacteria Are Different: Observations, Interpretations, Speculations and Opinions about the Mechanisms of Adaptive Evolution in Prokaryotes." *Proceedings of the National Academy of Sciences*, 97: 6981–6985.

Lewin, R. (1993) *Human Evolution: An Illustrated Introduction*, 3rd edn. Boston, MA: Blackwell Scientific Publications.

Lewin, R. (1994) "I Buzz Therefore I Think." *New Scientist*, 141: 29–33.

Lewin, R. (1997) *Bones of Contention*. Chicago, IL: University of Chicago Press.

Lewontin, R.C. (1978) "Adaptation." *Scientific American*, 239(3): 213–230.

Lewontin, R.C. (2000a) *The Triple Helix: Gene Organism and Environment*. Cambridge, MA: Harvard University Press.

Lewontin, R.C. (2000b) "Computing the Organism." *Natural History*, 109(3): 94.

Liddle, A.R. (1999) *An Introduction to Modern Cosmology*. New York: John Wiley & Sons.

Lieberman, P. (1998) *Eve Spoke: Human Language and Human Evolution*. New York: W.W. Norton.

Liebes, S., *et al.* (1998) *A Walk Through Time: From Stardust to Us*. New York: John Wiley & Sons.

Ligon, J.D., and S.H. Ligon (1982) "The Cooperative Breeding Behavior of the Green Woodhoopoe." *Scientific American*, 247(1): 126–134.

Lindley, R. (1988) "Is the Earth Alive or Dead?" *Nature*, 322: 483–484.

Linton, S. (1971) "Woman the Gatherer: Male Bias in Anthropology." In *Women in Cross-Cultural Perspective*, ed. S.-E. Jacob, pp. 9–21. Champaign: University of Illinois Press.

Little, P. (1995) "The Genome Directory: Navigational Progress." *Nature*, 377: 288.

Lloyd Morgan, C.H. (1891) *Animal Life and Intelligence*. Boston, MA: Ginn & Co.

Lloyd Morgan, C.H. (1896a) "On Modification and Variation." *Science*, 4: 733–740.

Lloyd Morgan, C.H. (1896b) *Habit and Instinct*. London: Edward Arnold.

Lloyd Morgan, C.H. (1900) *Animal Behaviour*. London: Edward Arnold.

Lloyd Morgan, C.H. (1923) *Emergent Evolution*. New York: Henry Holt.

Lloyd Morgan, C. (1926) *Life, Mind and Spirit*. London: Williams & Norgate.

Lloyd Morgan, C. (1933) *The Emergence of Novelty*. New York: Henry Holt.

Lorenz, K. (1966) *On Aggression*. London: Methuen.

Losick, R., and D. Kaiser (1997) "Why and How Bacteria Communicate." *Scientific American*, 276(2): 68–73.

Losos, J.B. (2001) "Evolution: A Lizard's Tale." *Scientific American*, 284(3): 64–69.

Lovejoy, C.O. (1981) "The Origin of Man." *Science*, 211: 341–350.

Lovelock, J.E. (1979) *Gaia: A New Look at Life on Earth*. Oxford, U.K.: Oxford University Press.

Lovelock, J.E. (1990) "Hands up for the Gaia Hypothesis." *Nature*, 344: 100–102.

Lovelock, J.E. (1991) *Healing Gaia*. New York: Crown Books.

Lovelock, J.E. (1995) *The Ages of Gaia: A Biography of Our Living Earth.* New York: W.W. Norton.

Lumsden, C.J., and E.O. Wilson (1981) *Genes, Mind and Culture: The Co-Evolutionary Process.* Cambridge, MA: Harvard University Press.

Lutzoni, F., *et al.* (2001) "A Maximum Likelihood Approach to Fitting Equilibrium Models of Microsatellite Evolution." *Molecular Biology and Evolution,* 18: 413–417.

Lux, K. (1990) *Adam Smith's Mistake.* Boston, MA: Shambhala Press.

Lynch, A. (1996) *Thought Contagion: How Belief Spreads through Society.* New York: Basic Books.

Lynch, D. (1992) *Titanic: An Illustrated History.* Toronto, Canada: Madison Press.

MacArthur, R.H., and E.R. Pianka (1966) "On Optimal Use of a Patchy Environment." *American Naturalist,* 100: 603–609.

Macdonald, D.W. (1986) "A Meerkat Volunteers for Guard Duty so its Comrades Can Live in Peace." *Smithsonian,* 17(11): 54.

MacMullen, R. (1988) *Corruption and the Decline of Rome.* New Haven, CT: Yale University Press.

MacPhee, R.D.E. (1999) *Extinctions in Near Time: Causes, Contexts, and Consequences.* New York: Plenum Press.

Maddox, J. (1990) "Cooperating Molecules in Biology." *Nature,* 345: 15.

Malone, J. (1997) *Predicting the Future: From Jules Verne to Bill Gates.* New York: M. Evans.

Malthus, T. (1789) *An Essay on the Principle of Population.* London: J. Johnson.

Mandoli, D.F. (1998) "Elaboration of Body Plan and Phase Change During Development of *Acetabularia*: How is the Complex Architecture of a Giant Unicell Built?" *Annual Review of Plant Physiology and Plant Molecular Biology,* 49: 173–198.

Mann, A. (1972) "Hominid and Cultural Origins." *Man,* 7: 379–386.

Mann, A. (1981) "Diet and Human Evolution." In *Omnivorous Primates: Gathering and Hunting in Human Evolution,* eds. R.S.O. Harding and G. Teleki, pp. 10–36. New York: Columbia University Press.

Mann, C. (1991) "Gaia: Myth or Mechanism?" *Science,* 252: 380.

Manson, J.H., and R.W. Wrangham (1991) "Intergroup Aggression in Chimpanzees and Humans." *Current Anthropology,* 32(4): 369–390.

Margulis, L. (1970) *Origin of Eukaryotic Cells.* New Haven, CT: Yale University Press.

Margulis, L. (1981) *Symbiosis in Cell Evolution.* San Francisco, CA: W.H. Freeman.

Margulis, L. (1990) "Words as Battle Cries – Symbiogenesis and the New Field of Endocytobiology." *BioScience,* 40(9): 673–677.

Margulis, L. (1993) *Symbiosis in Cell Evolution,* 2nd edn. San Francisco, CA: W.H. Freeman.

Margulis, L., and M. Dolan (1999) "Did Centrioles and Kinetosomes Evolve from Bacterial Symbionts? Report of the Henneguy-Lenhossek Theory Meeting." *Symbiosis,* 26: 199–204.

Margulis, L., and R. Fester (eds.) (1991) *Symbiosis as a Source of Evolutionary Innovation: Speciation and Morphogenesis.* Cambridge, MA: MIT Press.

Margulis, L., and R. Guerrero (1991) "Two Plus Three Equals One." In *Gaia 2: Emergence: The New Science of Becoming*, ed. W.I. Thompson, pp. 50–67. Hudson, NY: Lindisfarne Press.

Margulis, L., and M. McMenamin (1990) "Marriage of Convenience: The Motility of the Modern Cell May Reflect an Ancient Symbiotic Union." *Sciences*, 30(5): 30–38.

Margulis, L., and M. McMenamin (eds.) (1993) *Concepts of Symbiogenesis: A Historical and Critical Study of the Research of Russian Botanists*. New Haven, CT: Yale University Press.

Margulis, L., and D. Sagan (1986) *Microcosmos: Four Billion Years of Evolution from Our Microbial Ancestors*. New York: Summit Books.

Margulis, L., and D. Sagan (1995) *What is Life?* New York: Simon & Schuster.

Margulis, L., and D. Sagan (1997) *Slanted Truths: Essays on Gaia, Symbiosis, and Evolution*. New York: Springer-Verlag.

Margulis, L., and D. Sagan (2001) "The Beast with Five Genomes." *Natural History*, 110(5): 38–41.

Margulis, L., *et al.* (2000) "The Chimeric Eukaryote: Origin of the Nucleus from the Karyomastigont in Amitochondriate Protists." *Proceedings of the National Academy of Sciences*, 97: 6954–6959.

Markoff, J. (1999) "Illness Becomes an Art Metaphor for Computers." *The New York Times*, June 14, 1999, pp. 1ff.

Markoff, J. (2000) "Technologists Get a Warning and a Plea from One of Their Own." *The New York Times*, March, 13, 2000, pp. C1ff.

Markos, A. (1995) "The Ontogeny of Gaia: The Role of Microorganisms in Planetary Information Network." *Journal of Theoretical Biology*, 176: 175–180.

Martin, P.S. (1967) "Prehistoric Overkill." In *Pleistocene Extinctions*, eds. P.S. Martin and H.E. Wright, pp. 75–120. New Haven, CT: Yale University Press.

Martin, P.S., and R.G. Klein (eds.) (1984) *Quaternary Extinctions: A Prehistoric Revolution*. Tucson: University of Arizona Press.

Martin, R.D. (1981) "Relative Brainsize in Terrestrial Vertebrates." *Nature*, 293: 57–60.

Martin, R.D., and A.M. MacLarnon (1985) "Gestation Period, Neonatal Size, and Maternal Investment in Placental Mammals." *Nature*, 313: 220–223.

Martin, S.J. (2002) "Parasitic Cape Honeybee Workers, Apis mellifera capensis, Evade Policing." *Nature*, 415: 163–164.

Martin, W., and M. Müller (1998) "The Hydrogen Hypothesis for the First Eukaryote." *Nature*, 391: 37.

Marzluff, J.M., and B. Heinrich (1991) "Foraging by Common Ravens in the Presence and Absence of Territory Holders: An Experimental Analysis of Social Foraging." *Animal Behaviour*, 42: 755–770.

Masland, R. H. (1986) "The Functional Architecture of the Retina." *Scientific American*, 255(6): 102–111.

Maslow, A.H. (1964) "Synergy and the Society in the Individual." *Journal of Individual Psychology*, 29: 153–164.

Maslow, A.H., and J.J. Honigman (1970) "Synergy: Some Notes of Ruth Benedict." *American Anthropologist*, 72: 320–333.

Mason, G., *et al.* (1998a) "A Demanding Task: Using Economic Techniques to Assess Animal Priorities." *Animal Behaviour*, 55: 1071–1078.

Mason, G., *et al.* (1998b) "Assessing Animal Priorities: Future Directions." *Animal Behaviour*, 55: 1082–1083.

May, G.H. (1996) "The Future as a Learning Process." *The Futurist*, 30(4): 60.

May, R.M. (1992) "How Many Species Inhabit the Earth?" *Scientific American*, 267(4): 42–48.

Maynard Smith, J. (1974) "The Theory of Games and the Evolution of Animal Conflicts." *Journal of Theoretical Biology*, 47(1): 209–221.

Maynard Smith, J. (1975) *The Theory of Evolution*, 3rd edn. New York: Penguin Books.

Maynard Smith, J. (1982a) "The Evolution of Social Behavior: A Classification of Models." In *Current Problems in Sociobiology*, ed. King's College Sociobiology Group, pp. 28–44. Cambridge, U.K.: Cambridge University Press.

Maynard Smith, J. (1982b) *Evolution and the Theory of Games*. Cambridge, U.K.: Cambridge University Press.

Maynard Smith, J. (1983) "Models of Evolution." *Proceedings of the Royal Society of London B*, 219: 315–325.

Maynard Smith, J. (1984a) "Game Theory and the Evolution of Behaviour." *Behavioral and Brain Sciences*, 7: 95–125.

Maynard Smith, J. (1984b) "Paleontology at the High Table." *Nature*, 309: 401.

Maynard Smith, J. (1989) *Evolutionary Genetics*. Oxford, U.K.: Oxford University Press.

Maynard Smith, J. (1998) *Shaping Life: Genes, Embryos and Evolution*. New Haven, CT: Yale University Press.

Maynard Smith, J., and E. Szathmáry (1995) *The Major Transitions in Evolution*. New York: W.H. Freeman Spektrum.

Maynard Smith, J., and E. Szathmáry (1999) *The Origins of Life: From the Birth of Life to the Origin of Language*. Oxford, U.K.: Oxford University Press.

Mayr, E. (1960) "The Emergence of Evolutionary Novelties." In *Evolution after Darwin*, vol I., ed. S. Tax, pp. 349–380. Chicago: University of Chicago Press.

Mayr, E. (1965) "Cause and Effect in Biology." In *Cause and Effect*, ed. D. Lerner, pp. 33–49. New York: Free Press.

Mayr, E. (1974a) "Behavior Programs and Evolutionary Strategies." *American Scientist*, 62: 650–659.

Mayr, E. (1974b) "Teleological and Teleconomic: A New Analysis." In *Boston Studies in the Philosophy of Science*, vol. 14, eds. R.S. Cohen and M.W. Wartofsky, pp. 91–117. Boston, MA: D. Reidel.

Mayr, E. (1976) *Evolution and the Diversity of Life: Selected Essays*. Cambridge, MA: Harvard University Press.

Mayr, E. (1982) *The Growth of Biological Thought: Diversity, Evolution, and Inheritance*. Cambridge, MA: Harvard University Press.

Mayr, E. (1988) "The Limits of Reductionism." *Nature*, 333: 475.

Mayr, E. (2002) *What Evolution Is*. London: Weidenfeld & Nicolson.

McBrearty, S., and A.S. Brooks (2000) "The Revolution that Wasn't: A New Interpretation of the Origin of Modern Human Behavior." *Journal of Human Evolution*, 39: 453–563.

McFall, M.J. (2000) "Negotiations between Animals and Bacteria: The 'Diplomacy' of the Squid–*Vibrio* Symbiosis." *Comparative Biochemistry and Physiology*, 126(A): 471–480.

McFall-Ngai, M.J. (1994) "Animal–Bacterial Interactions in the Early Life History of Marine Invertebrates: The *Euprymna scolopes/Vibrio fischeri symbiosis*." *American Zoologist*, 34: 554–561.

McFall-Ngai, M.J. (1998) "The Development of Cooperative Associations Between Animals and Bacteria: Establishing Détente Among Domains." *American Zoologist*, 38: 593–608.

McFarland, D. (ed.) (1987) *The Oxford Companion to Animal Behavior*. Oxford, U.K.: Oxford University Press.

McGrew, W.C. (1992) *Chimpanzee Material Culture: Implications for Human Evolution*. Cambridge, U.K.: Cambridge University Press.

McGrew, W.C. (1993) "The Intelligent Use of Tools: Twenty Propositions." In *Tools, Language and Cognition in Human Evolution*, eds. K.R. Gibson and T. Ingold, pp. 151–170. Cambridge, U.K.: Cambridge University Press.

McGrew, W.C. (2001) "The Nature of Culture: Prospects and Pitfalls of Cultural Primatology." In *Tree of Origin: What Primate Behavior Can Tell Us about Human Social Evolution*, ed. F.B.M. de Waal, pp. 229–254. Cambridge, MA: Harvard University Press.

McHenry, H.M. (1992) "How Big Were Early Hominids?" *Evolutionary Anthropology*, 1: 15–20.

McKibben, B. (1999) "Indifferent to a Planet in Pain." *The New York Times*, September 4, 1999, p. A13.

McLaughlin, A. (1999) "New Midwest Crop Comes Sweeping Down the Plain." *Christian Science Monitor*, March 9, 1999, p. 2.

McMahon, T. (1973) "Size and Shape in Biology." *Science*, 179: 1201–1204.

McManus, J.F., *et al.* (1999) "A 0.5-million-year Record of Millenial-scale Climate Variability in the North Atlantic." *Science*, 283: 971.

McMenamin, M.A.S. (1998) *The Garden of Ediacara: Discovering the First Complex Life*. New York: Columbia University Press.

McNeil, D.G., Jr. (1998) "AIDS Stalling Africa's Struggling Economics." *The New York Times*, September 15, 1998, pp. 1, 20.

McNeill, J.R. (2000) *Something New under the Sun: An Environmental History of the Twentieth Century World*. New York: W.W. Norton.

McNeill, W.H. (1963) *The Rise of the West: A History of the Human Community*. Chicago, IL: University of Chicago Press.

Mech, L.D. (1970) *The Wolf: The Ecology and Behavior of an Endangered Species*, 1981 edn. Minneapolis: University of Minnesota Press.

Mellaart, J. (1970) "Neolithic Anatolia." In *The Cambridge Ancient History*, vol. 1, eds. I.E.S. Edwards, *et al.*, pp. 316–317. Cambridge, U.K.: Cambridge University Press.

Mellaart, J. (1972) "Anatolian Neolithic Settlement Patterns." In *Man, Settlement and Urbanism*, ed. P.J. Ucko *et al.*, pp. 279–285. London: Duckworth.

Mellars, P. (1999) "The Neanderthal Problem." *Current Anthropology*, 40(3): 341–364.

Merrell, D.J. (1949) "Selective Mating in *Drosophila melanogaster*." *Genetics*, 34: 370–389.

Merrell, D.J. (1953) "Selective Mating as a Cause of Gene Frequency Changes in Laboratory Populations of *D. melanogaster*." *Evolution*, 7: 287–296.

Mesterton-Gibbons, M., and L.A. Dugatkin (1992) "Cooperation among Unrelated Individuals: Evolutionary Factors." *Quarterly Review of Biology*, 67(3): 267–281.

Michod, R.E. (1996) "Cooperation and Conflict in the Evolution of Individuality. II. Conflict Mediation." *Proceedings of the Royal Society of London B*, 263: 813–822.

Michod, R.E. (1999) *Darwinian Dynamics: Evolutionary Transitions in Fitness and Individuality*. Princeton, NJ: Princeton University Press.

Miele, F. (1998) "The Ionian Instauration." *Skeptic*, 6(1): 76–85.

Milgram, S. (1965) "Some Conditions of Obedience and Disobedience to Authority." *Human Relations*, 18: 57–76.

Milgram, S. (1973) *Obedience to Authority: An Experimental View*. New York: Harper & Row.

Milius, S. (2000) "Overlooked Fossil Spread First Feathers." *Science News*, 157: 405.

Miller, J.G. (1978) *Living Systems*, 1995 edn. Niwot: University of Colorado Press.

Miller, K. (1999) *Finding Darwin's God: A Scientist's Search for Common Ground, Between God and Evolution*. New York: Cliff Street Books.

Miller, M. (2001) "Mammoth Mystery." *New Scientist*, 170: 32–35.

Møller, A.P. (1987) "Advantages and Disadvantages of Coloniality in the Swallow, *Hirundo rustica*." *Animal Behaviour*, 35: 819–832.

Monden, Y. (1998) *Toyota Production System: An Integrated Approach to Just-In-Time*. Norcross, GA: Industrial Engineering and Management Press.

Monod, J. (1971) *Chance and Necessity*, trans. A. Wainhouse. New York: Alfred A. Knopf.

Monod, J., *et al.* (1965) "On the Nature of Allosteric Transitions: A Plausible Model." *Journal of Molecular Biology*, 12: 88–118.

Moore, A.M.T. (1985) "The Development of Neolithic Societies in the Near East." *Advances in World Archaeology*, 4: 1–69.

Moore, A.M.T. (1992) " 'The Enduring Dead': Death and Ancestors on the Euphrates." *American Journal of Archaeology*, 96(2): 1–69.

Moore, A.M.T., and G.C. Hillman (1992) "The Pleistocene to the Holocene Transition and Human Economy in Southwest Asia: The Impact of the Younger Dryas." *American Antiquity*, 57: 482–494.

Moore, H., *et al.* (2002) "Exceptional Sperm Cooperation in the Wood Mouse." *Nature*, 418: 174–177.

Moore, J. (1984) "The Evolution of Reciprocal Sharing." *Ethology and Sociobiology*, 5: 5–14.

Morgan, E. (1982) *The Aquatic Ape*. New York: Stein & Day.

Morgan, E. (1997) *The Aquatic Ape Hypothesis*. London: Souvenir Press.

Morowitz, H.J. (1968) *Energy Flow in Biology*. New York: Academic Press.

Morowitz, H.J. (1978a) *Foundations of Bioenergetics*. New York: Academic Press.

Morowitz, H.J. (1978b) "Proton Semiconductors and Energy Transduction in Biological Systems." *American Journal of Physiology*, 235: R99–114.

Morowitz, H.J. (1981) "Phase Separation, Charge Separation and Biogenesis." *BioSystems*, 14: 41–47.

Morowitz, H.J. (1992) *Beginnings of Cellular Life: Metabolism Recapitulates Biogenesis*. New Haven, CT: Yale University Press.

Morris, D. (1967) *The Naked Ape*. New York: Dell.

Morris, D.R. (1965) *The Washing of Spears*. New York: Simon & Schuster.

Morton, D.J. (1927) "Human Origin, Correlation of Previous Studies on Private Fect and Posture with Other Morphological Evidence." *American Journal of Physical Anthropology*, 10: 173–203.

Moxon, E.R., and D.S. Thaler (1997) "The Tinkerer's Evolving Tool-box." *Nature*, 387: 659–662.

Moxon, E.R., and C. Wills (1999) "DNA Microsatellites: Agents of Evolution?" *Scientific American*, 280(1): 94–99.

Mueller, U.G., *et al.* (1998) "The Evolution of Agriculture in Ants." *Science*, 281: 2034.

Mueller, U.G., *et al.* (2001) "The Origin of the Attine Ant–Fungus Mutualism." *Quarterly Review of Biology*, 76(2): 169–197.

Müller, G.B., and G.P. Wagner (1991) "Novelty in Evolution: Restructuring the Concept." *Annual Review of Ecology and Systematics*, 22: 229–256.

Mumme, R.L. (1992) "Do Helpers Increase Reproductive Success?" *Behavioral Ecology and Sociobiology*, 31: 319–328.

Mumme, R.L., *et al.* (1988) "Costs and Benefits of Joint Nesting in the Acorn Woodpecker." *American Naturalist*, 131: 654–677.

Mumme, R.L., *et al.* (1989) "Helping Behaviour Reproductive Value, and the Future Component of Indirect Fitness." *Animal Behaviour*, 38: 331–343.

Murdock, G.P. (1967) *Ethnographic Atlas*. Pittsburgh, PA: University of Pittsburgh Press.

Napier, J. (1967) "The Antiquity of Human Walking." *Scientific American*, 216(4): 56–66.

Naroll, R. (1967) "Imperial Cycles and World Order." *Peace Research Society (International) Papers*, 7: 83–101.

Nash, T.H. III (ed.) (1996) *Lichen Biology*. Cambridge, U.K.: Cambridge University Press.

Needham, J. (1937) *Integrative Levels: A Reevaluation of the Idea of Progress*. Oxford, U.K.: Clarendon Press.

Nei, M. (1995) "Genetic Support for the Out-of-Africa Theory of Human Evolution." *Proceedings of the National Academy of Sciences*, 96: 6720–6722.

Newman, R.W. (1970) "Why Man Is Such a Sweaty and Thirsty Animal: A Speculative Review." *Human Biology*, 42: 12–27.

Nicholls, D.G., and S.J. Ferguson (1992) *Bioenergetics 2*. San Diego, CA: Academic Press.

Nicholls, J.G., *et al.* (1992) *From Neuron to Brain*, 3rd edn. Sunderland, MA: Sinauer Associates.

Nicholson, T.R. (1970) *Passenger Cars, 1863–1904*. New York: Macmillan.

Nitecki, M.H. (ed.) (1988) *Evolutionary Progress*. Chicago, IL: University of Chicago Press.

Nitecki, M.H. (ed.) (1990) *Evolutionary Innovations*. Chicago, IL: University of Chicago Press.

Noble, W., and I. Davidson (1994) *Human Evolution, Language and the Mind: A Psychological and Archeological Inquiry*. Cambridge, U.K.: Cambridge University Press.

Noë, R., *et al.* (eds.) (2001) *Economics in Nature: Social Dilemmas, Mate Choice and Biological Markets*. Cambridge, U.K.: Cambridge University Press.

Norell, M., and J. Clarke (2001) "Fossil that Fills a Critical Gap in Avian Evolution." *Nature*, 409: 181–183.

Nossal, G.J.V. (1993) "Life, Death and the Immune System." *Scientific American*, 269(3): 53–62.

Novikoff, A. (1945) "The Concept of Integrative Levels in Biology." *Science*, 101: 209–215.

Nowak, M.A. (2000) "*Homo grammaticus*." *Natural History*, 109(10): 36–45.

Nowak, M.A., and K. Sigmund (1993) "A Strategy of Win–Stay, Lose–Shift that Outperforms Tit-for-tat in the Prisoner's Dilemma Game." *Nature*, 364: 56–58.

Nowak, M.A., and K. Sigmund (1998a) "Evolution of Indirect Reciprocity by Image Scoring." *Nature*, 394: 573–578.

Nowak, M.A., and K. Sigmund (1998b) "The Dynamics of Indirect Reciprocity." *Journal of Theoretical Biology*, 194: 561–574.

O'Connell, J.F., *et al.* (1999) "Grandmothering and the Evolution of *Homo erectus*." *Journal of Human Evolution*, 36: 461–485.

O'Leary, T.J., and D. Levinson (eds.) (1991) *Encyclopedia of World Cultures*. Boston, MA: G.K. Hall & Co.

Ochert, A. (1999) "Transposons." *Discover*, 20(12): 59–66.

Ofek, H. (2001) *Second Nature: Economic Origins of Human Evolution*. Cambridge, U.K.: Cambridge University Press.

Okasha, S. (2001) "Why Won't the Group Selection Controversy Go Away?" *British Journal of the Philosophy of Science*, 52(1): 25–50.

Oldstone, M.B.A. (1998) *Viruses, Plagues, and History*. New York: Oxford University Press.

Oliver, S.C. (1962) "Ecology and Cultural Continuity as Contributing Factors in the Social Organization of the Plains Indians." In *Man in Adaptation: The Cultural Present*, ed. Y.A. Cohen, pp. 243–261. 1974 edn: Chicago, IL: Aldine Press.

Osborn, H.F. (1896a) "Discussion following Presentation of Paper by A. Graf, 'The Problem of the Transmission of Acquired Characters'." *Transactions of the New York Academy of Sciences*, 15: 141–143.

Osborn, H.F. (1896b) "Ontogenetic and Phylogenetic Variation." *Science*, 4: 786–789.

Osborn, H.F. (1897) "The Limits of Organic Selection." *American Naturalist*, 31: 944–951.

Oster, G.F., and E.O. Wilson (1978) *Caste and Ecology in the Social Insects*. Princeton, NJ: Princeton University Press.

Otterbein, K.F. (1970) *The Evolution of War*. New Haven, CT: Human Relations Area Files Press.

Otterbein, K.F. (1985) *The Evolution of War: A Cross-Cultural Study*, 2nd edn. New Haven, CT: Human Relations Area Files Press.

Otterbein, K.F. (ed.) (1994) *Feuding and Warfare: Selected Works of Keith F. Otterbein*, vol. 1, *War and Society*. Langhorne, PA: Gordon & Breach.

Ottoni, E., and M. Mannu (2001) "Semifree-Ranging Tufted Capuchins (*Cebus apella*) Spontaneously Use Tools to Crack Open Nuts." *International Journal of Primatology*, 22(3): 347–358.

Overpeck, J., and R. Webb. (2000) "Nonglacial Rapid Climate Events: Past and Future." *Proceedings of the National Academy of Sciences*, 1335–1338.

Pääbo, S. (1999) "Human Evolution." *Trends in Cell Biology*, 9(12): M13–M16.

Pääbo, S. (2001) "The Human Genome and Our View of Ourselves." *Science*, 291: 1219–1220.

Packer, C., and A.E. Pusey (1982) "Cooperation and Competition Within Coalitions of Male Lions: Kin Selection or Game Theory?" *Nature*, 296: 740–742.

Packer, C., and L. Ruttan (1988) "The Evolution of Cooperative Hunting." *American Naturalist*, 132(2): 159–198.

Packer, C., *et al.* (1990a) "Why Lions Form Groups: Food is Not Enough." *American Naturalist*, 136(1): 1–19.

Packer, C., *et al.* (1990b) "Reproductive Success of Lions." In *Reproductive Success*, ed. T.H. Clutton-Brock, pp. 363–383. Chicago, IL: University of Chicago Press.

Packer, C., *et al.* (2001) "Egalitarianism in Female African Lions." *Science*, 293: 690–693.

Padian, K., and L.M. Chiappe (1998) "The Origin of Birds and their Flight." *Scientific American*, 278(2): 38–47.

Page, R.E., and S.D. Mitchell (1998) "Self-Organization and the Evolution of Division of Labor." *Apidologie*, 29: 171–190.

Palameta, B., and L.K. Lefebvre (1985) "The Social Transmission of a Food-finding Technique in Pigeons: What is Learned?" *Animal Behaviour*, 33: 892–896.

Pankiw, P. (1967) "Studies of Honey Bees on Alfalfa Flowers." *Journal of Apicultural Research*, 6: 105–112.

Papaseit, C., *et al.* (2000) "Microtubule Self-organization is Gravity Dependent." *Proceedings of the National Academy of Sciences*, 97: 8364–8368.

Parish, A.R. (1996) "Female Relationships in Bonobos (*Pan paniscus*)." *Human Nature*, 7(1): 61–96.

Parker, P.G., *et al.* (1994) "Do Common Ravens Share Ephemeral Food Resources with Kin? DNA Fingerprinting Evidence." *Animal Behaviour*, 48: 1085–1093.

Partridge, B.L. (1982) "The Structure and Function of Fish Schools." *Scientific American*, 246(6): 114–123.

Pattee, H.H. (ed.) (1973) *Hierarchy Theory*. New York: George Braziller.

Pauly, D., *et al.* (1998) "Fishing Down Marine Food Webs." *Science*, 279: 860–863.

Peace, W.J.H., and P.J. Grubb (1982) "Interaction of Light and Mineral Nutrient Supply in the Growth of *Impatiens parviflora*." *New Phytologist*, 90: 127–150.

Peacock, J.A. (1999) *Cosmological Physics*. New York: Cambridge University Press.

Pendick, D. (1997) "Living Proof." *Earth*, 6(2): 24.

Penrose, R. (1989) *The Emperor's New Mind: Concerning Computers, Minds, and the Laws of Physics*. New York: Oxford University Press.

Penrose, R. (1994) *Shadows of the Mind: A Search for the Missing Science of Consciousness*. New York: Oxford University Press.

Perkins, S. (2000) "Fossil Birds Sport a New Kind of Feather." *Science News*, 158: 374.

Perkins, S. (2001) "Did Fibers and Filaments Become Feathers?" *Science News*, 159: 262.

Perkowitz, S. (1999) "The Rarest Element." *The Sciences*, 39(1): 34–37.

Perrow, C. (1984) *Normal Accidents: Living with High-Risk Technologies*. New York: Basic Books.

Peterson, I. (2000) "Calculating Swarms." *Science News*, 158: 314–316.

Pfeiffer, J.E. (1977) *The Emergence of Society: A Pre-history of the Establishment*. New York: McGraw-Hill.

Piaget, J. (1978) *Behavior and Evolution*. New York: Random House.

Pigliucci, M. (2001) *Phenotypic Plasticity: Beyond Nature and Nurture*. Baltimore, MD: Johns Hopkins University Press.

Pika, S., and M. Tomasello (2001) "Separating the Wheat from the Chaff: A Novel Food Processing Technique in Captive Gorillas (*Gorilla g. gorilla*)." *Primates*, 42(2): 167–170.

Pimentel, D., and M. Pimentel (eds.) (1996) *Food, Energy, and Society*. Denver: University of Colorado Press.

Pimentel, D., *et al.* (1997) "Water Resources: Agriculture, the Environment and Society." *BioScience*, 47: 97–106.

Pimentel, D., *et al.* (1998) "Ecology of Increasing Disease." *BioScience*, 48(10): 817–826.

Pinker, S. (1994) *The Language Instinct*. New York: William Morrow.

Pinker, S. (2002) *The Blank State: The Modern Denial of Human Nature*. New York: Viking.

Pirenne, J. (1962) *The Tides of History*. New York: Dutton.

Pirozynski, K.A., and D.W. Malloch (1975) "The Origins of Land Plants: A Matter of Mycotrophism." *BioSystems*, 6: 153–164.

Pittendrigh, C.S. (1958) "Adaptation, Natural Selection and Behavior." In *Behavior and Evolution*, eds. A. Roe and G.G. Simpson, pp. 390–416. New Haven, CT: Yale University Press.

Plato. *The Republic*, 1946 edn., trans. B. Jowett. Cleveland, OH: World Publishing Co.

Plotkin, H.C. (ed.) (1982) *Learning, Development, and Culture: Essays in Evolutionary Epistemology*. Chichester, U.K.: John Wiley & Sons.

Plotkin, H.C. (ed.) (1988) *The Role of Behavior in Evolution*. Cambridge, MA: MIT Press.

Plotkin, H.C. (1994) *Darwin Machines and the Nature of Knowledge*. Cambridge, MA: Harvard University Press.

Polanyi, M. (1968) "Life's Irreducible Structure." *Science*, 160: 1308–1312.

Polkinghorne, J.C. (1984) *One Word: The Interaction of Science and Theology*. London: Society for the Propagation of Christian Knowledge.

Polkinghorne, J.C. (1998) *Belief in God in an Age of Science*. New Haven, CT: Yale University Press.

Popper, K. (1957) *The Poverty of Historicism*. New York: Harper & Row.

Postel, S. (2000) "Troubled Waters." *The Sciences*, 40(2): 19–24.

Potts, R. (1984) "Home Bases and Early Hominids." *American Scientist*, 72(4): 338–347.

Potts, R. (1988) *Early Hominid Activities at Olduvai*. Chicago, IL: Aldine Press.

Potts, R. (1998a) "Environmental Hypothesis of Hominid Evolution." *Yearbook of Physical Anthropology*, 41: 93–136.

Potts, R. (1998b) "Variability Selection in Hominid Evolution." *Evolutionary Anthropology*, 7(3): 81–96.

Potts, R., and P. Shipman. (1981) "Cutmarks Made by Stone Tools on Bones from Olduvai Gorge, Tanzania." *Nature*, 291: 557–580.

Poulin, R., and A.S. Grutter (1996) "Cleaning Symbiosis: Proximate and Adaptive Explanations." *BioScience*, 46: 512–517.

Powers, W.T. (1973) *Behavior: The Control of Perception*. Chicago, IL: Aldine Press.

Price, P.W. (1991) "The Web of Life: Development Over 3.8 Billion Years of Trophic Relationship." In *Symbiosis as a Source of Evolutionary Innovation*, eds. L. Margulis and R. Fester, pp. 262–272. Cambridge, MA: MIT Press.

Price, P.W., *et al.* (1986) "Parasite Mediation in Ecological Interactions." *Annual Review of Ecology and Systematics*, 17: 487–505.

Price, P.W., *et al.* (1988) "Parasite-Mediated Competition: Some Predictions and Tests." *American Naturalist*, 131: 544–555.

Price, M.E. *et al.* (2002) "Punitive Sentiment as an Anti-Free Rider Psychological Device." *Evolution and Human Behavior*, 23: 203–231.

Prigogine, I., *et al.* (1972a) "Thermodynamics of Evolution, I." *Physics Today*, 25: 23–28.

Prigogine, I., *et al.* (1972b) "Thermodynamics of Evolution, II." *Physics Today*, 25: 38–44.

Prigogine, I., *et al.* (1977) "The Evolution of Complexity and the Laws of Nature." In *Goals in a Global Society*, eds. E. Laszlo and J. Bierman, pp. 1–63. New York: Pergamon.

Qiang, J., *et al.* (1998) "Two Feathered Dinosaurs from Northeastern China." *Nature*, 393: 753–761.

Queller, D.C., *et al.* (1988) "Genetic Relatedness in Colonies of Tropical Wasps with Multiple Queens." *Science*, 242: 1155–1157.

Queller, D.C., *et al.* (1990) "Wasps Fail to Make Distinctions." *Nature*, 344: 388.

Quiatt, D., and J. Kelso (1985) "Household Economics and Hominid Origins." *Current Anthropology*, 26(2): 207–222.

Rabin, M. (1993) "Incorporating Fairness into Game Theory and Economics." *American Economic Review*, 83: 1281–1302.

Raven, J.A. (1992) "Energy and Nutrient Acquisition by Autotrophic Symbiosis and Their Asymbiotic Ancestors." *Symbiosis,* 14: 33–60.

Raver, A. (1996) "Silent Spring after a Devastating Winter for Bees." *The New York Times,* June 16, 1996, p. 17.

Ravve, A. (2000) *Principles of Polymer Chemistry,* 2nd edn. New York: Kluwer.

Read, A.F., and P.H. Harvey (1993) "Evolving in a Dynamic World." *Science,* 260: 1760–1762.

Redfield, R. (1942) *Levels of Integration in Biological and Social Systems.* Lancaster, PA: J. Cattell Press.

Rees, M.J. (2000) *New Perspectives in Astrophysical Cosmology,* 2nd edn. New York: Cambridge University Press.

Regan, G. (1993) *The Guinness Book of Naval Blunders.* London: Guinness Publishing.

Reinhardt, J.F. (1952) "Responses of Honey Bees to Alfalfa Flowers." *American Naturalist,* 86: 257–275.

Relethford, J.H. (1995) "Genetics and Modern Human Origins." *Evolutionary Anthropology,* 4: 53–63.

Rensch, B. (1947) *Biophilosophy,* 1971 edn., trans. C.A.M. Sym. New York: Columbia University Press.

Revkin, A.C. (2000) "Warming's Effects to be Widespread." *The New York Times,* June 12, 2000, pp. A1, A27.

Reynolds, V., *et al.* (1987) *The Sociobiology of Ethnocentrism: Evolutionary Dimensions of Xenophobia, Discrimination, Racism and Nationalism.* London: Croom Helm.

Richardson, R.C. (2001) "Complexity, Self-Organization and Selection." *Biology and Philosophy,* 16: 655–683.

Richerson, P.J., and R. Boyd (1992) "Cultural Inheritance and Evolutionary Ecology." In *Evolutionary Ecology and Human Behavior,* eds. E.A. Smith and B. Winterhalder, pp. 61–94. New York: Aldine de Gruyter.

Richerson, P.J., and R. Boyd (1999) "Complex Societies: The Evolutionary Origins of a Crude Superorganism." *Human Nature,* 10: 253–289.

Richmond, B.G., and D.S. Strait (2000) "Evidence that Humans Evolved from a Knuckle-Walking Ancestor." *Nature,* 404: 382–385.

Richmond, B.G., and D.S. Strait (2001) "Knuckle-Walking Hominid Ancestors: A Reply to Corruccini and McHenry." *Journal of Human Evolution,* 40: 513–520.

Ricketts, E.F., *et al.* (1985) *Between Pacific Tides.* Stanford, CA: Stanford University Press.

Ricklefs, R.E. (1996) *The Economy of Nature.* New York: W.H. Freeman.

Ridley, M. (1997) *The Origins of Virtue: Human Instincts and the Evolution of Cooperation.* New York: Viking.

Ridley, M. (2001) *The Cooperative Gene: How Mendel's Demon Explains the Evolution of Complex Beings.* New York: Free Press.

Ridley, M. (2003) *Nature versus Nurture: The Origin of the Individual.* New York: HarperCollins.

Rilling, J.K., and T.R. Insel (1999) "The Primate Neocortex in Comparative Perspective using Magnetic Resonance Imaging." *Journal of Human Evolution*, 37:
191–223.

Rinehart, J., *et al.* (1997) *Just Another Car Factory? Lean Production and Its Discon-
tents.* Ithaca, NY: ILR Press.

Rissing, S.W., and G.B. Pollack (1991) "An Experimental Analysis of Pleometrotic
Advantage in the Desert Seed-harvester Ant *Messor pergandei* (Hymenoptera:
Formicidae)." *Insect Society,* 38: 205–211.

Rissing, S.W., *et al.* (1989) "Foraging Specialization without Relatedness or Dominance among Co-founding Ant Queens." *Nature,* 338: 420–422.

Robinson, B.W., and R. Dukas (1999) "The Influence of Phenotypic Modifications
on Evolution: The Baldwin Effect and Modern Perspectives." *Oikos,* 85(3): 582–
589.

Rock, I. (1984) *Perception.* New York: W.H. Freeman.

Rock, I., and S. Palmer (1990) "The Legacy of Gestalt Psychology." *Scientific
American,* 263(6): 84–90.

Rodman, P.S., and H.M. McHenry (1980) "Bioenergetics and the Origins of
Bipedalism." *American Journal of Physical Anthropology,* 52: 103–106.

Roe, A., and G.G. Simpson (eds.) (1958) *Behavior and Evolution.* New Haven, CT:
Yale University Press.

Roes, F. (1998) "A Conversation with George C. Williams." *Natural History,* 107(4):
10–15.

Rogers, L.J. (1998) *Minds of Their Own: Thinking and Awareness in Animals.*
St. Leonards, Australia: Allen & Unwin.

Roitt, I.M. (1988) *Essential Immunology,* 6th edn. Oxford, U.K.: Blackwell Scientific
Publications.

Romanes, G.J. (1883) *Animal Intelligence.* New York: Appleton.

Rose, L., and F. Marshall (1996) "Meat Eating, Hominid Sociality, and Home Bases
Revisited." *Current Anthropology,* 37(2): 307–338.

Rothschild, M. (1990) *Bionomics: Economy as Ecosystem.* New York: Henry Holt.

Roulin, A. (2002) "Why Do Lactating Females Nurse Alien Offspring? A Review of Hypotheses and Empirical Evidence." *Animal Behaviour,* 63: 201–
208.

Rouvray, D.H. (1986) "Predicting Chemistry from Topology." *Scientific American,*
255(3): 40–47.

Runciman, W.G., *et al.* (eds.) (1996) *Evolution of Social Behaviour Patterns in Primates
and Man.* Oxford, U.K.: Oxford University Press.

Ruse, M. (1988) *Philosophy of Biology Today.* Albany: State University of New York
Press.

Ruse, M. (1999) *The Darwinian Revolution: Science Red in Tooth and Claw.* Chicago,
IL: University of Chicago Press.

Ruse, M. (2000) *The Evolution Wars: A Guide to the Debates.* Santa Barbara, CA:
ABC-Clio.

Ryan, P.B. (1985) *The Iranian Rescue Mission.* Annapolis, MD: Naval Institute
Press.

Sabine, G.H. (1961) *A History of Political Theory*, 3rd edn. New York: Holt, Rinehart & Winston.

Saffre, F., *et al.* (1999) "Collective Decision-making in Social Spiders: Dragline-mediated Amplification Process Acts as a Recruitment Mechanism." *Journal of Theoretical Biology*, 198: 507–517.

Sagan, C. (1997) *Billions and Billions: Thoughts on Life and Death at the Brink of the Millennium*. New York: Random House.

Salisbury, R.F. (1973) "Economic Anthropology." *Annual Review of Anthropology*, 2: 85–94.

Salthe, S.N. (1985) *Evolving Hierarchical Systems*. New York: Columbia University Press.

Savage, J.M. (1977) *Evolution*. New York: Holt, Rinehart & Winston.

Schaller, G.B. (1972) *The Serengeti Lion: A Study of Predator–Prey Relations*. Chicago, IL: University of Chicago Press.

Scheel, D., and C. Packer (1991) "Group Hunting Behaviour of Lions: A Search for Cooperation." *Animal Behaviour*, 41(4): 697–710.

Schick, K.D., and N. Toth (1993) *Making Silent Stones Speak*. New York: Simon & Schuster.

Schlesinger, A.M. (1939) "The Tides of American Politics." *Yale Review*, 29: 217–230.

Schmidt-Neilsen, K.S. (1972) *How Animals Work*. Cambridge, U.K.: Cambridge University Press.

Schmitz, O.J. (1997) "Commemorating 30 Years of Optimal Foraging Theory." *Evolutionary Ecology*, 11: 631–632.

Schneider, S.H., and P.E. Boston (eds.) (1991) *Scientists on Gaia*. Cambridge, MA: MIT Press.

Schoener, T.W. (1971) "Theory of Feeding Strategies." *Annual Review of Ecology and Systematics*, 2: 369–404.

Schopf, J.W. (1993) "Microfossils of the Early Archean Apex Chert: New Evidence of the Antiquity of Life." *Science*, 260: 640–645.

Schrödinger, E. (1945) *What is Life? The Physical Aspect of the Living Cell*. New York: Macmillan.

Schumpeter, J. (1911) *The Theory of Economic Development*, 1934 edn. Cambridge, MA: Harvard University Press.

Schwartz, M.W., and J.D. Hoeksema (1998) "Specialization and Resource Trade: Biological Markets as a Model of Mutualisms." *Ecology*, 79(3): 1029–1038.

Scott, J.P. (1958) *Animal Behavior*. Chicago, IL: University of Chicago Press.

Scott, M.P. (1994) "Competition with Flies Promotes Communal Breeding in the Burying Beetle (*Necrophorus tomentosus*)." *Behavioral Ecology and Sociobiology*, 34: 367–373.

Scriver, C.R., *et al.* (eds.) (2001) *The Metabolic and Molecular Bases of Inherited Disease*, 8th edn. New York: McGraw-Hill.

Searle, J.R. (1992) *The Rediscovery of the Mind*. Cambridge, MA: MIT Press.

Searle, J.R., *et al.* (1997) *The Mystery of Consciousness*. London: Granta Books.

Seeley, T.D. (1995) *The Wisdom of the Hive: The Social Physiology of Honey Bee Colonies*. Cambridge, MA: Harvard University Press.

Seeley, T.D., and S.C. Buhrman (1999) "Group Decision Making in Swarms of Honey Bees." *Behavioral Ecology and Sociobiology,* 45(1): 19–31.

Seeley, T.D., and R.A. Levien (1987) "A Colony of Mind: The Beehive as Thinking Machine." *The Sciences,* 27(4): 38–43.

Sekuler, R., and R. Blake (1990) *Perception,* 2nd edn. New York: McGraw-Hill.

Selosse, M.-A., and F. Le Tacon (1998) "The Land Flora: A Phototroph–Fungus Partnership?" *Trends in Ecology and Evolution,* 13(1): 15–20.

Semino, O., *et al.* (2000) "The Genetic Legacy of *Homo sapiens sapiens* in Extant Europeans: A Y-Chromosome Perspective." *Science,* 290: 1155.

Service, E.R. (1971) *Cultural Evolutionism: Theory in Practice.* New York: Holt, Rinehart & Winston.

Service, E.R. (1975) *Origins of the State and Civilization: The Process of Cultural Evolution.* New York: W.W. Norton.

Sethi, R. and E. Somanathan (2001) "Preference Evolution and Reciprocity." *Journal of Economic Theory,* 97: 273–297.

Shapiro, J.A. (1988) "Bacteria as Multicellular Organisms." *Scientific American,* 258(6): 82–89.

Shapiro, J.A. (1991) "Genomes as Smart Systems." *Genetica,* 84: 3–4.

Shapiro, J.A. (1992) "Natural Genetic Engineering in Evolution." *Genetica,* 86: 99–111.

Shapiro, J.A., and M. Dworkin (eds.) (1997) *Bacteria as Multicellular Organisms.* New York: Oxford University Press.

Shaw, P., and Y. Wong (1989) *Genetic Seeds of Warfare: Evolution, Nationalism and Patriotism.* London: Unwin Hyman.

Shepherd, G.M. (ed.) (1990) *The Synaptic Organization of the Brain,* 3rd edn. New York: Oxford University Press.

Sherman, P.W., *et al.* (1988) "Parasites, Pathogens, and Polyandry in Social Hymenoptera." *American Naturalist,* 131(4): 602–610.

Sherman, P.W., *et al.* (1992) "Naked Mole Rats." *Scientific American,* 267(2): 72–78.

Sherman, P.W., *et al.* (eds.) (1991) *The Biology of the Naked Mole-Rat.* Princeton, NJ: Princeton University Press.

Sherrington, C. (1940) *Man on his Nature.* Cambridge, U.K.: Cambridge University Press.

Shields, W.M., and J.R. Crook (1987) Barn Swallow Coloniality: A Net Cost for Group Breeding in the Adirondacks? *Ecology,* 68(5): 1373–1386.

Shingleton, A.W., and W.A. Foster (2000) "Behaviour, Morphology and the Division of Labour in Two Soldier-Producing Aphids." *Animal Behaviour,* 62: 671–679.

Shipman, P. (1983) "Early Hominid Lifestyle: Hunting and Gathering or Foraging and Scavenging." In *Animals and Archaeology,* eds. J. Clutton-Brock and C. Grigson, pp. 31–49. Oxford, U.K.: B.A.R.

Shipman, P. (1986) "Studies of Hominid–Faunal Interactions at Olduvai Gorge." *Journal of Human Evolution,* 15(8): 691–706.

Shipman, P., and A. Walker (1989) "The Costs of Becoming a Predator." *Journal of Human Evolution,* 18: 373–392.

Sick, G. (1985) *America's Tragic Encounter with Iran.* New York: Random House.

Sigmund, K. (1993) *Games of Life: Explorations in Ecology, Evolution, and Behaviour.* Oxford, U.K.: Oxford University Press.

Silk, J. (1994) *A Short History of the Universe.* New York: W.H. Freeman.

Silverstein, A., *et al.* (1994) *Diabetes.* Hillside, NJ: Enslow Publishers.

Simpson, G.G. (1953) "The Baldwin Effect." *Evolution,* 2: 110–117.

Simpson, G.G. (1967) *The Meaning of Evolution,* rev. edn. New Haven, CT: Yale University Press.

Singer, W. (1999) "Striving for Coherence." *Nature,* 397: 391.

Sirower, M.L. (1998) *The Synergy Trap: How Companies Lose the Acquisition Game.* New York: Free Press.

Skinner, B.F. (1974) *About Behaviorism.* New York: Alfred A. Knopf.

Sleigh, M.A. (1989) *Protozoa and Other Protists.* London: Edward Arnold.

Slurink, P. (1993) "Ecological Dominance and the Final Sprint in Hominid Evolution." *Human Evolution,* 8: 265–273.

Small, M. (2000) "Aping Culture." *Discover,* 21(5): 53–57.

Smillie, D. (1993) "Darwin's Tangled Bank: The Role of Social Environments." In *Perspectives in Ethology,* vol. 10, *Behavior and Evolution,* ed. P.P.G. Bateson *et al.,* pp. 119–141. New York: Plenum Press.

Smillie, D. (1995) "Darwin's Two Paradigms: An Opportunistic Approach to Natural Selection Theory." *Journal of Social and Evolutionary Systems,* 18: 231–255.

Smith, A. (1776) *The Wealth of Nations,* 1964 edn. 2 vols. London: J.M. Dent.

Smith, D.C., and A.E. Douglas (1987) *The Biology of Symbiosis.* Baltimore, MD: Edward Arnold.

Smith, E.A. (1985) "Inuit Foraging Groups: Some Simple Models Incorporating Conflicts of Interest, Relatedness and Central-Place Sharing." *Ethology and Sociobiology,* 6: 27–47.

Smith, M.R., and L. Marx (eds.) (1994) *Does Technology Drive History? The Dilemma of Technological Determinism.* Cambridge, MA: MIT Press.

Smith, N.G. (1968) "The Advantage of Being Parasitized." *Nature,* 219: 690–694.

Smolin, L. (1997) *The Life of the Cosmos.* New York: Oxford University Press.

Smuts, B., *et al.* (eds.) (1987) *Primate Societies.* Chicago, IL: University of Chicago Press.

Smuts, J.C. (1926) *Holism and Evolution.* New York: Macmillan.

Snowdon, C.T. (2001) "From Primate Communication to Human Language." In *Tree of Origin: What Primate Behavior Can Tell Us about Human Social Evolution,* ed. F.B.M. de Waal, pp. 193–228. Cambridge, MA: Harvard University Press.

Sober, E. (ed.) (1984) *Conceptual Issues in Evolutionary Biology.* Cambridge, MA: MIT Press.

Sober, E., and D.S. Wilson (1998) *Unto Others: The Evolution and Psychology of Unselfish Behavior.* Cambridge, MA: Harvard University Press.

Sonea, S. (1991) "Bacterial Evolution Without Speciation." In *Symbiosis as a Source of Evolutionary Innovation,* eds. L. Margulis and L. Fester, pp. 95–105. Cambridge, MA: MIT Press.

Sorokin, P. (1937) *Social and Cultural Dynamics,* 4 vols. New York: American Book Co.

Sosis, R. (2000) "The Emergence and Stability of Cooperative Fishing on Ifaluk Atoll." In *Human Behavior and Adaptation. An Anthropological Perspective*, eds. L. Cronk, N. Chagnon, and B. Irons, pp. 437–472. New York: Aldine de Gruyter.

Spencer, H. (1852a) "The Development Hypothesis." In *Essays: Scientific, Political and Speculative*. 1892 edn: New York: Appleton.

Spencer, H. (1852b) "A Theory of Population Deduced from the General Law of Animal Fertility." *Westminster Review*, 57: 468–501.

Spencer, H. (1862) *First Principles*. London: Watts.

Spencer, R. (1959) *The North Alaskan Eskimo: A Study in Ecology and Society*. Washington, D.C.: U.S. Government Printing Office.

Sperry, R.W. (1969) "A Modified Concept of Consciousness." *Psychological Review*, 76: 532–536.

Sperry, R.W. (1991) "In Defense of Mentalism and Emergent Interaction." *Journal of Mind and Behavior*, 12(2): 221–246.

Speth, J.D. (1987) "Early Hominid Subsistence Strategies in Seasonal Habitats." *Journal of Archaeological Science*, 14: 13–29.

Speth, J.D. (1989) "Early Hominid Hunting and Scavenging: The Role of Meat as an Energy Source." *Journal of Human Evolution*, 18(4): 329–343.

Stamps, J.A. (1991) "Why Evolutionary Issues Are Reviving Interest in Proximate Behavioral Mechanisms." *American Zoologist*, 31: 338–348.

Stander, P.E. (1992) "Cooperative Hunting in Lions: The Role of the Individual." *Behavioral Ecology and Sociobiology*, 29: 445–454.

Stanford, C.B. (1992) "Costs and Benefits of Allomothering in Wild Capped Langurs *(Presbytis pileata)*." *Behavioral Ecology and Sociobiology*, 30: 29–34.

Stanford, C.B (1999) *The Hunting Apes: Meat Eating and the Origins of Human Behavior*. Princeton, NJ: Princeton University Press.

Stanford, C.B (2000) "The Cultured Ape?" *The Sciences*, 40(3): 38–43.

Stanley, S.M. (1992) "Ecological Theory for the Origin of *Homo*." *Paleobiology*, 18: 237–257.

Stanley, S.M. (2000) "The Past Climate Change Heats Up." *Proceedings of the National Academy of Sciences*, 97(4): 1319.

Starr, C.G. (1991) "*A History of the Ancient World*. Oxford, U.K.: Oxford University Press.

Stearns, S.C. (1989) "The Evolutionary Significance of Phenotypic Plasticity: Phenotypic Sources of Variation Can Be Described by Devlopmental Switches and Reaction Norms." *BioScience*, 39(7): 436.

Steele, J. (1999) " Stone Legacy of Skilled Hands." *Nature*, 399: 24–25.

Steudel, K.L. (1994) "Locomotor Energetics and Hominid Evolution." *Evolutionary Anthropology*, 3: 42–48.

Steudel, K.L. (1996) "Limb Morphology, Bipedal Gait, and the Energetics of Hominid Locomotion." *Amercian Journal of Physical Anthropology*, 99: 345–355.

Stevens, W.K. (1999a) "Scientists Warn Against Ignoring Climate Change." *The New York Times*, January 30, 1999, pp. F1, F9.

Stevens, W.K. (1999b) "Surveys Uncover Substantial Melting of Greenland Ice Sheet." *The New York Times*, March 5, 1999, p. A13.

Stevens, W.K. (1999c) "Human Imprint on Climate Change Grows Clearer." *The New York Times*, June 29, 1999, pp. F1, F9.

Stevens, W.K. (2000a) "New Survey Shows Growing Loss of Arctic Atmosphere's Ozone." *The New York Times*, April 6, 2000, p. A79.

Stevens, W.K. (2000b) "Megadrought Appears to Loom in Africa." *The New York Times*, February 8, 2000, p. F3.

Steward, J.H. (1938) *Basin-Plateau Aboriginal Sociopolitical Groups*. Washington, D.C.: U.S. Government Printing Office.

Steward, J.H. (1941) *Nevada Shoshoni*. Berkeley: University of California Press.

Steward, J.H. (1943) *Northern and Gosiute Shoshoni*. Berkeley: University of California Press.

Steward, J.H. (1955) *Theory of Cultural Change: The Methodology of Multilinear Evolution*, 1963 edn. Urbana: University of Illinois Press.

Stiassny, M.L.J., and A. Meyer (1999) "Cichlids of the Rift Lakes." *Scientific American*, 280(2): 64–69.

Stoinski, T.S., and B.B. Beck (2001) "Spontaneous Tool Use in Captive, Free-Ranging Golden Tamarins (*Leontopithecus rosalia rosalia*)." *Primates*, 42(4): 319–326.

Stoneking, M. (1993) "DNA and Recent Human Evolution." *Evolutionary Anthropology*, 2: 60–73.

Stoner, J.A.F. (1982) *Management*, 2nd edn. Englewood Cliffs, NJ: Prentice-Hall.

Strassmann, J.E., and D.C. Queller (1989) "Ecological Determinants of Social Evolution." In *The Genetics of Social Evolution*, eds. M.D. Breed and R.E. Page, pp. 81–101. Boulder, CO: Westview Press.

Strassman, J.E. (1993) "Weak Queen or Social Contract?" *Nature*, 363: 502–503.

Strassmann, J.E., *et al.* (1997) "Absence of Within-Colony Kin Discrimination in Behavioural Interactions of Swarm-Founding Wasps." *Proceedings of the Royal Society of London B*, 264: 1565–1570.

Strum, S.C. (1975a) "Primate Predation: Interim Report on the Development of a Tradition in a Troop of Olive Baboons." *Science*, 187: 755–757.

Strum, S.C. (1975b) "Life with the Pumphouse Gang." *National Geographic Magazine*, 147: 687–691.

Strum, S.C. (1987) *Almost Human: A Journey into the World of Baboons*. New York: Random House.

Su, B., *et al.* (1999) "Y-Chromosome Evidence for a Northward Migration of Modern Humans into Eastern Asia during the Last Ice Age." *American Journal of Human Genetics*, 65(6): 1718.

Sumner, W.G. (1934) *Essays of William Graham Sumner*. New Haven, CT: Yale University Press.

Swimme, B. (1996) *The Hidden Heart of the Cosmos: Humanity and the New Story*. Mary Knoll, NY: Orbis Books.

Swimme, B., and T. Berry (1992) *The Universe Story*. San Francisco, CA: Harper & Row.

Swinburne, R. (1990) "Argument from the Fine-Tuning of the Universe." In *Physical Cosmology and Philosophy*, ed. J. Leslie, pp. 154–173. New York: Macmillan.

Switzer, P.V., and D.A. Cristol (1999) "Avian Prey-dropping Behavior, I. The Effects of Prey Characteristics and Prey Loss." *Behavioral Ecology*, 10(3): 213–219.

Szathmáry, E., F. Jordán, and C. Pál (2001) "Can Genes Explain Biological Complexity?" *Science,* 292: 1315–1316.

Taagepera, R. (1968) "Growth Curves of Empires." *General Systems,* 13: 171–175.

Taagepera, R. (1978a) "Size and Duration of Empires: Systematics of Size." *Social Science Research,* 7: 108–127.

Taagepera, R. (1978b) "Size and Duration of Empires: Growth–Decline Curves, 3000 to 600 B.C." *Social Science Research,* 7: 180–196.

Taagepera, R. (1979) "Size and Duration of Empires: Growth–Decline Curves, 600 B.C. to 600 A.D." *Social Science History,* 3: 115–138.

Tainter, J. A. (1988) *The Collapse of Complex Societies.* Cambridge, U.K.: Cambridge University Press.

Tanner, N. (1987) "Gathering by Females: The Chimpanzee Model Revisited and the Gathering Hypothesis." In *The Evolution of Human Behavior: Primate Models,* ed. W. Kinzey, pp. 3–37. New York: Cambridge University Press.

Tattersall, I. (1998) *Becoming Human: Evolution and Human Uniqueness.* New York: Harcourt Brace.

Tattersall, I. (2000) "Once We Were Not Alone." *Scientific American,* 282(1): 56–62.

Tattersall, I. (2002) *The Monkey in the Mirror.* New York: Harcourt Brace.

Taylor, K. (1999) "Rapid Climate Change." *American Scientist,* 87: 320–327.

Teaford, M.F., and P.S. Ungar (2000) "Diet and the Evolution of the Earliest Human Ancestors." *Proceedings of the National Academy of Sciences,* 97(5): 13506–13511.

Tedeschi, B. (2000) "Construction Heads into the Internet Age." *The New York Times,* February 21, 2000, pp. C1, C9.

Teilhard de Chardin, P. (1959) *The Phenomenon of Man.* New York: Harper & Row.

Temple, D. (1994) "The New Design Argument." In *Science, Technology, and Religious Ideas,* eds. M.H. Shale and G.W. Shields, p. 135. Lanham, MD: University Press of America.

ten Cate, C. (2000) "How Learning Mechanisms Might Affect Evolutionary Processes." *Trends in Ecology and Evolution,* 15(5): 179–181.

Thaler, D.S. (1994) "The Evolution of Genetic Intelligence." *Science,* 264: 224–225.

Thieme, H. (1999) "Lower Paleolithic Hunting Spears from Germany." *Nature,* 285, 807–810.

Thomas, J.H. (1994) "The Mind of a Worm." *Science,* 264: 1698–1699.

Thompson, D.W. (1917) *On Growth and Form.* Cambridge, U.K.: Cambridge University Press.

Thompson, D.W. (1942) *On Growth and Form,* 2nd edn, 2 vols. Cambridge, U.K.: Cambridge University Press.

Thompson, E.A. (1948) *A History of Attila and the Huns.* Oxford, U.K.: Clarendon Press.

Thorndike, E.L. (1911) *Animal Intelligence: Experimental Studies,* 1965 edn. New York: Hafner.

Thorne-Miller, B., and J.G. Catena (1991) *The Living Ocean.* Washington, D.C.: Island Press.

Thorpe, W.H. (1956) *Learning and Instinct in Animals.* London: Methuen.

Thuan, T.X. (1995) *The Secret Melody.* New York: Oxford University Press.

Tiger, L., and R. Fox (1971) *The Imperial Animal.* New York: Holt, Rinehart & Winston.

Tinbergen, N. (1965) *Social Behaviour in Animals.* London: Methuen.

Tobias, P.V. (1971) *The Brain in Hominid Evolution.* New York: Columbia University Press.

Tobias, P.V. (ed.) (1985) *Hominid Evolution: Past, Present and Future*, Proceedings of the Taung Diamond Jubilee International Symposium. New York: Alan R. Liss, Inc.

Toigo, J.W. (2000) "Avoiding a Data Crunch." *Scientific American*, 282(5): 58–74.

Tooby, J., and I. DeVore (1987) "The Reconstruction of Hominid Behavioral Evolution Through Strategic Modelling." In *The Evolution of Human Behavior: Primate Models*, ed. G. Kinzey, pp. 183–237. Albany: State University of New York Press.

Topoff, H. (1999) "Slave-making Queens." *Scientific American*, 281(5): 84–90.

Toth, N. (1987a) "Behavioral Influences from Early Stone Artifact Assemblages: An Experimental Model." *Journal of Human Evolution*, 16(7/8): 763–787.

Toth, N. (1987b) "The First Technology." *Scientific American*, 255(4): 112–121.

Toth, N. *et al.* (1993) "Pan the Tool-maker: Investigations into the Stone Tool-making and Tool Using Capabilities of a Bonobo (*Pan Paniscus*)." *Journal of Archaeological Science*, 20: 81–91.

Travis, J. (1998) "The Hydrogen Hypothesis." *Science News*, 153(16): 253–255.

Trefil, J.S. (1983) *The Moment of Creation.* New York: Charles Scribner's Sons.

Trefil, J.S. (1997) "Was the Universe Designed for Life? Coming to Terms with the Anthropic Principle." *Astronomy*, June 1997(25): 54–57.

Treisman, A. (1986) "Features and Objects in Visual Processing." *Scientific American*, 255(5): 114–125.

Trinkaus, E. (1987) "Bodies, Brawn, Brains and Noses: Human Ancestors and Human Predation." In *The Evolution of Human Hunting*, eds. M.A. Nitecki and D.V. Nitecki, pp. 107–145. New York: Plenum Press.

Trivers, R.L. (1971) "The Evolution of Reciprocal Altruism." *Quarterly Review of Biology*, 46: 35–57.

Trivers, R.L. (1985) *Social Evolution.* Menlo Park, CA: Benjamin/Cummings.

Tschinkel, W.R. (1998) "An Experimental Study of Pleometrotic Colony Founding in the Fire Ant, *Solenopsis invicta*: What Is the Basis for Association?" *Behavioral Ecology and Sociobiology*, 43: 247–257.

Tuchman, B.W. (1985) *The March of Folly: From Troy to Vietnam.* New York: Ballantine.

Tullock, G. (1971) "The Coal Tit as a Careful Shopper." *American Naturalist*, 105: 77–80.

Tullock, G. (1979) "Sociobiology and Economics." *Atlantic Economic Journal*, September, 1–10.

Tullock, G. (1994) *The Economics of Non-Human Societies.* Tucson, AZ: Pallas Press.

Turke, P. (1984) "Effects of Ovulary Concealment and Synchrony on Proto-hominid Mating Systems and Parental Roles." *Ethology and Sociobiology*, 53: 33–44.

Turner, J.C., *et al.* (1987) *Rediscovering the Social Group: A Self-Categorization Theory.* Oxford, U.K.: Blackwell Scientific Publications.

Turner, J.S. (2000) *The Extended Organism: The Physiology of Animal-Built Structures.* Cambridge, MA: Harvard University Press.

Turner, M.S. (2000) "More than Meets the Eye." *The Sciences,* 40(6): 32–37.

Turney-High, H.H. (1949) *Primitive War: Its Practice and Concepts.* Columbia: University of South Carolina Press.

Tuttle, R.H. (2001) "On Culture and Traditional Chimpanzees." *Current Anthropology,* 42(3): 407–408.

Tylor, E.B. (1871) *Primitive Culture: Researches into the Development of Mythology, Philosophy, Religion, Art and Customs,* 2 vols. 1889 edn: New York: Henry Holt.

U.S. Department of Agriculture (1999) *Crop Production.* National Agricultural Statistics Service. Washington, D.C.: U.S. Government Printing Office.

U.S. Department of Agriculture (2000) *Non-Citrus Fruits and Nuts 1999 Summary.* National Agricultural Statistics Service, July 2000. Washington, D.C.: U.S. Government Printing Office.

U.S. Department of Commerce (2001) *The Statistical Abstract of the United States.* Washington, D.C.: U.S. Government Printing Office.

Umpleby, S. (2000a) "Why the Century Date Change Went So Smoothly." *16th Annual Washington Consortium of Business Schools,* Faculty Research Forum, Gallaudet University, April 22, 2000.

Umpleby, S. (2000b) "Y2K Revisited: Why the Sky Didn't Fall." *Centerpiece,* School of Business Public Management, George Washington University, Fall 2000, pp. 32–35.

Umpleby, S. (2000c) "Coping with an Error in the Knowledge Society: The Case of the Year 2000 Computer Crisis." *By George!,* September 5, 2000, pp. 2 and 11.

van den Ent, F., L.A. Amos, and J. Löwe (2001) "Prokaryotic Origin of the Actin Cytoskeleton." *Nature,* 413: 39–44.

van der Dennen, J.M.G. (1995) *The Origin of War: The Evolution of a Male-Coalitional Reproductive Strategy.* Groningen, The Netherlands: Origin Press.

van der Dennen, J.M.G. (1999) "Human Evolution and the Origin of War: A Darwinian Heritage." In *The Darwinian Heritage and Sociobiology,* eds. J.M.G. van der Dennen *et al.,* pp. 159–192. Westport, CT: Praeger.

van der Dennen, J.M.G., and V.S.E. Falger (eds.) (1990) *Sociobiology and Conflict: Evolutionary Perspectives on Competition, Cooperation, Violence and Warfare.* London: Chapman & Hall.

van Hooff, J.A.R.A.M. (2001) "Conflict, Reconciliation and Negotiation in Non-human Primates: The Value of Long–Term Relationships." In *Economics in Nature: Social Dilemmas, Mate Choices and Biological Markets,* eds. R. Noë *et al.,* pp. 67–90. Cambridge, U.K.: Cambridge University Press.

van Schaik, C.P. (2002) "Orangutan Tool Use and the Evolution of Technology and Language." Paper presented at the Plenary Session of the 16th Biennial Meeting, International Society of Human Ethology, Montreal, Canada, August 7–10, 2002.

van Schaik, C.P., and A. Paul (1996) "Male Care in Primates: Does It Ever Reflect Paternity?" *Evolutionary Anthropolgy,* 5: 152–156.

van Schaik, C.P., *et al.* (1999) "The Conditions for Tool Use in Primates: Implications for the Evolution of Material Culture." *Journal of Human Evolution*, 36: 719–741.

van Staaden, M.J. (1994) *"Suricata suricatta." Mammalian Species*, 483: 1–8.

Van Helden, A. (1985) *Measuring the Universe: Cosmic Dimensions from Aristarchus to Halley*. Chicago, IL: University of Chicago Press.

Vaneechoutte, M. (2000) "Report of the Symposium 'Water and Human Evolution'." *Human Evolution*, 15(3–4): 243–251.

Vaughan, C. W., *et al.* (1997) "How Opioids Inhibit GABA-mediated Neurotransmission." *Nature*, 390: 611–614.

Vernadsky, V.I. (1986) *The Biosphere*. Oracle, AZ: Synergetic Press.

Vetter, R.D. (1991) "Symbiosis and the Evolution of Novel Trophic Strategies: Thiotrophic Organisms at Hydrothermal Vents." In *Symbiosis as a Source of Evolutionary Innovation*, eds. L. Margulis and L. Fester, pp. 219–245. Cambridge, MA: MIT Press.

Vevers, G. (1971) *The Underwater World*. New York: St. Martin's Press.

Vogel, C. (2000) "Guggenheim in Pact with Hermitage." *The New York Times*, June 20, 2000, pp. B1, B6.

Vogel, G. (1998) "Did the First Complex Cell Eat Hydrogen?" *Science*, 270: 1633–1634.

Vogel, S. (1988) *Life's Devices: The Physical World of Animals and Plants*. Princeton, NJ: Princeton University Press.

Vogel, S. (1998) *Cats' Paws and Catapults: Mechanical Worlds of Nature and People*. New York: W.W. Norton.

Volk, T. (1997) *Gaia's Body: Toward a Physiology of Earth*. New York: Springer-Verlag.

von Frisch, K. (1967) *The Dance Language and Orientation of Bees*, trans. L. Chadwick. Cambridge, MA: Harvard University Press.

von Wagner, H.O. (1954) "Massenansammlungen von Weberknechten in Mexiko." *Zeitschrift für Tierpsychologie*, 11: 349–352.

Vrba, E.S., *et al.* (eds.) (1995) *Paleoclimate and Evolution, with Emphasis on Human Origins*. New Haven, CT: Yale University Press.

Waddington, C.H. (1942) "Canalization of Development and the Inheritance of Acquired Characters." *Nature*, 150: 563–565.

Waddington, C.H. (1952) "Selection of the Genetic Basis for an Acquired Character." *Nature*, 169: 278.

Waddington, C.H. (1957) *The Strategy of the Genes: A Discussion of Some Aspects of Theoretical Biology*. New York: Macmillan.

Waddington, C.H. (1961) *The Nature of Life*. New York: Harper & Row.

Waddington, C.H. (1975) *The Evolution of An Evolutionist*. Ithaca, NY: Cornell University Press.

Wade, N. (1999) "For Leaf-Cutter Ants, Farm Life Isn't So Simple." *The New York Times*, August 3, 1999, pp. D1, D4.

Wald, G. (1994) "The Cosmology of Life and Mind." In *New Metaphysical Foundations of Modern Science*, eds. W. Harman with J. Clark, pp. 123–131. Sausalito, CA: Institute of Noetic Sciences.

Wall, J.F. (ed.) (1992) *The Andrew Carnegie Reader*. Pittsburgh, PA: University of Pittsburgh Press.

Weinberg, S. (2002) "Is the Universe a Computer?" *The New York Review of Books*, October 24, 2002, pp. 79–89.

Wallin, I.E. (1927) *Symbionticism and the Origin of Species*. Baltimore, MD: Williams & Wilkins.

Ward, P., and Rockman (2002) *Future Evolution: An Illuminated History of Life to Come*. New York: Henry Holt.

Ward, P., and A. Zahavi (1973) "The Importance of Certain Assemblages of Birds as 'Information Centers' for Food-finding." *Ibis*, 115: 517–534.

Washburn, S.L., and C.S. Lancaster (1968) "The Evolution of Hunting." In *Man the Hunter*, eds. R.B. Lee and I. DeVore, pp. 293–303. Chicago, IL: Aldine Press.

Waterman, T.H. (1968) "Systems Theory and Biology – View of a Biologist." In *Systems Theory and Biology: Proceedings of the 3rd Systems Symposium at Case Institute of Technology*, ed. M.D. Mesarović, pp. 1–37. New York: Springer-Verlag.

Watson, A.J., and J.E. Lovelock (1983) "Biological Homeostasis of the Global Environment: The Parable of Daisyworld." *Tellus*, 35B: 284–289.

Wcislo, W.T. (1989) "Behavioral Environments and Evolutionary Change." *Annual Review of Ecology and Systematics*, 20: 137–169.

Webster, O.W. (1991) "Living Polymerization Methods." *Science*, 251: 887.

Weigl, P.D., and E.V. Hanson (1980) "Observational Learning and the Feeding Behavior of the Red Squirrel *Tamiasciurus hudsonicus*: The Ontogeny of Optimization." *Ecology*, 61(2): 213–218.

Weinberg, S. (1987) "Newtonianism, Reductionism and the Art of Congressional Testimony." *Nature*, 330: 433–437.

Weinberg, S. (1988) *The First Three Minutes: A Modern View of the Origin of the Universe*. New York: Basic Books.

Weinberg, S. (1992) *Dreams of a Final Theory*. New York: Pantheon Books.

Weinberg, S. (2002) "Is the Universe a Computer?" *The New York Review of Books*, October 24, 2002, pp. 79–89.

Weiner, J. (1994) *The Beak of the Finch*. New York: Vintage Books.

Weismann, A. (1904) *The Evolution Theory*, trans. J.A. Thomson and M.R. Thomson. London: Edward Arnold.

Weiss, H. (1996) "Desert Storm: Weather Brought Destruction to the First Ancient Civilization." *Sciences*, 36(3): 30–37.

Weiss, H. (2000) "Beyond the Younger Dryas: Collapse as Adaptation to Abrupt Climate Change in Ancient West Asia and the Eastern Mediterranean." In *Environmental Disasters and the Archaeology of Human Response*, eds. G. Bawden and R. Reycraft, pp. 75–98. Albuquerque: University of New Mexico Press.

Weiss, H. and R. S. Bradley (2001) "What Drives Societal Collapse?" *Science*, 291: 609–610.

Weiss, H., *et al.* (1993) "The Genesis and Collapse of Third Millennium North Mesopotamian Civilization." *Science*, 261: 995–1004.

Weiss, P.A. (ed.) (1971) *Hierarchically Organized Systems in Theory and Practice*. New York: Hafner.

West, S.A., *et al.* (2002) "Cooperation and Competition between Relatives." *Science,* 296: 72–75.

West-Eberhard, M.J. (1989) "Phenotypic Plasticity and the Origins of Diversity." *Annual Review of Ecology and Systematics,* 20: 249–278.

West-Eberhard, M.J. (1992) "Adaptation: Current Usages." In *Keywords in Evolutionary Biology,* eds. E.F. Keller and E.A. Lloyd. Cambridge, MA: Harvard University Press.

Wheeler, P.E. (1985) "The Evolution of Bipedalism and the Loss of Functional Body Hair in Hominids." *Journal of Human Evoltion,* 14: 23–28.

Wheeler, P.E. (1991) "The Influence of Bipedalism on the Energy and Water Budgets of Early Hominids." *Journal of Human Evolution,* 21: 117–136.

Wheeler, W.M. (1927a) *Emergent Evolution and the Social.* London: Kegan Paul, Trench, Trubner.

Wheeler, W.M. (1927b) "Emergent Evolution of the Social." In *Proceedings of the 6th International Congress of Philosophy,* ed. E.S. Brightman, pp. 33–46. London: Longmans, Green & Co.

White, L.A. (1949) *The Science of Culture: A Study of Man and Civilization.* New York: Grove Press.

White, L.A. (1959) *The Evolution of Culture.* New York: McGraw-Hill.

White, L.A. (1975) *The Concept of Cultural Systems: A Key to Understanding Tribes and Nations.* New York: Columbia University Press.

White, T.D. (1995) "African Omnivores: Global Climatic Change and Plio-Pleistocene Hominids and Suids." In *Paleo Climate and Evolution: With Emphasis on Human Origins,* eds. E.S. Vrba *et al.,* pp. 369–384. New Haven, CT: Yale University Press.

Whiten, A., and C. Boesch (2001) "The Cultures of Chimpanzees." *Scientific American,* 284(1): 61–67.

Whiten, A., and R.W. Byrne (eds.) (1997) *Machiavellian Intelligence,* vol. 2, *Extensions and Evaluations.* Cambridge, U.K.: Cambridge University Press.

Whiten, A., *et al.* (1999) "Cultures in Chimpanzees." *Nature,* 399: 682–685.

Whitesides, G.M., and B. Grzybowski (2002) "Self-Assembly at All Scales." *Science,* 295: 2418–2421.

Whyte, L.L. (1965) *Internal Factors in Evolution.* New York: Braziller.

Wilford, J.N. (2001) "Fossil Found in China Is Seen as Linking Birds to Dinosaurs." *The New York Times,* April 26, 2001, pp. A1, A9.

Wilford, J.N. (2002) "When Humans Became Human." *The New York Times,* February 26, 2002, pp. F1, F5.

Wilkinson, D.M. (1999) "Is Gaia Really Conventional Ecology?" *Oikos,* 84: 533–536.

Wilkinson, G.S. (1984) "Reciprocal Food Sharing in the Vampire Bat." *Nature,* 308: 181–184.

Wilkinson, G.S. (1988) "Reciprocal Altruism in Bats and Other Mammals." *Ethology and Sociobiology,* 9: 85–100.

Wilkinson, G.S. (1990) "Food Sharing in Vampire Bats." *Scientific American,* 262(2): 76–82.

Wilkinson, G.S. (1992) "Communal Nursing in the Evening Bat, *Nycticeius humeralis.*" *Behavioral Ecology and Sociobiology,* 31: 225–235.

Williams, G.C. (1966) *Adaptation and Natural Selection: A Critique of Some Current Evolutionary Thought*. Princeton, NJ: Princeton University Press.

Williams, G.C. (1993) "Mother Nature Is a Wicked Old Witch." In *Evolutionary Ethics*, eds. M.H. Nitecki and D.V. Nitecki, pp. 217–223. Albany: State University of New York Press.

Williams, G.R. (1996) *The Molecular Biology of Gaia*. New York: Columbia University Press.

Williams, J.T. (1997) "The Painless Synergism of Aspirin and Opium." *Nature,* 390: 557–559.

Wills, C. (1993) *The Runaway Brain: The Evolution of Human Uniqueness*. New York: Basic Books.

Wills, C. (1998) *Children of Prometheus: The Accelerating Pace of Human Evolution*. Reading, MA: Perseus Books.

Wilson, D.S. (1975) "A General Theory of Group Selection." *Proceedings of the National Academy of Sciences,* 72: 143–146.

Wilson, D.S. (1980) *The Natural Selection of Populations and Communities*. Menlo Park, CA: Benjamin/Cummings.

Wilson, D.S. (2001) "Evolutionary Biology: Struggling to Escape Exclusively Individual Selection." *Quarterly Review of Biology,* 76(2): 199–205.

Wilson, D.S. (2002) *Darwin's Cathedral: Evolution, Religion, and the Nature of Society*. Chicago, IL: University of Chicago Press.

Wilson, D.S., and L.A. Dugatkin (1997) "Group Selection and Assortative Interactions." *American Naturalist,* 149: 336–351.

Wilson, D.S., and E. Sober (1989) "Reviving the Superorganism." *Journal of Theoretical Biology,* 136: 337–356.

Wilson, D.S., and E. Sober (1994) "Reintroducing Group Selection to the Human Behavioral Sciences." *Behavioral and Brain Sciences,* 17: 585–608.

Wilson, E.O. (1975) *Sociobiology: The New Synthesis*. Cambridge, MA: Harvard University Press.

Wilson, E.O. (1998) *Consilience: The Unity of Knowledge*. New York: Alfred A. Knopf.

Wilson, J.Q. (1993) *The Moral Sense*. New York: Free Press.

Wimsatt, W.C. (1999) "Genes, Memes and Cultural Heredity." *Biology and Philosophy,* 14: 279–310.

Winter, S.G., Jr. (1964) "Economic 'Natural Selection' and the Theory of the Firm." *Yale Economic Essays,* 4(1): 225–272.

Witt, U. (1991a) *Individualistic Foundations of Evolutionary Economics*. Cambridge, U.K.: Cambridge University Press.

Witt, U. (1991b) "Economics, Sociobiology, and Behavioral Psychology on Preferences." *Journal of Economic Psychology,* 12: 557–573.

Woese, C.R. (2000) "Interpreting the Universal Phylogenetic Tree." *Proceedings of the National Academy of Sciences,* 97: 8392–8396.

Wolf, J.B. *et al.* (1999) "Interacting Phenotypes and the Evolutionary Process, II. Selection Resulting from Social Interaction." *American Naturalist,* 153(3): 254–266.

Wolfe, D.W. (2001a) *Tales from the Underground: A Natural History of Subterranean Life*. Reading, MA: Perseus Books.

Wolfe, D.W. (2001b) "Out of Thin Air." *Natural History,* 110(7): 44–51.

Wolfram, S. (2002) *A New Kind of Science.* Champaign, IL: Wolfram Media, Inc.

Wolpoff, M.H. (1999a) *Paleoanthropology,* 2nd edn. New York: McGraw-Hill.

Wolpoff, M.H. (1999b) "The Systematics of *Homo.*" *Science,* 284: 1774–1775.

Wolpoff, M.H., *et al.* (2001) "Modern Human Ancestry at the Peripheries: A Test of the Replacement Theory." *Science,* 291: 293–297.

Wood, B., and M. Collard (1999) "The Human Genus." *Science,* 284: 65–71.

Woodier, O. (1998) "How to Protect our Imperiled Pollinators." *National Wildlife,* 36(2): 36–46.

World Bank (2002) *Improving Air Quality in Mexico City: An Economic Evaluation.* Washington, D.C.: World Bank Policy Research Dissemination Center.

Wrangham, R.W. (1987) "The Significance of African Apes for Reconstructing Human Evolution." In *The Evolution of Human Behavior: Primate Models,* ed. W.G. Kinzey, pp. 51–71. Albany: State University of New York Press.

Wrangham, R.W. (2001) "Out of the *Pan,* Into the Fire: How Our Ancestors' Evolution Depended on What They Ate." In *Tree of Origin: What Primate Behavior Can Tell Us about Human Social Evolution,* ed. F.B.M. de Waal, pp. 121–143. Cambridge, MA: Harvard University Press.

Wrangham, R.W., and D. Peterson (1996) *Demonic Males: Apes and the Origins of Human Violence.* Boston, MA: Houghton Mifflin.

Wrangham, R.W., *et al.* (eds.) (1994) *Chimpanzee Cultures.* Cambridge, MA: Harvard University Press.

Wrangham, R.W., *et al.* (1999) "The Raw and the Stolen: Cooking and the Ecology of Human Origins." *Current Anthropology,* 40(5): 567–594.

Wright, R. (1999) "The Accidental Creationist: Why Stephen Jay Gould Is Bad for Evolution." *New Yorker,* 75(38): 56–65.

Wright, R. (2000) *Nonzero: The Logic of Human Destiny.* New York: Pantheon Books.

Wright, S. (1964) "Biology and the Philosophy of Science." In *Process and Divinity: Philosophical Essays Presented to Charles Hartshorne,* eds. W.R. Reese and E. Freeman, pp. 101–125. LaSalle, IL: Open Court Press.

Wynne, C.D.L. (2001) *Animal Cognition: The Mental Lives of Animals.* London: Macmillan.

Wynne-Edwards, V.C. (1962) *Animal Dispersion in Relation to Social Behaviour.* New York: Hafner.

Wynne-Edwards, V.C. (1963) "Intergroup Selection in the Evolution of Social Systems." *Nature,* 200: 623.

Xing, X., X.-L. Wang, and X.-C. Wu (1999) "A Dromaesaurid Dinosaur with a Filamentous Integument from the Yixian Formation of China." *Nature,* 401: 262–266.

Yates, F.E., *et al.* (1987) *Self-Organizing Systems: The Emergence of Order.* New York: Plenum Press.

Yellen, J.E., *et al.* (1995) "A Middle Stone Age Worked Bone Industry from Katanda, Upper Semliki Valley, Zaire." *Science,* 268: 553.

Yoerg, S.I. (2001) *Clever as a Fox: Animal Intelligence and What It Can Teach Us about Ourselves.* New York: Bloomsbury.

Yoffee, N., and G.L. Cowgill (eds.) (1988) *The Collapse of Ancient States and Civilizations*. Tucson: University of Arizona Press.

Zajonc, R.B. (1965) "Social Facilitation." *Science,* 149: 269–274.

Zentall, T.R., and B.G. Galef, Jr. (eds.) (1988) *Social Learning: Psychological and Biological Perspectives*. Hillsdale, NJ: Lawrence Erlbaum.

Zhang, F., and Z. Zhou (2000) "A Primitive Enantiornithine Bird and the Origin of Feathers." *Science*, 290: 1955.

Zihlman, A.L. (1981) "Women as Shapers of Human Adaptation." In *Woman the Gatherer*, ed. F. Dahlberg, pp. 75–120. New Haven, CT: Yale University Press.

Zihlman, A.L., and N. Tanner (1978) "Gathering and the Hominid Adaptation." In *Female Hierarchies*, eds. L. Tiger and H. Fowler, pp. 163–194. Chicago, IL: Beresford Books.

Zihlman, A.L., *et al.* (1978) "Pygmy Chimpanzees as a Possible Prototype for the Common Ancestor of Humans, Chimpanzees and Gorillas." *Nature*, 275: 744–746.

Ziman, J. (ed.) (2000) *Technological Innovation as an Evolutionary Process*. New York: Cambridge University Press.

Zimmer, C. (1998) "A Sickle in the Clouds." *Discover*, 19(6): 32.

Zimmer, C. (2000) *Parasite Rex: Inside the Bizarre World of Nature's Most Dangerous Creatures*. New York: Free Press.

Copyright Notices

Index